AF361674

Recent Technical Developments in Energy-Efficient 5G Mobile Cells

Recent Technical Developments in Energy-Efficient 5G Mobile Cells

Special Issue Editors

Raed A. Abd-Alhameed
Issa Elfergani
Jonathan Rodriguez

MDPI • Basel • Beijing • Wuhan • Barcelona • Belgrade • Manchester • Tokyo • Cluj • Tianjin

Special Issue Editors

Raed A. Abd-Alhameed
University of Bradford
UK

Issa Elfergani
Campus Universitário
de Santiago
Portugal

Jonathan Rodriguez
Campus Universitário
de Santiago
Portugal

Editorial Office
MDPI
St. Alban-Anlage 66
4052 Basel, Switzerland

This is a reprint of articles from the Special Issue published online in the open access journal *Electronics* (ISSN 2079-9292) (available at: https://www.mdpi.com/journal/electronics/special_issues/Energy_5G).

For citation purposes, cite each article independently as indicated on the article page online and as indicated below:

LastName, A.A.; LastName, B.B.; LastName, C.C. Article Title. *Journal Name* **Year**, *Article Number*, Page Range.

ISBN 978-3-03936-212-7 (Pbk)
ISBN 978-3-03936-213-4 (PDF)

Contents

About the Special Issue Editors

Raed A. Abd-Alhameed (M′02–SM′13) received B.Sc. and M.Sc. degrees from Basrah University, Basrah, Iraq, in 1982 and 1985, respectively, and the PhD degree from the University of Bradford, West Yorkshire, U.K., in 1997. Raed Abd-Alhameed is Professor of Electromagnetic and Radio Frequency Engineering at the University of Bradford, UK. He has long years′ research experience in the areas of Radio Frequency, Signal Processing, propagations, antennas and electromagnetic computational techniques, and has published over 600 academic journal and conference papers; in addition, he is co-authors of four books and several book chapters. At present, he is the leader of Radio Frequency, Propagation, sensor design and Signal Processing in addition to leading the Communications research group for years within the School of Engineering and Informatics, Bradford University, UK. He is the Principal Investigator for several funded applications to EPSRCs and leader of several successful knowledge Transfer Programmes, such as with Arris (previously known as Pace plc), Yorkshire Water plc, Harvard Engineering plc, IETG ltd, Seven Technologies Group, Emkay ltd, and Two World ltd including many Research Development Projects awards supported by Regional European funds. He has also been a co-investigator in several funded research projects, including 1) H2020 MARIE Skłodowska-CURIE ACTIONS: Innovative Training Networks (ITN) "Secure Network Coding for Next Generation Mobile Small Cells 5G-US", 2) Nonlinear and demodulation mechanisms in biological tissue (Dept. of Health, Mobile Telecommunications & Health Research Programme), and 3) Assessment of the Potential Direct Effects of Cellular Phones on the Nervous System (EU: collaboration with six other major research organizations across Europe). He was awarded the Business Innovation Award for his successful KTP with Pace and Datong companies on the design and implementation of MIMO sensor systems and antenna array design for service localizations. He is the chair of several successful workshops on Energy Efficient and Reconfigurable Transceivers (EERT): Approach towards Energy Conservation and CO2 Reduction that addresses the biggest challenges for future wireless systems. He has also been appointed as a guest editor for the IET Science, Measurements and Technology Journal since 2009 and 2012. He is also a research visitor for Wrexham University, Wales, since Sept 2009, covering the wireless and communications research areas. His interests lie in 5G Green Communications Systems, computational methods and optimizations, wireless and mobile communications, sensor design, EMC, MIMO systems, beam steering antennas, energy-efficient PAs, RF predistorter design applications. He is a fellow of the Institution of Engineering and Technology, a fellow of the Higher Education Academy and a chartered engineer.

Issa Elfergani received M.Sc. and Ph.D. degrees in electrical and electronic engineering from the University of Bradford, U.K., in 2008 and 2012, respectively, with a specialization in tunable antenna design for mobile handset and UWB applications. He is currently a Senior Researcher with the Instituto de Telecomunicações, Aveiro, Portugal, working with several national and international research funded projects, such as ENIAC ARTEMIS from 2011 to 2014; EUREKA BENEFIC from 2014 to 2017; CORTIF from 2014 to 2017; GREEN-T from 2011 to 2014; VALUE from 2016 to 2016; H2020-SECRET Innovative Training Network from 2017 to 2020. Since his Ph.D. graduation, he has successfully completed the supervision of several Master and Ph.D. students. He has around 100 high-impact publications in academic journals and international conferences; in addition, he is the author of two book editorial and nine book chapters. He has been on the technical program committee of a large number of IEEE conferences. He has several years of experience in 3G/4G and 5G radio frequency systems research with particular expertise on several and different antenna structures along with novel approaches in accomplishing a size reduction, low cost, improved bandwidth, and gain and efficiency. His expertise includes research in various antenna designs, such as MIMO, UWB, balanced and unbalanced mobile phone antennas, RF tunable filter technologies, and power amplifier designs. In 2014, he received prestigious FCT fellowship for his postdoctoral research. He is also the IEEE and an American Association for Science and Technology (AASCIT) member. He reviewed several highly ranked journals, such as the IEEE ANTENNAS AND WIRELESS PROPAGATION LETTERS, the IEEE TRANSACTIONS ON VEHICULAR TECHNOLOGY, IET MICROWAVES, ANTENNAS AND PROPAGATION, IEEE ACCESS, Transactions on Emerging Telecommunications Technologies, Radio Engineering Journal, IET-SMT, the IET Journal of Engineering. He was the Chair of both 4th and 5th International Workshop on Energy Efficient and Reconfigurable Transceivers (EERT). He is also a Guest Editor of the Hindawi Special Issue "Antenna Design Techniques for 5G Mobile Communications and Electronics", a Special Issue on "Recent Technical Developments in Energy-Efficient 5G Mobile Cells", and a Special Issue on " Recent Advances in Antenna Design for 5G Heterogeneous Networks"

Jonathan Rodriguez received his Master's degree in Electronic and Electrical Engineering and Ph.D. from the University of Surrey (UK), in 1998 and 2004, respectively. In 2005, he became a researcher at the Instituto de Telecomunicações (Portugal) where he was a member of the Wireless Communications Scientific Area. In 2008, he became a Senior Researcher where he established the 4TELL Research Group targeting next-generation mobile systems. He has served as a project coordinator for major international research projects, including Eureka LOOP and FP7 C2POWER whilst serving as technical manager for FP7 COGEU and FP7 SALUS. He is currently the coordinator of the H2020-SECRET Innovative Training Network. Since 2009, he has served as Invited Assistant Professor at the University of Aveiro (Portugal) and attained Associate Level in 2015. In 2017, he was appointed the Professor of Mobile Communications at the University of South Wales (UK). He is author of more than 400 scientific works, including 10 book editorials. His professional affiliations include Senior Member of the IEEE and Chartered Engineer (CEng) since 2013 and Fellow of the IET (2015).

Editorial

Recent Technical Developments in Energy-Efficient 5G Mobile Cells: Present and Future

Raed A. Abd-Alhameed [1,*]**, Issa Elfergani** [2] **and Jonathan Rodriguez** [2]

[1]   Faculty of Engineering and Informatics, University of Bradford, Bradford BD7 1DP, UK
[2]   Mobile Systems Group, Instituto de Telecomunicações, 3810-193 Aveiro, Portugal;
      i.t.e.elfergani@av.it.pt (I.E.); Jonathan@av.it.pt (J.R.)
*   Correspondence: R.A.A.Abd@bradford.ac.uk

Received: 13 April 2020; Accepted: 14 April 2020; Published: 20 April 2020

---

1. Introduction

The chapter of 4G (4th Generation) mobile systems is finally coming to an end, with waves of 4G systems having been deployed throughout Europe and worldwide. These systems provide a universal platform for broadband mobile services at any time and anywhere. However, mobile traffic is still growing at an unprecedented rate and the need for more sophisticated broadband services is further pushing the limits of the current standards to provide even tighter integration between wireless technologies and higher speeds [1]. The increasing number of mobile devices and traffic, the change in the nature of service and devices, along with the pressure on the operation, costs, and energy efficiency are all continuously putting stringent limits on the requirements of the designs of mobile networks. It is widely accepted that incremental enhancements of the current networking paradigm will not come close to meeting the requirements of networking by 2020 [2]. This has led to the need for a new generation of mobile communications: so-called 5G. The interests of stakeholders and academic researchers are now focused on the 5G paradigm. Although 5G systems are not expected to penetrate the market until 2020, the evolution towards 5G is widely accepted to manifest in the convergence of internet services with existing mobile networking standards, leading to the commonly used term "mobile internet" over heterogeneous networks (HetNets), with very high connectivity speeds.

The envisaged plan is to narrow the gap between current networking technologies and the foreseen requirements of 2020 networking and beyond, providing higher network capacities, the ability to support more users, lower cost per bit, better energy efficiency, and finally, adaptability to the new nature of services and devices, such as support of smart cities and the Internet of Things (IoT).

Certain technology trends, properties, and offered services have been widely envisioned to form part of the highly anticipated 5G [3,4]. It is almost globally accepted that the densification of mobile networks is the way to go for 5G. It is expected that mobile networks will become hugely densified with the adoption of multitier heterogeneous networks, including macrocells, a huge number of small cells, remote radio units (RRUs), and device-to-device communications [5]. Additionally, cooperation and network virtualization are expected to play the main roles in 5G systems [5]. Small cells are envisaged as the vehicle for ubiquitous densified 5G services, providing cost-effective, energy-efficient, high-speed communication. Small cells were partly adopted in the 4G revolution in the form of the femtocell, and the outdoor version, the picocell; however, femtocells are confined to indoor use, and picocells require radio networking infrastructure and planning, representing a significant cost for operators. Yet, small-cell technology is here to stay, with its desirable energy rating making it a winning candidate for a basic building block upon which the mobile networks of the future will evolve.

2. The Present Issue

This Special Issue features 15 articles which addresses the main aspects of technology trends which are widely accepted to form part of 5G, by providing a virtual cooperative wireless network of small cells. The main aim of these investigations goes beyond the current vision of densification and small-cell 5G through disruptive, new "femtocell"-like paradigms, where end-users play the role of prosumers of wireless connectivity, i.e., "Mobile Small Cells".

The 15 articles within this special issue illustrate the true innovation in engineering design that can occur by blending models and methodologies from different disciplines. In this special issue, the target was to follow this approach to deliver a new disruptive architecture to deliver next-generation mobile small-cell technologies. According to this design philosophy, the novelty of these articles resides in the intersection of engineering paradigms that include cooperation, network coding, and smart energy-aware frontends. These technologies will not only be considered as individual building blocks, but will be re-engineered according to an interdesign approach, serving as enablers for energy-efficient femtocell-like services on the move.

Next-generation handsets will need to be green, or in other words, "energy-aware", so as to support emerging smart services that are likely to be bandwidth-hungry, as well as to support multimode operation (5G, LTE, LTE-A, HSDPA, 3G among others) in HetNet environments. This vision gives way to stringent design requirements in the RF system design that, in today's handset, are the key consumers of power. To address the RF frontend and propose multi-standard flexible transceivers, the power consumption must be considered as a key design metric. This will include investigating RF building blocks such as energy-efficient power amplifiers (PAs) and antenna techniques, and tuneable RF bandpass filters.

Seven articles in this special issue propose novel and efficient antenna designs that employ both single and MIMO synthesis for use in heterogeneous networks [6–12]. Some of these designs operate on fixed single/multiband and radiations, as in [6–10], while some have the feature of reconfigurability that allows them to operate in a tuned manner in which the resonant frequencies and patterns can be shifted/reconfigured even after the designs have been made [11,12]. The other two articles present recent work on a highly efficient power amplifier which will provide hardware solutions to the growing RF front-end integration challenges with additional design requirements towards energy efficiency for Pas [13,14]. A paper presenting the recent progress of 4G/5G reconfigurable filters for multimode operation with potential energy efficiency traits, good linearity, and potentially low-cost manufacturing over a variety of substrates is also included in this issue [15].

The other three papers study Non-Orthogonal Multiple Access (NOMA) schemes. The first paper addresses an investigation of two transmission scenarios for the base station (BS) in cellular networks to serve users who are located at the cell-edge area [16]. In this study, it was shown that wireless-powered NOMA and the cell-center user can harvest energy from the BS in such a model. Moreover, the problem of the cell-edge user, i.e., due to the weak received signal, has been solved by fabricating a far NOMA user with multiple antennae to achieve improved performance. A similar work [17] proposes NOMA as a promising technology that could be used in next-generation networks in the near future. Within this work, a multipoints cooperative relay (MPCR) NOMA model, instead of just a relay, as suggested in previous studies, was proposed. The third paper [18] introduces the Power Domain-based Multiple Access (PDMA) scheme as a kind of NOMA that can be used in green communications and which can support energy-limited devices by employing wireless power transfer. Such a technique is known as a lifetime-expanding solution for operations in future access policy, especially in the deployment of power-constrained relays for three-node, dual-hop systems.

To equip the network with small cells, parameters such as cell size, interference in the network, and deployment strategies to maximize the network's performance gains expected from small cells are important, as stated in reference [19]. Furthermore, the network performance was evaluated for different Pmax values for small-cell uplink. Various deployment scenarios for furnishing the existing macro layer in LTE networks with small cells were considered within this work.

The last work in this special issue presents a theoretical study of electromagnetic propagation in a complex medium suspended multilayer coplanar waveguide (CPW). This work is based on the generalized exponential matrix technique (GEMT) that was joined with Galerkin's spectral method of moments, and then applied to a CPW printed on a bianisotropic medium. The analytical formulation is based on a Full-GEMT, a method that avoids the usual procedure of heavy and tedious mathematical expressions, using matrix-based mathematics instead [20].

3. Future

From a future perspective, to help current mobile standards to move forward a cooperative approach, a more user-network centric approach is desirable, i.e., where all devices are seen as a "pool of resources" to be used by the network as a vehicle, leading to enhanced spectral and energy efficiency. It is essential to break the femto-barrier and reduce the energy consumption in the network by at least a factor of 10, while providing higher data rates, higher capacities, and ubiquitous service through reduced-cost solutions for future 5G systems.

Thus, initially, to support reliability, throughput, coverage, and the coexistence requirements of 5G wireless systems in a cost-effective and energy-efficient manner, some vital issues should be considered, such as analyses, design, and optimization of NCC communications for mobile small cells and of small-cell overlay deployment for HetNets, thereby enabling the potential of 5G systems.

Moreover, in order to accomplish secure network coding for 5G cooperative mobile small cells, we must go beyond the previously proposed mechanisms by using random linear network coding, as well as modifying and adapting the proposed protocols to multihop secret key distribution in highly dynamic wireless networks. The use of random linear network coding is expected to boost performance.

In terms of a frontend that can meet the requirements of 5G systems, it is apparent that reliance on a single technology will no longer have a place in the mobile communication paradigm; rather, the very careful integration of diverse radio technologies in a cost-effective way will be required. Forthcoming 5G systems comprise a truly mobile multimedia platform that constitutes a convergent networking arena, that not only includes legacy heterogeneous mobile networks, but also advanced radio interfaces and the possibility of operating at mm-wave frequencies to capitalize on the large swathe of available bandwidth. This provides the impetus for a new breed of handset designs that, in principle, should be multimode in nature, energy efficient, and above all, able to operate at the mm-wave band, placing new design drivers on antenna design. Therefore, the target in future is to investigate advanced 5G massive array/MIMO antennas for 5G smart future applications that can operate in the mm- range, i.e., above 30 GHz, and meet the essential requirements of 5G systems such as large bandwidth (>1 GHz) and gain and efficiencies up to 15 dBi and 95% respectively. Also, it should extend the current Doherty amplifier implementation towards a three-step approach to promote the concept of efficiency enhancement and linearity compensation in PA design. Also, new reconfigurable switchable 5G filters should be designed using tuning technology with an emphasis on low-loss, low-power consumption, reduced size, and high-Q, which would also give rise to easy integration with the CMOS PA.

Author Contributions: R.A.A.-A., I.E. and J.R. worked together during the whole editorial process of the special issue, "Recent Technical Developments in Energy-Efficient 5G Mobile Cells", published in the MDPI journal *Electronics*. I.E. drafted this editorial summary. R.A.A.-A. and J.R. reviewed, edited and finalized the manuscript. All authors have read and agreed to the published version of the manuscript.

Acknowledgments: The editors would like to thank not only the authors who have contributed with submitting excellent work to this special issue but also the reviewers for their fruitful and valuable comments and feedback, which improve the quality of the published work within this special issue. A special appreciation also goes to the editorial board of MDPI *Electronics* journal for the opportunity to guest edit this special issue, and to the *Electronics* Editorial Office staff for the hard and precise work to keep a rigorous peer-review schedule and timely publication.

Conflicts of Interest: The authors declare no conflict of interest.

References

1. Cisco Visual Networking Index. Global Mobile Data Traffic Forecast Update, 2012–2017. Available online: http://www.cisco.com/en/US/solutions/collateral/ns341/ns525/ns537/ns705/ns827/white_paper_c11-520862.html (accessed on 3 April 2020).

2.   Andrews, J.G.; Buzzi, S.; Choi, W.; Hanly, S.V.; Lozano, A.; Soong, A.C.K. What will 5G Be? *IEEE J. Sel. Areas Commun.* **2014**, *32*, 1065–1082. [CrossRef]

3.   Hossain, E.; Hasan, M. 5G cellular: Key enabling technologies and research challenges. *IEEE Instrum. Meas. Mag.* **2015**, *18*, 11–21. [CrossRef]

4.   llJoyn. Proximity-Based Peer-to-Peer Mobile Application Development Framework. Available online: https://www.alljoyn.org/ (accessed on 3 April 2020).

5.   Heath, R.W.; Kountouris, M. Modeling heterogeneous network interference. *IEEE Inf. Theory Appl. Workshop* **2012**, *61*, 17–22.

6.   Elfergani, I.; Iqbal, A.; Zebiri, C.; Basir, A.; Rodriguez, J.; Sajedin, M.; De Oliveira Pereira, A.; Mshwat, W.; Raed, A.-A.; Ullah, A. Low-profile and closely spaced four elements mimo antenna for wireless body area networks. *Electronics* **2020**, *9*, 258. [CrossRef]

7.   Zebiri, C.; Sayad, D.; Elfergani, I.; Iqbal, A.; Widad, F.A.; Mshwat, J.K.; Rodriguez, J.; Raed, A.-A. Compact semi-circular and arc-shaped slot antenna for heterogeneous rf front-ends. *Electronics* **2019**, *8*, 1123. [CrossRef]

8.   Ojaroudi Parchin, N.; Jahanbakhsh Basherlou, H.; Al-Yasir, Y.I.A.M.; Abdulkhaleq, A.; Patwary, M.A.; Abd-Alhameed, R.A. New CPW-fed diversity antenna for MIMO 5G smartphones. *Electronics* **2020**, *9*, 261. [CrossRef]

9.   Abdullah, M.; Kiani, S.H.; Abdulrazak, L.F.; Iqbal, A.; Bashir, M.A.; Khan, S.; Kim, S. High-performance multiple-input multiple-output antenna system for 5G mobile terminals. *Electronics* **2019**, *8*, 1090. [CrossRef]

10.   Ojaroudi Parchin, N.; Jahanbakhsh Basherlou, H.; Al-Yasir, Y.I.A.; Abd-Alhameed, R.A.; Abdulkhaleq, A.M.; Noras, J.M. Recent developments of reconfigurable antennas for current and future wireless communication systems. *Electronics* **2019**, *8*, 128. [CrossRef]

11.   Iqbal, A.; Smida, A.; Abdulrazak, L.F.; Omar, A.; Saraereh, N.K.M.; Elfergani, I.; Sunghwan, K. Low-profile frequency reconfigurable antenna for heterogeneous wireless systems. *Electronics* **2019**, *8*, 976. [CrossRef]

12.   Iqbal, A.; Smida, A.; Mallat, N.; Ghayoula, R.; Elfergani, I.T.E.; Rodriguez, J.; Kim, S. Frequency and pattern reconfigurable antenna for emerging wireless communication systems. *Electronics* **2019**, *4*, 407. [CrossRef]

13.   Sajedin, M.; Elfergani, I.; Rodriguez, J.; Abd-Alhameed, R.; Barciela, M.F. A survey on rf and microwave doherty power amplifier for mobile handset applications. *Electronics* **2019**, *8*, 717. [CrossRef]

14.   Abdulkhaleq, A.M.; Yahya, M.A.; McEwan, N.; Rayit, A.; Abd-Alhameed, R.A.; Ojaroudi Parchin, N.; Al-Yasir, Y.I.A.; Noras, J. Recent developments of dual-band doherty power amplifiers for upcoming mobile communications systems. *Electronics* **2019**, *8*, 638. [CrossRef]

15.   Al-Yasir, Y.I.A.; Ojaroudi Parchin, N.; Abd-Alhameed, R.A.; Abdulkhaleq, A.M.; Noras, J.M. Recent progress in the design of 4G/5G reconfigurable filters. *Electronics* **2019**, *8*, 114. [CrossRef]

16.   Le, C.-B.; Do, D.-T.; Voznak, M. Wireless-powered cooperative MIMO NOMA networks: Design and performance improvement for cell-edge users. *Electronics* **2019**, *8*, 328. [CrossRef]

17.   Ran, T.-N.; Voznak, M. Multi-points cooperative relay in NOMA system with N-1 DF relaying nodes in HD/FD mode for N user equipments with energy harvesting. *Electronics* **2019**, *8*, 167.

18.   Nguyen, T.-L.; Nguyen, M.-S.V.; Do, D.-T.; Voznak, M. Enabling non-linear energy harvesting in power domain based multiple access in relaying networks: Outage and ergodic capacity performance analysis. *Electronics* **2019**, *8*, 817. [CrossRef]

19.   Haider, A.; Hwang, S.-H. Maximum transmit power for UE in an LTE small cell uplink. *Electronics* **2019**, *8*, 796. [CrossRef]

20.   Sayad, D.; Zebiri, C.; Elfergani, I.; Rodriguez, J.; Abobaker, H.; Ullah, A.; Abd-Alhameed, R.; Otung, I.; Benabdelaziz, F. Complex bianisotropy effect on the propagation constant of a shielded multilayered coplanar waveguide using improved full generalized exponential matrix technique. *Electronics* **2020**, *9*, 243. [CrossRef]

Review

Recent Progress in the Design of 4G/5G Reconfigurable Filters

Yasir I. A. Al-Yasir [1,*], Naser Ojaroudi Parchin [1], Raed A. Abd-Alhameed [1], Ahmed M. Abdulkhaleq [1,2] and James M. Noras [1]

[1] School of Electrical Engineering and Computer Science, Faculty of Engineering and Informatics, University of Bradford, Bradford BD7 1DP, UK; N.OjaroudiParchin@bradford.ac.uk (N.O.P.); R.A.A.Abd@bradford.ac.uk (R.A.A.-A); A.ABD@sarastech.co.uk (A.M.A.); jmnoras@bradford.ac.uk (J.M.N.)

[2] SARAS Technology Limited, Leeds LS12 4NQ, UK

* Correspondence: Y.I.A.AL-YASIR@bradford.ac.uk; Tel.: +44-127-423-4033

Received: 22 December 2018; Accepted: 16 January 2019; Published: 20 January 2019

Abstract: Currently, several microwave filter designs contend for use in wireless communications. Among various microstrip filter designs, the reconfigurable planar filter presents more advantages and better prospects for communication applications, being compact in size, light-weight and cost-effective. Tuneable microwave filters can reduce the number of switches between electronic components. This paper presents a review of recent reconfigurable microwave filter designs, specifically on current advances in tuneable filters that involve high-quality factor resonator filters to control frequency, bandwidth and selectivity. The most important materials required for this field are also highlighted and surveyed. In addition, the main references for several types of tuneable microstrip filters are reported, especially related to new design technologies. Topics surveyed include microwave and millimetre wave designs for 4G and 5G applications, which use varactors and MEMSs technologies.

Keywords: microstrip; tuneable filter; microwave filter; 5G; MEMSs; varactor

---

1. Introduction

Reconfigurable microwave filters are vital in wireless communications. Many applications require diversity in filter performance. Traditional filter banks occupy much space on circuit boards, fuelling interest in replacing them with compact tuneable filters, saving space and improving performance. The centre frequency is the only tunable parameter in most reconfigurable filters and relatively few filter designs offer other tunable parameters such as bandwidth, poles, zeros and quality factors. To select the suitable technique of reconfiguration for a given application, researchers should take into account the following parameters: operating frequency, physical size, performance and power handling. Microwave filters can be categorized in terms of the position of the poles and their effect on the insertion loss and the effect of the zeros on the characteristics of the passband. The zeros are usually distributed within the passband to give equiripple or Chebyshev characteristics. From the other side, when the poles are analysed, this kind of filter has all these positioned at DC or infinity and it is usually called an all-pole Chebyshev filter or simply a Chebyshev filter. It is worth mentioning that it is highly recommended to place poles where they are most required and also to minimise their number; each extra pole complicates systems and increases cost [1–3].

Some researchers have designed tuneable microwave filters using varactor diodes [4–27]. In these articles, most designs are focussed on bandpass tuneable resonators [4–20] and tuneable band-stop resonators using varactor diodes [21–25]. Only a few designs of microwave low-pass tuneable resonators and high-pass tuneable resonators are presented [26,27]. That is because of the deficiency of practical monolithic reconfigurable inductor solutions that increase the complexity of realizing a good performance for the design. In general, research into reconfigurable bandpass and band-stop resonator filters generally investigated reconfigurable frequency and bandwidth. Among a variety of prototype designs, $\lambda/4$ and $\lambda/2$ tuneable filters with varactor diodes, as well as multi-mode filters, are mostly used because of their compact size and the simplicity of the tuning circuit. For example, by using a $\lambda/4$ resonator, Hunter and Rhodes [4] presented a microstrip second-order combline filter at 3450–5000 MHz with a 3.2–5.2 dB insertion loss using striplines and varactor diodes as switches. To achieve constant impedance bandwidth, the filter was required to have electrical length at the mid-point of the frequency band. This technique also used in the design of a tuneable microstrip combline filter using stepped impedance resonators with varactor diodes [5].

Sanchez et al. presented a reconfigurable bandpass combline filter resonating at 470 MHz, adjusting the mutual coupling between the resonating elements [6]. Wang and et al. presented a planar reconfigurable combline resonator filter using varactor diodes [7]. In this design, the short-circuited end of the resonators was replaced by lumped series lines. According to this technique, the slope parameter of the introduced lines can be adjusted to achieve a constant fractional bandwidth covering a wide tuning range. Park et al. reported a second-order reconfigurable filter using varactor diodes [8].

Three different types of bandwidth responses have been achieved: constant absolute bandwidth, constant fractional bandwidth and decreasing fractional bandwidth. By using the independent electric and magnetic mutual coupling technique, designs can cover a wide tuning range of 845–1500 MHz. In addition, by utilizing the concept of the $\lambda/2$ resonator, Zhang et al. presented a second-order reconfigurable microstrip bandpass and band-stop filters by using varactor diodes [9] and [22], respectively. In these designs, a constant absolute bandwidth had been achieved by utilizing a mixed electric and magnetic mutual coupling technique. Similarly, a second-order microstrip tuneable filter using varactor diodes was presented in Reference [10]. By utilizing a corrugated coupled lines, the design covered the frequency band 1.4–2.0 GHz. On the other hand, other recent designs were reported in Reference [11–13] with different kinds of multi-mode filters, such as multi-mode open-loop planar tuneable filter [11], multi-mode microstrip ring resonator tuneable filter [12] and multi-mode triangular-microstrip resonator tuneable filter [13]. These filters are designed using varactor diodes to achieve reconfigurability for both the resonance frequency and absolute bandwidth.

It has been shown that multi-mode resonators have separately coupled degenerate modes that result between tuning elements and can be adjusted so as to affect each resonating mode independently. Microstrip bandpass and band-stop tuneable filters using varactor diodes were studied in Reference [14] and [21]. The main benefit of these reconfigurable filters was their compact size as compared with other prototypes.

Our paper aims to provide a survey of some important materials and designs for reconfigurable microwave filters. Different important designs and techniques used to accomplish reconfigurable filters are discussed in the following sections. In addition, the paper provides a common review of recent development in the design and implementation of tuneable RF, microwave and mmWave filters. Wireless communication applications driven by tuneable filters have shown a continuous development in both theoretical concepts and in the technology applied to realize them. This is surveyed in this paper, highlighting major design improvements.

This paper is organized as follows: Section 2 is a general literature review, highlighting the main books and review papers in the field of microwave filters. Section 3 surveys the tunable filter designs and simulation tools required by the 5G applications. Section 4 discuses BAW, SAW and active reconfigurable filters. Section 5 focuses mainly on recent microstrip tunable filter designs and gives a comparison summary. Finally, Section 6 presents our conclusions.

2. Literature Review and Highlighting Key Sources

In this section, we review and highlight the most important reference tools for researchers in the field of tuneable filters, especially key books and references on this topic.

In 2001, I.C. Hunter published the book entitled "Theory and Design of Microwave Filters" [1]. This book is valuable to researchers of the topic as well as to practitioners of the art and science of tuneable microwave resonator filters. Designing of tuneable microwave filters is unusual because it requires network synthesis, a technique requiring systematic processes to go ahead with the requirement of the last prototype model. This way is convenient for engineering regulations that aim to apply the model theory according to the design concepts. Circuit synthesis can be understood in terms of the circuit theory of passive elements. This scope has been deeply investigated in recent electrical engineering research. Accordingly, a prerequisite for the design and implementation of tuneable filters a knowledge of network synthesis. Synthesis enables an engineer to be familiar with the prototype circuit which requires to be converted into different microwave circuit modes like TEM, waveguides and dielectric resonators. Therefore, researchers need good information about the electromagnetic characteristics of such networks. The advantages of the book in Reference [1] are to afford a good reference for the designer including the basic concepts of microwave filters. Network synthesis models of many microwave filter designs were surveyed by specific structures with numerical analysis and simulations.

In 2002, Jia-Sheng Hong and M. J. Lancaster published another important book entitled "Microstrip Filters for RF/Microwave Applications" [2]. It provides a good and comprehensive study of RF and microwave filters based on the theory of microstrip design, as well as a link to the software of computer-aided structure tools and the techniques of advanced materials. Many results using the computer-aided tool were reported, from fundamental theory to practical implementation. This reference is not just a valuable academic reference but also a manageable resource for students, researchers and engineers in the scope of tuneable microwave filters. This source covers the designing of different new planar filter structures with progressive filtering properties, novel design concepts and miniaturization techniques for microstrip filters. Commercial systems are presented with design theory and methodology, which not only apply to planar filter but also to other types of filters, such as 3D designs and transmission line circuits.

In 2007, Richard J. Cameron, Raafat Mansour and Chandra M. Kudsia authored the book entitled "Microwave Filters for Communication Systems: Fundamentals, Design and Applications" [3]. This reference presents important developments in network synthesis and practical implementations of microwave resonator circuits over previous years. It delivers a handy and clear explanation of system characteristics and focuses on microwave resonator filters, basic requirements in the concepts and theory of microwave resonator filters, up to recent techniques of network synthesis. This review is the most comprehensive available with important design techniques concerning the study of coupling matrices. Thus, it is a very useful reference for every microwave filter design researcher.

The three books mentioned above constitute the most important sources for the designer of microwave tuneable filters. In addition, some useful review papers are also available [28–30].

In 1948, Ralph Levy and Seymour B Cohn presented a survey paper entitled "A History of Microwave Filter Research, Design and Development" [28]. This paper provides developments in the historical perspective of microwave resonator filters. The reference may resemble a review paper but this was not the main object of the paper. Therefore, the authors did not include comparatively new subjects such as millimetre waves.

In 2002, Ralph Levy, Richard Snyder and George Matthaei introduced a review paper for the main techniques used in the structure of microwave filters. The article explained the basic theory of important microwave resonator filters by using lumped-elements, adopted practically straightaway for different systems with frequencies up to 18 GHz. Several kinds of microwave filters were reported in this paper by referring to the most valuable resources, particularly the new designs for this topic [29].

In the same year, Ian C. Hunter, Laurent Billonet, Bernard Jarry and Pierre Guillon presented a review paper explaining the development of microwave filter technology from the viewpoint of its applications [30]. It is an interesting paper in the scope of filter theory, investigating designs of numerous types of passive elements: couplers, power dividers and phase shifters. For example, it shows that military communication systems require a wide-spectrum and reconfigurable performance for microelectronic maintenance receivers. That led to improving the highly discerning wide-spectrum cavity filter, coaxial resonator, suspended substrate diplexers and reconfigurable microwave filter with DC biasing. Moreover, it shows that satellite applications require low-mass and narrow-band filters with low-loss, selective amplitude and linear phase properties, which led to the growth of the multi-mode cavity and dielectric resonator filter with developments in the structure of diplexers and multiplexers. Finally, it shows that mobile phone base station applications require selective filter characteristics with low loss, compact size and handling of high power and that they should be able to be designed, implemented and manufactured in hundreds of millions at low cost, while mobile phone handset applications demand the construction of tens of thousands of filters of very compact size, low cost, low loss and high selectivity. These requirements stimulated the development of reconfigurable filters [30].

3. Reconfigurable Filters for 5G Applications

New wireless applications like LTE, LTE advanced and fifth generation use several bands of radio frequency instantaneously to assign the bandwidth necessary to increase data rates. Accordingly, the need for reconfigurable filters is enhanced. RF noise is an increasingly serious issue in modern wireless communication applications such as 5G and wide-band radar systems. Many recent books and articles discuss the use of reconfigurable antennas for "green" flexible RF in 5G applications [31–34]. Nowadays, 5G wireless communication technology is being considered for use in 700 MHz, 3.6 GHz and 26 GHz bands [35]. Band-pass filters are useful units in many 5G systems for rejecting unwanted signals. In addition, there are particular requirements for band-pass filters in such systems [36,37].

In Reference [38], a compact microstrip band-pass filter (BPF) covering the 3.4–3.8 GHz spectrum bandwidth for 5G wireless communications is presented. The planar filter uses three resonators, each terminated by a via to hole ground at one end and a capacitor at the other end with 50 Ω transmission line impedances for input and output terminals. The coupling between the lines is adjusted to resonate at the centre frequency with third-order band-pass Butterworth properties. The proposed combline filter is designed on an alumina substrate with a relative dielectric constant of 9.8 and a very small size of $9 \times 5 \times 1.2$ mm^3. This filter can be easily tuned in frequency by adjusting the capacitors as shown in Figure 1.

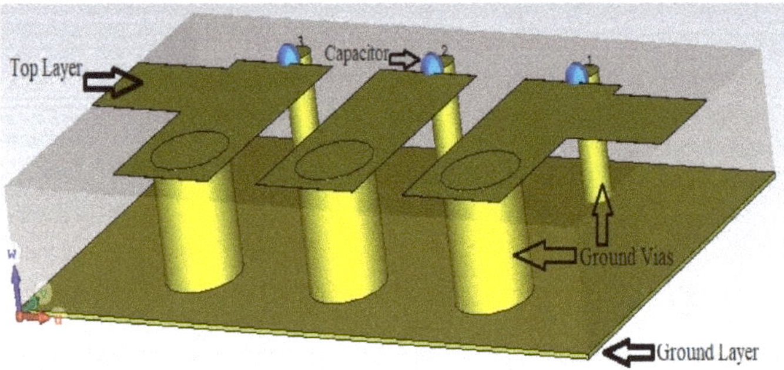

Figure 1. 3D configuration of the Combline BPF filter [38].

In Reference [39], the authors reported the design and optimization of reconfigurable cavity bandpass filters using MEMS switches with continuously tuneable resonant frequency over a wide tuning band. The tuneable filters are implemented by using silicon micromachining methods, which enables them to be tuned with the desired frequency in the millimetre wave (28–90 GHz) band. Besides, a novel feeding technique is used with fully passive input and output impedance matching over the total tuning range. The filters are tested through electromagnetic software and practically implemented.

Figure 2 explains the system installation. An Agilent E8361A network analyser is used to measure the return/insertion loss via a pair of SMA coaxial connectors, which are fixed under the layer of the Au-metallization. The frequency characteristics of the filters are separately adjusted with suitable DC biasing circuits.

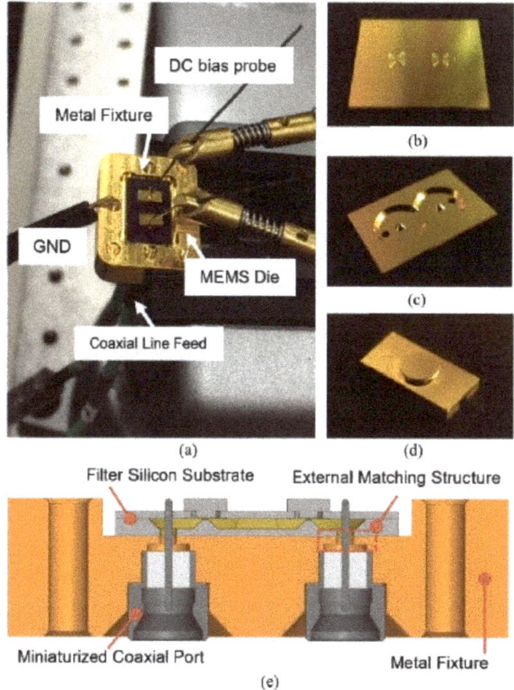

Figure 2. (**a**) RF system installation of the two BPF reconfigurable filters; (**b**) The front side of the diaphragm; (**c**) The front side of the cavity; (**d**) Side view of the DC bias circuit; (**e**) A cross-sectional view of the installed package system [39].

In Reference [40], a novel concept in coplanar waveguide (CPW) reconfigurable band-stop resonator filter was suggested and investigated, with reconfigurability achieved by adjusting the tuning frequency with shortcutting a defected ground (DG) layer. The microstrip reconfigurable filter is implemented based on CMOS techniques with a high resistivity silicon substrate of 300 nm thickness and by exploiting a 1 μm × 10 μm area with a vanadium oxide switch. The tuneable microstrip filter operates in the Ka-band and the reconfigurable resonant frequency covers the mmWave spectrum from 28.2–35 GHz. The paper proposed the use of vanadium oxide in a coplanar waveguide band-stop reconfigurable microstrip filter in the Ka-band. The design was more compact in physical size compared with other coplanar waveguide defected ground reconfigurable filters. The microstrip layer is divided into the amount of the square of the free space wavelength. It is implemented with the highest tuned frequency while showing a reconfigurability of 20%. Figure 3 shows a photo of the filter.

It is noteworthy that a new electronic system developer stage is important to design the compact size, low power handling components that are necessary for fifth generation applications [41] as shown in Figure 4. Recently, many filter companies have established some tools that are not yet developed to get optimal characteristics for filters to be used in 5G applications and there is a challenge to get the optimum designs with the required accuracy while reducing both the cost and time needed. As a result, there is a need for new tools and improvements in current ones to design whole boards and to take into account the following aspects: 1) Modern filter theory. 2) Finite element modelling. 3) New optimization techniques.

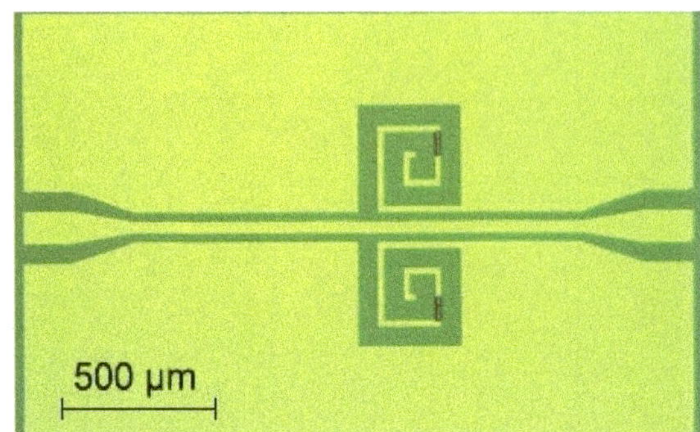

Figure 3. A photo of the reconfigurable microstrip filter [40].

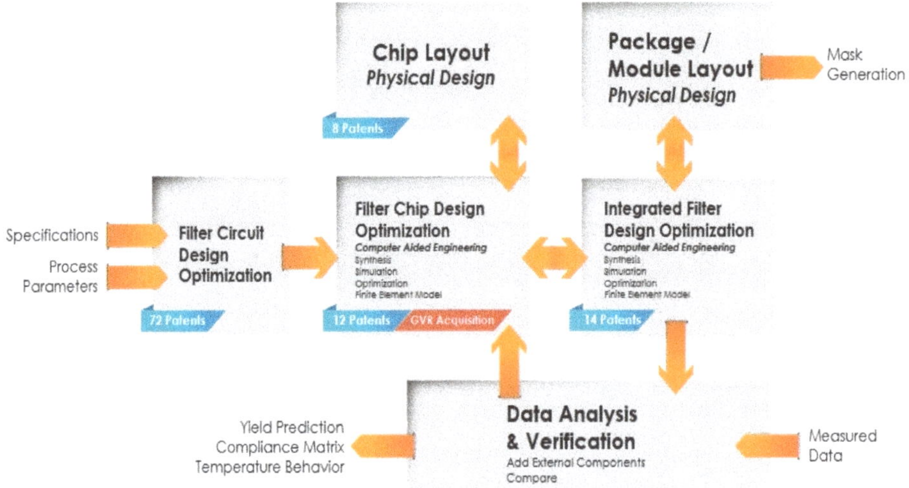

Figure 4. The design procedure for synthesized networks [41].

4. BAW, SAW and Active Reconfigurable Filters

In this section, we survey and compare reconfigurable surface acoustic wave (SAW), bulk acoustic wave (BAW) and active filters. Firstly, a discussion introduces SAW/BAW reconfigurable filters using varactor diodes integrated with microelectromechanical system (MEMS) technologies and some key techniques that can be employed, showing that high performance can be expected and delivered [42–44]. Secondly, active reconfigurable filters based on semiconductor materials and their analysis by hybrid technologies will be discussed [45–47].

4.1. BAW and SAW Reconfigurable Filters

In recent wireless communications, some applications, excluding BPFs with duplexers and power amplifiers, are combined into a single CMOS chip. That is because a high performance is achieved by using SAW/BAW filters for current wireless communication, so providing tuneability without degrading performance is essential. When a very small loss and a very low nonlinearity is preserved and wide range of reconfigurability is realized, reconfigurable SAW/BAW filters can be introduced in various areas. A combination of varactors with wideband SAW filters is presented in Reference [48]. The configurations of this technique are shown in Figure 5a. The advantage of this topology is that the passband characteristic is not sensitive to the quality factors of the varactors but the disadvantage of this approach is the limited reconfigurability due to the availability of a frequency band that is always involved in the passband of the designed filter. With the approach explained in Figure 5b [48], the resonance frequency and the bandwidth of the passband region could be controlled more easily than with previous techniques. The main disadvantage of this topology is the sensitivity of performance of the designed filters to the quality factors of the varactors.

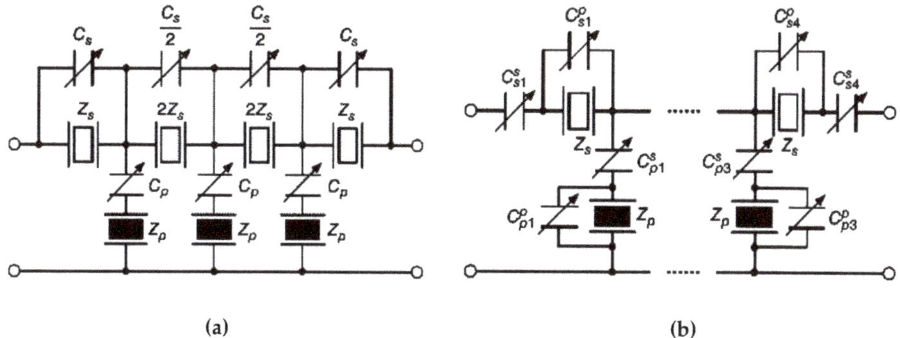

(a)   (b)

Figure 5. A reconfigurable filter topology using: (**a**) Acoustic wave resonators and variable capacitors; (**b**) SAW / BAW resonators and varactors [48].

In Reference [49], a novel reconfigurable wideband SAW filter using π type topology is presented for the LTE and UMTS low-band frequency range. Figure 6a shows the configuration of a proposed π type reconfigurable SAW filter circuit. Two reconfigurable filters are connected in parallel with a series arm which includes an inductor L_0 and a capacitor C_0 selected using a RF switch. Figure 6b shows a top view of the implemented π type reconfigurable filter. Measurements show an isolation value of more than 50 dB with a reconfigurability exceeding 30%.

(a)   (b)

Figure 6. π type reconfigurable SAW filter: (**a**) Circuit configuration; (**b**) a top view of the fabricated filter [49].

In Reference [50], the authors presented a new reconfigurable BAW filter with negative capacitors. Some similar previous designs were introduced with inductances [51,52]. The disadvantage of these proposed designs is that a second-order parasitic resonant frequency will be generated, leading to a low quality factor of the inductance that reduces the high quality factor of the BAW filter. In Reference [50], the paper presented a solution with a negative capacitance which adjusts the shunt resonance frequency of the BAW filter without generating a parasitic effect. The design includes flip-chip elements of a

BAW solidly mounted resonator (SMR) placed on the top of a BiCMOS 0.25 μm chip. Figure 7a shows the configuration of the negative capacitance circuit and the filter layout is depicted in Figure 7b. The designed tunable filter presents a very good level of rejection with less than -50 dB at 1 GHz, improving on previous filters using inductor technologies: the disadvantages of this filter are its power consumption and excess noise.

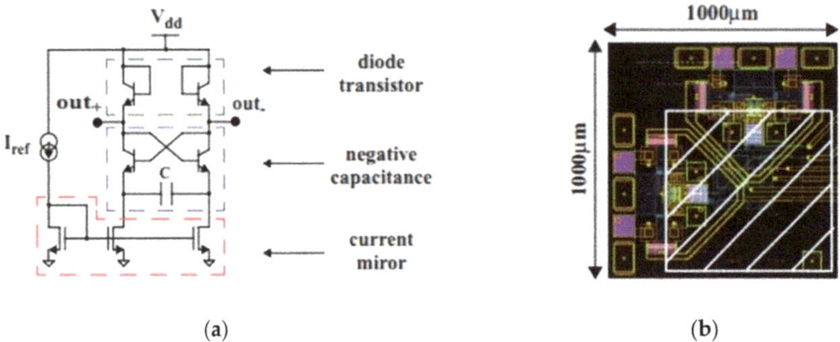

(a)   (b)

Figure 7. Reconfigurable BAW filter: (**a**) Negative capacitance circuit; (**b**) Filter layout [50].

Generally, both SAW and BAW filters have specific advantages and drawbacks. Mostly, they complement each other. In a few, very limited systems and applications, they compete against each other. As a comparison, BAW tuneable filters have more ability to operate with high frequency bands, high power circuits and high performance compared with SAW tuneable filters.

4.2. Active Reconfigurable Filters

Active tunable filters are also surveyed in this sub-suction. Generally, there are two categories of active tunable filters. Firstly, a resonator tunable filter based on gyrator active inductor [53] as shown in Figure 8a. The design produces a parallel active inductor with C_c and with a voltage gain e_o/e_i. The centre frequency and the quality factor are adjustable by controlling the values of C_L and the transconductance g_m, respectively. Secondly, an active filter based a positive feedback loop [45] as shown in Figure 8b. In this filter, the centre frequency and the quality factor are adjustable by varying the values of the phase shift K and the gain G of the amplifier, respectively. These filters can be realized by using the integrated circuits (ICs). The device is consisted of two filter cells coupled in shunt to generate a two pole responses. It should be noted that the centre frequency and the quality factor of each filter cells can be controlled automatically. In these filters, the resonance frequency and the bandwidth can be controlled with a certain renege with a constant filter performance. However, the drawback of these kinds of filters is the nonlinearity.

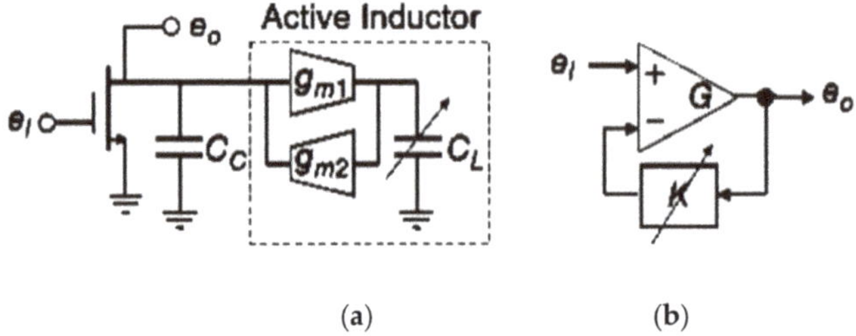

(a)   (b)

Figure 8. Active reconfigurable filter: (**a**) Active inductor-based filter [53]; (**b**) Recursive filter [45].

In Reference [54], the paper presented a reconfigurable microwave photonic filter with a wide tuning range, based on a semiconductor optical amplifier. The filter was experimentally implemented and measured with a new configuration, producing a reconfigurable microwave photonic notch filter with a high tuning range about 100%. Figure 9 shows the laboratory measurement set-up. The reconfigurability is realized by controlling the resonance wavelength of the device and the characteristics of the frequency response are fixed throughout the entire tuning mechanism. Additionally, the presented tunable filter has a very compact size, which can be easily combined with the techniques of photonic integrated devices. The filter has a configuration layout similar to that of the filter presented in Reference [55], except that this design incorporates filter detuning.

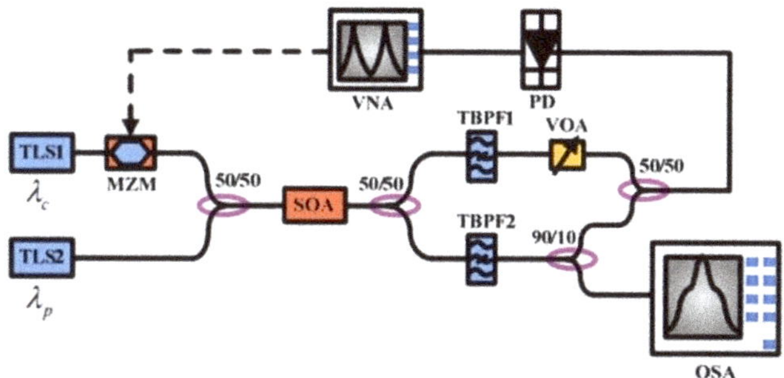

Figure 9. Laboratory install of the proposed filter [54]. VNA: Vector Network Analyser; PD: Photon Detector; TBPF: Tunable Band-Pass Filter; MZM: Mach-Zehnder Modulator; OSA: Optical Spectrum Analyzer; TLS: Tunable Laser Source.

Another semiconductor-tuned filter by using a cross-phase modulation (XPM) is presented in Reference [56]. The reconfigurability of the proposed filter is realized by varying the fibre delay lines or by dispersive optical fibre. In the designed tunable notch filter, both negative and positive coefficients are achieved by the cross-phase modulation, which leads to a constant frequency response. This scheme prevents further degradation of the system transfer function, unlike the cross-gain modulation (XGM) scheme presented in Reference [57] where only negative coefficients are employed over the amplifier. The XPM configuration also provides a higher RF bandwidth and a lower semiconductor optical amplifier-induced chirp than the XGM configuration in Reference [58].

5. Important Recent Microstrip Tuneable Filter Designs

In this section, we present the most important designs published recently in the microstrip tuneable filter field. Electrically tuneable or reconfigurable RF and microwave resonator filters are increasingly attractive to researchers and developers of RF and microwave circuit, because this technique is necessary to achieve compact and profitable electronic systems for future generations of wireless communication applications. These filters have different applications in wireless communication systems, such as in mixers and receiver preselection. To include electrical reconfigurability in resonator filters, reconfiguration components and switching elements like RF MEMS, semiconductor diodes, transistors and optical switches are used as shown in Figure 10 [59,60].

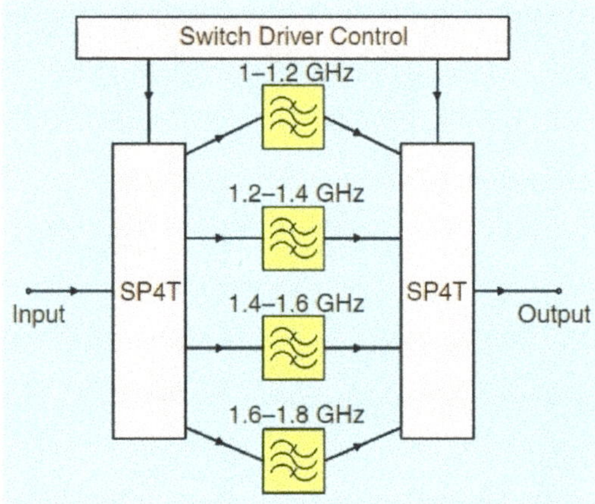

Figure 10. Electronically-tuned reconfigurable filters [60].

A design for a compact 5G reconfigurable-microstrip bandpass filter with third-order and Butterworth filter properties is presented in Reference [61]. The filter is reconfigurable in both resonant frequency and bandwidth to cover 3.4–3.8 GHz under the control of a single varactor diode switch. The design exhibits a 50–130 MHz bandwidth with return loss between 20–37 dB and insertion loss around 1 dB. The filter covers the 5G frequency spectrum for possible use in stationary terminals of both wireless communication and cognitive radio systems. Figure 11 shows a 3D structure of the filter. The biasing circuit, with the SPICE model for the varactor, is modelled as shown in Figure 12. For implementation, Skyworks Solutions SMV1234, size 1.5×0.7 mm^2, is used as the varactor switch.

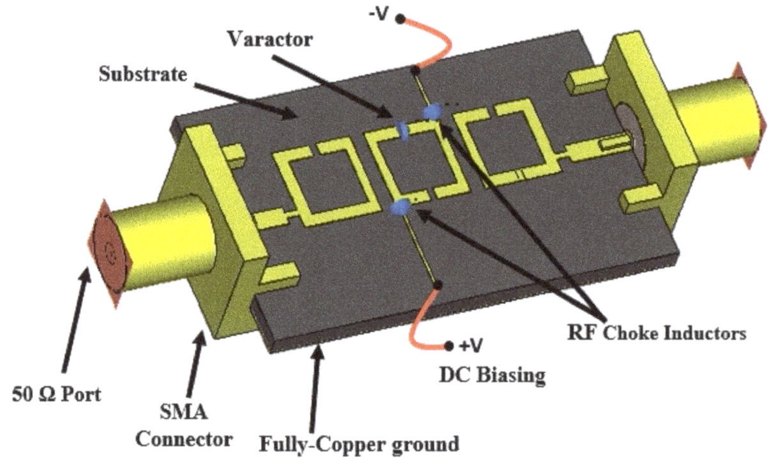

Figure 11. 3D Layout of the tuneable microstrip filter [61].

Another novel tuneable low-pass filter (LPF) using varactor diodes with high selectivity within the tuning range and wide stopband has been reported in Reference [62]. Stepped impedance resonators (SIR) and low impedance stubs of stepped-impedance resonators are used to achieve a tuneable resonator filter by means of varactor switches arranged in parallel and reconfigurable coupling lines arranged in series.

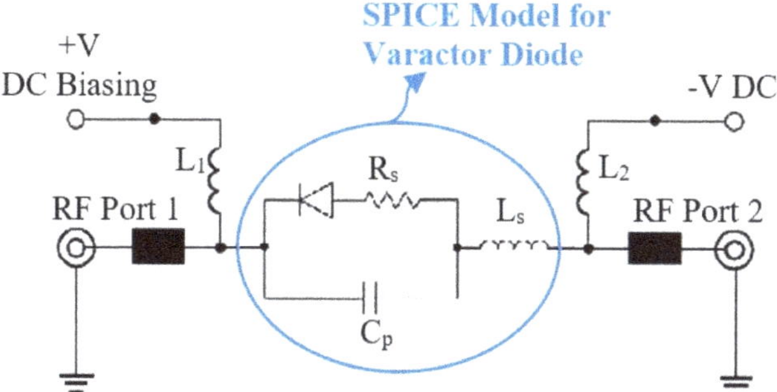

Figure 12. Biasing circuit with SPICE model of varactor switch [61].

Two semi-circular slots are adopted in the ground layer, while the microstrip layers have been connected with feed lines at both terminals to achieve a wide stopband for the filter. A stopband is maintained up to 22 GHz for the lower frequency with a rejection level of less than 10 dB. The presented tuneable microstrip filter has a very good performance compared with other LPF filter designs. Figure 13 shows the structure of the tuneable microstrip LPF.

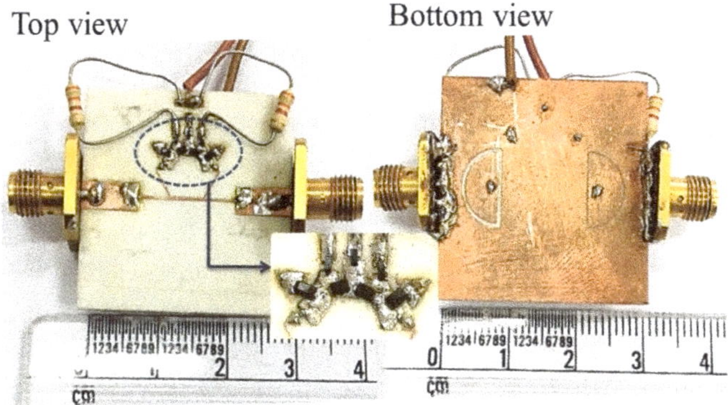

Figure 13. The tuneable microstrip LPF [62].

Modern wireless communication applications usually work with multi-bands. A novel structure of the miniaturized ring resonator tuneable filter is proposed in Reference [63]. The filter is designed with a single element, multi-mode, adjustable line impedance and microstrip material. The bandwidth of the resonator is enhanced significantly to include the entire range of the ultra-wideband (UWB) frequency with a wide area of reconfigurability. The filter has a high stopband and good performance. As long as the design of the filter is symmetrical, even and odd modes analysis have been used to study the required performance for the structure. The filter can be adapted to fulfil the required characteristics for various wireless applications. In addition, PIN diodes and varactor switches are used to achieve a high-level capacity for tunability with different performances are achieved for bandwidths, resonant frequencies and stopband frequencies. Figure 14 shows a photograph of the filter [63].

Another interesting design is introduced in Reference [64] to operate with a wideband covering the frequency range 3.5–10 GHz. Most referenced works in this paper were analogue and digital MEMS resonator filters. Open-stub ring resonators and advanced microelectromechanical system (MEMS) switches were used, giving high selectivity for the passband and adjusted DC biasing voltage up to 25 V.

Figure 14. The UWB reconfigurable filter [63].

A hybrid tuneable structure employing both series and shunt resonators is employed to achieve the desired characteristics. The switches are used for the biasing circuit to drive the required voltage levels. The design delivers the exact expected performance with the aid of the equivalent transmission line circuit of the microstrip filter. Compared with the PIN diode tuneable filters in Reference [65–68], the MEMS tuneable filter has a high quality factor and a good rejection in the stopband. Figure 15 shows the design with the installed layout.

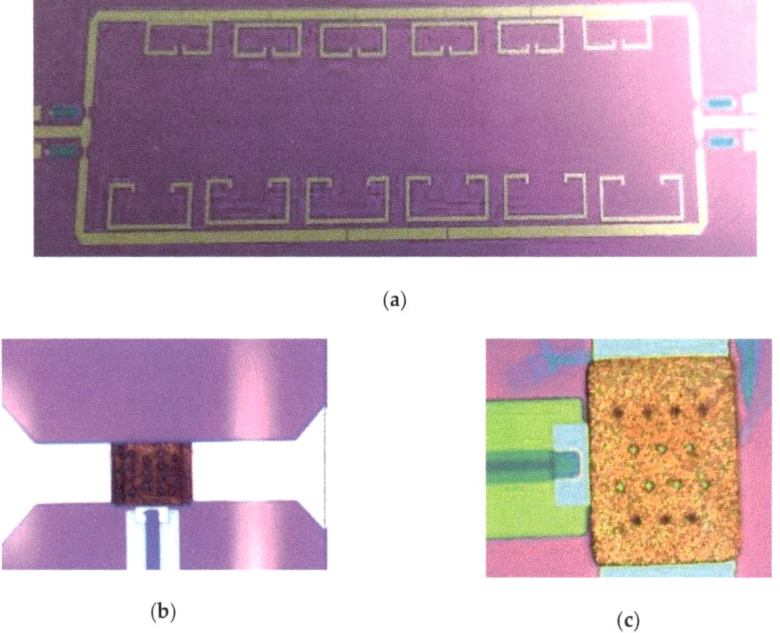

Figure 15. Design and layout: (a) Filter structure; (b,c) MEMS switch [64].

A multi-band reconfigurable filter is proposed in Reference [69]. The filter has a tunable resonance frequency and selection of bandpass and band-stop characteristics by using a PIN diode switch.

Two varactor diodes are used at the input and output of the transmission lines to adjust the quality factor. The centre frequency can be tuned continuously from 1.6–2.6 GHz during the bandpass configuration or from 1.6–3 GHz during the stopband configuration. The insertion loss in the bandpass configuration is adjusted by controlling the biasing voltage across the varactor diode, taking into account the loss of the PIN switch [70]. Figure 16 shows the filter.

Figure 16. The tuneable filter [69].

Table 1 Summarises performance comparisons for main recent and important references in the literature with the scope of microstrip reconfigurable filters.

Table 1. Performance comparisons for some recent microstrip tuneable filter designs.

Ref.	Filter Type	Freq. (GHz)	BW (MHz)	Reconfiguration	No. of Switches	RL (dB)	IL (dB)	Filter Size (mm³)
[62]	LPF	1–2.2	—	Freq.	4	20	0.6	30 × 30 × 1.52
[69]	BPF/ BSF	1.7–2.9	40	Freq.	7	16	4	36 × 35 × 0.8
[71]	BPF	1.1–2.1	40	Freq.	7	15–25	6	12.5 × 52 × 1.5
[72]	BSF	0.66–0.99	80	Freq.	2	0.8	27	41 × 55 × 1.5
[73]	BPF/BSF	0.8–1.5	215–535	Freq./BW/BS/BP	3	15/0.5	0.5/15	35 × 12 × 1.6
[74]	BSF	2.8–3.4	0–96	Freq./BW	2	12–25	4	26 × 26 × 3.1
[75]	BPF	1.8/2.9	20	Passband	—	0.5	22/38	35 × 35 × 0.5
[76]	BPF	0.5–1.1	60–230	Freq./BW	6	15	1.4–4.5	15 × 4.6 × 1.27
[77]	BSF	1.25–2.5	184	Freq.	4	2	50	100 × 20 × 0.7
[78]	BPF	0.76–2.6	75–285	Freq./BW/Selct.	2	15–30	1.2–4.2	100 × 8 × 0.5
[79]	BPF	2.4	900–1500	BW	4	15	1.1	64 × 64 × 0.81
[80]	BPF	6–11	400	Freq.	1	15	2	14 × 14 × 0.8

Varactor and PIN diodes are used as switches for all these designs. Low-pass filter (LPF), bandpass filter (BPF) and band-stop filter (BSF) are detailed in this comparison with different kinds of reconfiguration such as frequency, bandwidth and selectivity reconfigurations. The number of switches, filter size with resulted return loss (RL) and insertion loss (IL) are also summarized in this table, giving readers and researchers a general overview of the latest designs and achievements in the field of microstrip tuneable filters.

6. Conclusions

In this paper, a brief overview in the development of RF microwave tuneable filter is presented. During more than a decade of wide research, there has been a tremendous improvement in the design and implementation of reconfigurable filters and their applications in the RF microwave field.

The paper reviews recent designs of microwave tuneable filter topology and it gives the general background to the important source materials needed by researchers in the field of tuneable filter technology. It reviews recent developments in the field of tuneable RF, microwave and mmWave filters. Reconfigurable filters for the next generation of wireless communication applications are surveyed, with a selection from the most important recently published articles. The main recent design references are summarised with performance comparisons of achievements in this field.

Author Contributions: Conceptualization, Y.I.A.A.-Y.; methodology, Y.I.A.A.-Y. and N.O.P.; investigation, Y.I.A.A.-Y., A.M.A. and R.A.A.-A.; resources, Y.I.A.A.-Y., N.O.P, A.M.A. and R.A.A.-A.; writing—original draft preparation, Y.I.A.A.-Y., N.O.P., A.M.A., R.A.A.-A. and J.M.N.; writing—review and editing, Y.I.A.A.-Y., A.M.A., R.A.A.-A. and J.M.N.; visualization, Y.I.A.A.-Y., A.M.A., R.A.A.-A. and J.M.N.

Funding: This project has received funding from the European Union's Horizon 2020 research and innovation programme under grant agreement H2020-MSCA-ITN-2016 SECRET-722424.

Acknowledgments: This project has received funding from the European Union's Horizon 2020 research and innovation programme under grant agreement H2020-MSCA-ITN-2016 SECRET-722424.

Conflicts of Interest: The authors declare no conflict of interest.

References

1. Hunter, I. *Theory and Design of Microwave Filters*; The Institution of Engineering and Technology: Stevenage, UK, 2001. [CrossRef]

2. Hong, J.-S.; Lancaster, M.J. *Microstrip Filters for RF/Microwave*; John Wiley and Sons: Hoboken, NJ, USA, 2004; Volume 167.

3. Cameron, R.J.; Kudsia, C.M.; Mansour, R.R. *Microwave Filters for Communication Systems: Fundamentals, Design and Applications*, 2nd ed.; Wiley: Hoboken, NJ, USA, 2018.

4. Hunter, I.C.; Rhodes, J.D. Electronically Tuneable Microwave Bandpass Filters. *IEEE Trans. Microw. Theory Tech.* **1982**, *30*, 1354–1360. [CrossRef]

5. Byung-Wook, K.; Sang-Won, Y. Varactor-tuned combline bandpass filter using step-impedance microstrip lines. *IEEE Trans. Microw. Theory Tech.* **2004**, *52*, 1279–1283. [CrossRef]

6. Sanchez-Renedo, M.; Gomez-Garcia, R.; Alonso, J.I.; Briso-Rodriguez, C. Tuneable combline filter with continuous control of center frequency and bandwidth. *IEEE Trans. Microw. Theory Tech.* **2005**, *53*, 191–199. [CrossRef]

7. Wang, X.; Cho, Y.; Yun, S. A Tuneable Combline Bandpass Filter Loaded With Series Resonator. *IEEE Trans. Microw. Theory Tech.* **2012**, *60*, 1569–1576. [CrossRef]

8. Park, S.; Rebeiz, G.M. Low-Loss Two-Pole Tuneable Filters With Three Different Predefined Bandwidth Characteristics. *IEEE Trans. Microw. Theory Tech.* **2008**, *56*, 1137–1148. [CrossRef]

9. Zhang, X.Y.; Xue, Q.; Chan, C.H.; Hu, B. Low-Loss Frequency-Agile Bandpass Filters With Controllable Bandwidth and Suppressed Second Harmonic. *IEEE Trans. Microw. Theory Tech.* **2010**, *58*, 1557–1564. [CrossRef]

10. El-Tanani, M.A.; Rebeiz, G.M. Corrugated Microstrip Coupled Lines for Constant Absolute Bandwidth Tuneable Filters. *IEEE Trans. Microw. Theory Tech.* **2010**, *58*, 956–963. [CrossRef]

11. Tang, W.; Hong, J. Varactor-Tuned Dual-Mode Bandpass Filters. *IEEE Trans. Microw. Theory Tech.* **2010**, *58*, 2213–2219. [CrossRef]

12. Tsai, H.; Chen, N.; Jeng, S. Center Frequency and Bandwidth Controllable Microstrip Bandpass Filter Design Using Loop-Shaped Dual-Mode Resonator. *IEEE Trans. Microw. Theory Tech.* **2013**, *61*, 3590–3600. [CrossRef]

13. Serrano, A.L.C.; Correra, F.S.; Vuong, T.; Ferrari, P. Synthesis Methodology Applied to a Tuneable Patch Filter With Independent Frequency and Bandwidth Control. *IEEE Trans. Microw. Theory Tech.* **2012**, *60*, 484–493. [CrossRef]

14. Xiang, Q.; Feng, Q.; Huang, X.; Jia, D. Electrical Tuneable Microstrip LC Bandpass Filters With Constant Bandwidth. *IEEE Trans. Microw. Theory Tech.* **2013**, *61*, 1124–1130. [CrossRef]

15. Chiou, Y.; Rebeiz, G.M. A Tuneable Three-Pole 1.5–2.2-GHz Bandpass Filter With Bandwidth and Transmission Zero Control. *IEEE Trans. Microw. Theory Tech.* **2011**, *59*, 2872–2878. [CrossRef]

16. Chiou, Y.; Rebeiz, G.M. A Quasi Elliptic Function 1.75–2.25 GHz 3-Pole Bandpass Filter With Bandwidth Control. *IEEE Trans. Microw. Theory Tech.* **2012**, *60*, 244–249. [CrossRef]

17. Chiou, Y.; Rebeiz, G.M. Tuneable 1.55-2.1 GHz 4-Pole Elliptic Bandpass Filter With Bandwidth Control and >50 dB Rejection for Wireless Systems. *IEEE Trans. Microw. Theory Tech.* **2013**, *61*, 117–124. [CrossRef]

18. Carey-Smith, B.E.; Warr, P.A. Distortion Mechanisms in Varactor Diode-Tuned Microwave Filters. *IEEE Trans. Microw. Theory Tech.* **2006**, *54*, 3492–3500. [CrossRef]

19. El-Tanani, M.A.; Rebeiz, G.M. A Two-Pole Two-Zero Tuneable Filter With Improved Linearity. *IEEE Trans. Microw. Theory Tech.* **2009**, *57*, 830–839. [CrossRef]

20. Athukorala, L.; Budimir, D. Compact Second-Order Highly Linear Varactor-Tuned Dual-Mode Filters With Constant Bandwidth. *IEEE Trans. Microw. Theory Tech.* **2011**, *59*, 2214–2220. [CrossRef]

21. Ou, Y.; Rebeiz, G.M. Lumped-Element Fully Tuneable Band-stop Filters for Cognitive Radio Applications. *IEEE Trans. Microw. Theory Tech.* **2011**, *59*, 2461–2468. [CrossRef]

22. Zhang, X.Y.; Chan, C.H.; Xue, Q.; Hu, B. RF Tuneable Band-stop Filters With Constant Bandwidth Based on a Doublet Configuration. *IEEE Trans. Ind. Electron.* **2012**, *59*, 1257–1265. [CrossRef]

23. Wang, X.; Wang, B.; Zhang, H.; Chen, K.J. A Tuneable Band-stop Resonator Based on a Compact Slotted Ground Structure. *IEEE Trans. Microw. Theory Tech.* **2007**, *55*, 1912–1918. [CrossRef]

24. Wang, Z.P.; Kelly, J.; Hall, P.S. Reconfigurable band-stop filter with wide tuning range. *Electron. Lett.* **2010**, *46*, 771–772. [CrossRef]

25. Guyette, A.C. Design of fixed- and varactor-tuned band-stop filters with spurious suppression. In Proceedings of the 40th European Microwave Conference, Paris, France, 28–30 September 2010; pp. 288–291.

26. Huang, C.; Chen, N.; Tsai, H.; Chen, J. A coplanar waveguide bandwidth-tuneable low-pass filter with broadband rejection. *IEEE Microw. Wirel. Compon. Lett.* **2013**, *23*, 134–136. [CrossRef]

27. Abbosh, A. Compact tuneable low-pass filter using variable mode impedance of coupled structure. *IET Microw. Antennas Propag.* **2012**, *6*, 1306–1310. [CrossRef]

28. Levy, R.; Cohn, S.B. A History of Microwave Filter Research, Design and Development. *IEEE Trans. Microw. Theory Tech.* **1984**, *32*, 1055–1067. [CrossRef]

29. Levy, R.; Snyder, R.V.; Matthaei, G. Design of microwave filters. *IEEE Trans. Microw. Theory Tech.* **2002**, *50*, 783–793. [CrossRef]

30. Hunter, I.C.; Billonet, L.; Jarry, B.; Guillon, P. Microwave filters-applications and technology. *IEEE Trans. Microw. Theory Tech.* **2002**, *50*, 794–805. [CrossRef]

31. Hussaini, A.; Al-Yasir, Y.I.A.; Voudouris, K.; Mohammed, B.; Abd-Alhameed, R.; Mohammed, H.; Elfergani, I.; Abdullah, A.; Makris, D.; Rodriguez, J.; et al. Green Flexible RF for 5G. In *Fundamentals of 5G Mobile Networks*, 1st ed.; Rodriguez, J., Ed.; John Wiley and Sons: Hoboken, NJ, USA, 2015.

32. Al-Yasir, Y.I.A.; Abdullah, A.; Mohammed, H.; Mohammedand, B.; Abd-Alhameed, R. Design of Radiation Pattern-Reconfigurable 60-GHz Antenna for 5G Applications. *J. Telecommun.* **2014**, *27*, 1–6.

33. Abdulraheem, Y.I.; Oguntala, G.A.; Abdullah, A.S.; Mohammed, H.J.; Ali, R.A.; Abd-Alhameed, R.A.; Noras, J.M. Design of frequency reconfigurable multiband compact antenna using two PIN diodes for WLAN/WiMAX applications. *IET Microw. Antennas Propag.* **2017**, *11*, 1098–1105. [CrossRef]

34. Al-Yasir, Y.; Abdullah, A.; Ojaroudi Parchin, N.; Abd-Alhameed, R.; Noras, J. A New Polarization-Reconfigurable Antenna for 5G Applications. *Electronics* **2018**, *7*, 293. [CrossRef]

35. Statement: Improving consumer access to mobile services at 3.6 GHz to 3.8 GHz. Available online: https://www.ofcom.org.uk/consultations-and-statements/category-1/future-use-at-3.6-3.8-ghz (accessed on 28 July 2017).

36. Jeon, J.S.; Kang, S.T.; Kim, H.S. GA-optimized compact broadband CRLH band-pass filter using stub-inserted interdigital coupled lines. *J. Electromagn. Eng. Sci.* **2015**, *15*, 31–36. [CrossRef]

37. Wang, C.; Haider, F.; Gao, X.; You, X.; Yang, Y.; Yuan, D.; Aggoune, H.M.; Haas, H.; Fletcher, S.; Hepsaydir, E. Cellular architecture and key technologies for 5G wireless communication networks. *IEEE Commun. Mag.* **2014**, *52*, 122–130. [CrossRef]

38. Al-Yasir, Y.; Abd-Alhameed, R.A.; Noras, J.M.; Abdulkhaleq, A.M.; Ojaroudi, N. Design of Very Compact Combline Band-Pass Filter for 5G Applications. In Proceedings of the 2018 Loughborough Antennas & Propagation Conference (LAPC), Loughborough, UK, 12–13 November 2018; pp. 1–4.

39. Yang, Z.; Psychogiou, D.; Peroulis, D. Design and Optimization of Tuneable Silicon-Integrated Evanescent-Mode Bandpass Filters. *IEEE Trans. Microw. Theory Tech.* **2018**, *66*, 1790–1803. [CrossRef]

40. Casu, E.A.; Müller, A.A.; Fernández-Bolaños, M.; Fumarola, A.; Krammer, A.; Schüler, A.; Ionescu, A.M. Vanadium Oxide Band-stop Tuneable Filter for Ka Frequency Bands Based on a Novel Reconfigurable Spiral Shape Defected Ground Plane CPW. *IEEE Access* **2018**, *6*, 12206–12212. [CrossRef]

41. Infinite Synthesized Networks. Available online: https://www.resonant.com/technology/infinite-synthesized-networks (accessed on 8 February 2012).

42. Weigel, R.; Morgan, D.P.; Owens, J.M.; Ballato, A.; Lakin, K.M.; Hashimoto, K.; Ruppel, C.C.W. Microwave acoustic materials, devices and applications. *IEEE Trans. Microw. Theory Tech.* **2002**, *50*, 738–749. [CrossRef]

43. Tokihiro, N.; Masafumi, I.; Go, E.; Xiaoyu, M.; Shinji, T.; Masanori, U.; Yoshio, S. BAW/SAW/IPD hybrid type duplexer with Rx balanced output for WCDMA Band I. In Proceedings of the 2008 IEEE MTT-S International Microwave Symposium Digest, Atlanta, GA, USA, 15–20 June 2008; pp. 831–834.

44. Yang, J.; Jiao, X.; Zhang, R.; Zhong, H.; Shi, Y. Fabrication of bulk acoustic wave resonator based on AlN thin film. In Proceedings of the 2012 Symposium on Piezoelectricity, Acoustic Waves and Device Applications (SPAWDA), Shanghai, China, 23–25 November 2012; pp. 191–194.

45. Omori, T.; Seo, K.; Ahn, T.F.C.; Hashimoto, K. Flexible RF one-chip active filter based on recursive architecture in UHF range. In Proceedings of the 2014 Asia-Pacific Microwave Conference, Sendai, Japan, 4–7 November 2014; pp. 1309–1311.

46. Darfeuille, S.; Gomez-Garcia, R.; Lintignat, J.; Sassi, Z.; Barelaud, B.; Billonnet, L.; Jarry, B.; Marie, H.; Gamand, P. Silicon-Integrated 2-GHz Fully-Differential Tunable Recursive Filter for MMIC Three-Branch Channelized Bandpass Filter Design. In Proceedings of the 2006 IEEE MTT-S International Microwave Symposium Digest, San Francisco, CA, USA, 11–16 June 2006; pp. 776–779.

47. Omori, T.; Nishiuma, S.; Seo, K.; Ahn, C.; Hashimoto, K.; Kamada, M. Integrated RF tunable filter based on recursive architecture and its application. In Proceedings of the 2013 European Microwave Integrated Circuit Conference, Nuremberg, Germany, 6–8 October 2013; pp. 548–551.

48. Tomoya, K.; Ken-ya, H.; Tatsuya, O.; Masatsune, Y. Tunable Radio-Frequency Filters Using Acoustic Wave Resonators and Variable Capacitors. *Jpn. J. Appl. Phys.* **2010**, *49*, 07HD24.

49. Wada, T.; Ogami, T.; Horita, A.; Obiya, H.; Koshino, M.; Kawashima, M.; Nakajima, N. A new tunable SAW filter circuit for reconfigurable RF. In Proceedings of the 2016 IEEE MTT-S International Microwave Symposium (IMS), San Francisco, CA, USA, 22–27 May 2016; pp. 1–4.

50. Tilhac, C.; Razafimandimby, S.; Cathelin, A.; Bila, S.; Madrangeas, V.; Belot, D. A tunable bandpass BAW-filter architecture using negative capacitance circuitry. In Proceedings of the 2008 IEEE Radio Frequency Integrated Circuits Symposium, Atlanta, GA, USA, 15–17 June 2008; pp. 605–608.

51. Razafimandimby, S.; Tilhac, C.; Cathelin, A.; Kaiser, A.; Belot, D. A novel architecture of a tunable bandpass BAW-filter for a WCDMA transceiver. *Analog Integr. Circuits Signal Process.* **2006**, *49*, 237–247. [CrossRef]

52. Razafimandimby, S.; Tilhac, C.; Cathelin, A.; Kaiser, A.; Belot, D. An Electronically Tunable Bandpass BAW-Filter for a Zero-IF WCDMA Receiver. In Proceedings of the 32nd European Solid-State Circuits Conference, Montreux, Switzerland, 19–21 September 2006; pp. 142–145.

53. Haiqiao, X.; Schaumann, R.; Daasch, W.R.; Wong, P.K.; Pejcinovic, B. A radio-frequency CMOS active inductor and its application in designing high-Q filters. In Proceedings of the 2004 IEEE International Symposium on Circuits and Systems (IEEE Cat. No.04CH37512), Vancouver, BC, Canada, 23–26 May 2004; p. IV-197.

54. Li, X.; Yu, Y.; Dong, J.; Zhang, X. Widely tunable microwave photonic filter based on semiconductor optical amplifier. In Proceedings of the Asia Communications and Photonics Conference and Exhibition, Shanghai, China, 8–12 December 2010; pp. 122–123.

55. Coppinger, F.; Yegnanarayanan, S.; Trinh, P.D.; Jalali, B. All-optical RF filter using amplitude inversion in a semiconductor optical amplifier. *IEEE Trans. Microw. Theory Tech.* **1997**, *45*, 1473–1477. [CrossRef]

56. Manzanedo, M.D.; Mora, J.; Ortega, B.; Capmany, J. Tunable all-optical microwave filter using Cross-Phase Modulation in Semiconductor Optical Amplifier Mach-Zehnder interferometer. In Proceedings of the 2006 International Topical Meeting on Microwave Photonics, Grenoble, France, 3–6 October 2006; pp. 1–4.

57. Mora, J.; Martinez, A.; Manzanedo, M.D.; Capmany, J.; Ortega, B.; Pastor, D. Microwave photonic filters with arbitrary positive and negative coefficients using multiple phase inversion in SOA based XGM wavelength converter. *Electron. Lett.* **2005**, *41*, 921–922. [CrossRef]

58. Capmany, J.; Pastor, D.; Martinez, A.; Ortega, B.; Sales, S. Microwave photonic filters with negative coefficients based on phase inversion in an electro-optic modulator. *Opt. Lett.* **2003**, *28*, 1415–1417. [CrossRef]
59. Hong, J. Reconfigurable planar filters. *IEEE Microw. Mag.* **2009**, *10*, 73–83. [CrossRef]
60. Wong, P.W.; Hunter, I. Electronically Tuneable Filters. *IEEE Microw. Mag.* **2009**, *10*, 46–54. [CrossRef]
61. Al-Yasir, Y.I.A.; Parchin, N.O.; Abd-Alhameed, R.A.; Ali, A.H.; Noras, J.M.; Abdulkhaleq, A.M. Design of Bandpass Reconfigurable Filter for 5G Applications. In Proceedings of the Submitted to the 49th European Microwave Conference, Paris, France, 29 September–4 October 2019; pp. 1–4.
62. Kumar, L.; Parihar, M.S. A Compact Reconfigurable Low-Pass Filter with Wide-Stopband Rejection Bandwidth. *IEEE Microw. Wirel. Components Lett.* **2018**, *28*, 401–403. [CrossRef]
63. Kheir, M.; Kröger, T.; Höft, M. A New Class of Highly-Miniaturized Reconfigurable UWB Filters for Multi-Band Multi-Standard Transceiver Architectures. *IEEE Access* **2017**, *5*, 1714–1723. [CrossRef]
64. Zhang, N.; Mei, L.; Wang, C.; Deng, Z.; Yang, J.; Guo, Q. A Switchable Bandpass Filter Employing RF MEMS Switches and Open-Ring Resonators. *IEEE Trans. Electron Devices* **2017**, *64*, 3377–3383. [CrossRef]
65. Xu, J. Compact Switchable Bandpass Filter and Its Application to Switchable Diplexer Design. *IEEE Microw. Wirel. Components Lett.* **2016**, *26*, 13–15. [CrossRef]
66. Chao, S.; Wu, C.; Tsai, Z.; Wang, H.; Chen, C.H. Electronically Switchable Bandpass Filters Using Loaded Stepped-Impedance Resonators. *IEEE Trans. Microw. Theory Tech.* **2006**, *54*, 4193–4201. [CrossRef]
67. Song, X.; Wei, B.; Cao, B.; Guo, X.; Zhang, X. UHF band switchable superconducting filter with pin diode switches. *Electron. Lett.* **2014**, *50*, 775–777. [CrossRef]
68. Chuang, M.; Wu, M. Switchable Dual-Band Filter With Common Quarter-Wavelength Resonators. *IEEE Trans. Circuits Syst. II* **2015**, *62*, 347–351. [CrossRef]
69. Chen, F.; Li, R.; Chen, J. A Tuneable Dual-Band Bandpass-to-Band-stop Filter Using p-i-n Diodes and Varactors. *IEEE Access* **2018**, *6*, 46058–46065. [CrossRef]
70. Cho, Y.; Rebeiz, G.M. Two- and Four-Pole Tuneable 0.7–1.1-GHz Bandpass-to-Band-stop Filters With Bandwidth Control. *IEEE Trans. Microw. Theory Tech.* **2014**, *62*, 457–463. [CrossRef]
71. Chen, C.; Wang, G.; Li, J. Microstrip Switchable and Fully Tuneable Bandpass Filter With Continuous Frequency Tuning Range. *IEEE Microw. Wirel. Components Lett.* **2018**, *28*, 500–502. [CrossRef]
72. Ebrahimi, A.; Baum, T.; Scott, J.; Ghorbani, K. Continuously Tuneable Dual-Mode Band-stop Filter. *IEEE Microw. Wirel. Components Lett.* **2018**, *28*, 419–421. [CrossRef]
73. Kingsly, S.; Kanagasabai, M.; Alsath, M.G.N.; Shrivastav, A.K.; Subbaraj, S.; Selvam, Y.P.; Sivasamy, R.; Ramanarao, Y.V. Compact Frequency and Bandwidth Tuneable Bandpass–Band-stop Microstrip Filter. *IEEE Microw. Wirel. Components Lett.* **2018**, *28*, 786–788. [CrossRef]
74. Jeong, S.; Lee, J. Frequency- and Bandwidth-Tuneable Band-stop Filter Containing Variable Coupling Between Transmission Line and Resonator. *IEEE Trans. Microw. Theory Tech.* **2018**, *66*, 943–953. [CrossRef]
75. Ieu, W.; Zhang, D.; Lv, D.; Wu, Y. Dual-band microstrip bandpass filter with independently-tuneable passbands using patch resonator. *Electron. Lett.* **2018**, *54*, 665–667. [CrossRef]
76. Zhang, G.; Xu, Y.; Wang, X. Compact Tuneable Bandpass Filter With Wide Tuning Range of Centre Frequency and Bandwidth Using Short Coupled Lines. *IEEE Access* **2018**, *6*, 2962–2969. [CrossRef]
77. Hickle, M.D.; Peroulis, D. Theory and Design of Frequency-Tuneable Absorptive Band-stop Filters. *IEEE Trans. Circuits Syst. I* **2018**, *65*, 1862–1874. [CrossRef]
78. Lu, D.; Tang, X.; Barker, N.S.; Feng, Y. Single-Band and Switchable Dual-/Single-Band Tuneable BPFs With Predefined Tuning Range, Bandwidth and Selectivity. *IEEE Trans. Microw. Theory Tech.* **2018**, *66*, 1215–1227. [CrossRef]
79. Arain, S.; Vryonides, P.; Abbasi, M.A.B.; Quddious, A.; Antoniades, M.A.; Nikolaou, S. Reconfigurable Bandwidth Bandpass Filter With Enhanced Out-of-Band Rejection Using pi-Section-Loaded Ring Resonator. *IEEE Microw. Wirel. Components Lett.* **2018**, *28*, 28–30. [CrossRef]
80. Masood, M.H.; Suseela, S.B. Compact bandpass filter with reconfigurable X-band using stepped impedance resonator and folded structure. *J. Eng.* **2018**, *2018*, 162–165. [CrossRef]

 *electronics*

Review

Recent Developments of Reconfigurable Antennas for Current and Future Wireless Communication Systems

Naser Ojaroudi Parchin [1,*], Haleh Jahanbakhsh Basherlou [2], Yasir I. A. Al-Yasir [1], Raed A. Abd-Alhameed [1], Ahmed M. Abdulkhaleq [1,3] and James M. Noras [1]

[1] Faculty of Engineering and Informatics, School of Electrical Engineering and Computer Science, University of Bradford, Bradford BD7 1DP, UK; Y.I.A.Al-Yasir@bradford.ac.uk (Y.I.A.A.-Y.); R.A.A.Abd@bradford.ac.uk (R.A.A.-A.); A.Abd@sarastech.co.uk (A.M.A.); jmnoras@bradford.ac.uk (J.M.N.)
[2] Microwave Technology Company, P 14, Sa'adi Avenue, Vezarat Kar Street, Azadi Street, Tehran, Iran; Hale.Jahanbakhsh@gmail.com
[3] SARAS Technology Limited, Leeds LS12 4NQ, UK
*   Correspondence: N.OjaroudiParchin@Bradford.ac.uk; Tel.: +44-734-143-6156

Received: 24 December 2018; Accepted: 18 January 2019; Published: 26 January 2019

Abstract: Reconfigurable antennas play important roles in smart and adaptive systems and are the subject of many research studies. They offer several advantages such as multifunctional capabilities, minimized volume requirements, low front-end processing efforts with no need for a filtering element, good isolation, and sufficient out-of-band rejection; these make them well suited for use in wireless applications such as fourth generation (4G) and fifth generation (5G) mobile terminals. With the use of active materials such as microelectromechanical systems (MEMS), varactor or p-i-n (PIN) diodes, an antenna's characteristics can be changed through altering the current flow on the antenna structure. If an antenna is to be reconfigurable into many different states, it needs to have an adequate number of active elements. However, a large number of high-quality active elements increases cost, and necessitates complex biasing networks and control circuitry. We review some recently proposed reconfigurable antenna designs suitable for use in wireless communications such as cognitive-ratio (CR), multiple-input multiple-output (MIMO), ultra-wideband (UWB), and 4G/5G mobile terminals. Several examples of antennas with different reconfigurability functions are analyzed and their performances are compared. Characteristics and fundamental properties of reconfigurable antennas with single and multiple reconfigurability modes are investigated.

Keywords: 4G; 5G; CR; MIMO; reconfigurable antenna; switch; UWB; WiMAX; WLAN; wireless communications

---

1. Introduction

Due to their attractive advantages such as multi-band function, steerable radiation patterns and polarization diversity, which can reduce the size, complexity, and cost of an antenna while improving the total performance of a radio frequency (RF) system, reconfigurable antennas have been studied intensively in recent years [1,2]. Such antennas can facilitate multiple services in a compact structure and are good candidates for use in the future generation of mobile and wireless communication systems [3]. Active elements such as switches or capacitors enable an antenna to change its operation frequency, radiation pattern or polarization by using different techniques [4,5]. This is a wireless system incorporating a reconfigurable antenna that is able to change its operation frequency, radiation pattern, and polarization, to cope with extendable and reconfigurable services, multiple standards, and multi-mode operations. Reconfigurable antennas can offer single or multiple reconfiguration features. However, the number of reconfigurable features is strongly dependent on the number of employed active elements and increasing as the number of switches increases [6,7].

To change antenna operation, switches like RF microelectromechanical systems (MEMS), varactor diodes, and PIN diodes, as well as the optically activated switches of fiber-optic cables, are being used [8–10]. Reconfigurable antennas with RF MEMS switches have a switching speed ranging from 1–200 µsec. This is normally considered to be low for most applications [8]. An antenna combined with a varactor diode, the capacitance of which is controllable by a varying bias voltage, can have a wide tuning range. The design of the biasing network with varactor diodes has also been employed for the reconfigurable antenna design [9]. Hence, PIN diodes are preferred in reconfigurable antenna designs as they provide fast switching, and thus fast dynamic reconfiguration, with switching speeds ranging from 1–100 nsec [10]. In this paper, we present a comprehensive study of different types of reconfigurable antennas for mobile and wireless systems with simple or multiple reconfigurability functions. The paper is organized as follows: Section 2 discusses single reconfigurability in antennas. Different types of reconfigurable antennas with multiple reconfigurable features are studied in Section 3. Section 4 concludes the study.

2. Antennas with a Single Reconfigure Feature

Frequency switching, bandwidth switching, radiation pattern switching, and polarization switching are the different operations which can be performed by the reconfigurable antennas through changing the antenna size and shape of the radiators using active elements. In this section, different types of reconfigurable antennas with a single reconfigurable function are studied.

2.1. Frequency-Reconfigurable Antennas

Due to limited space available in modern wireless devices such as smartphones and tablets, the design of frequency-reconfigurable antennas requires much attention in terms of fundamental characteristics. These antennas can move dynamically from one frequency band to another or vary continuously in a range of frequencies [11–20]. They are mostly used in multi-band/multi-service systems requiring efficient reallocation of the dynamic spectrum: different types of switches are used. The most common method is to tune the electrical length of the antenna radiator using PIN diodes [11–13]. Varactor diode switches can tune the operation band in a wideband or ultra-wideband (UWB) mode and choose a desired narrowband mode [14,15]. Alternatively, the matching network has been used to set the resonance frequency [16,17]. Some other methods such as adding a shorting post on the antenna configuration or changing the input-impedance have been also proposed [18,19]. Hence, the antenna operational spectrum can be switched to the desired frequency in a variety of ways.

To help explain the theory of the frequency reconfigurable antenna, the configuration and performance of the antenna in [20] are represented in Figure 1. As illustrated, the antenna configuration is based on a composite right/left-handed transmission line (CRLH-TL), containing a shunt capacitance/inductance and series capacitance/inductance. The operation band of the antenna can be tuned by loading the varactor on the meander-line slot. The design exhibits frequency scanning in the range of 4.13–4.50 GHz at 0–36 V. The antenna provides very good frequency-reconfigurable characteristics and might be suitable for use in a phased array form for fifth generation (5G) base station applications.

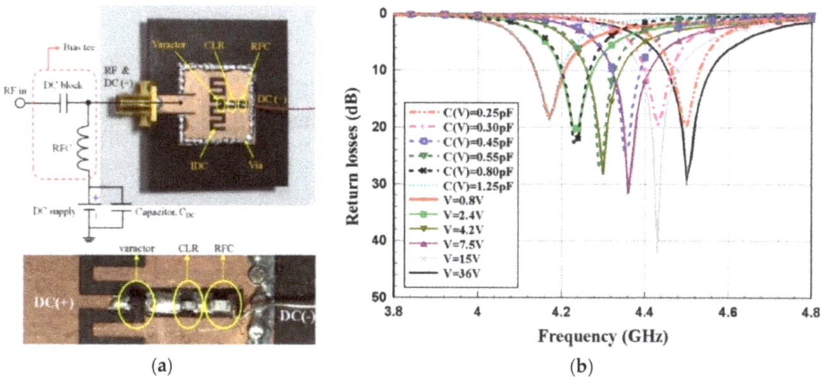

(a)   (b)

Figure 1. (**a**) Fabricated prototype of the reconfigurable antenna and (**b**) its frequency response for different voltage values [20]. Reproduced with permission from [20], Copyright IEEE, 2014.

2.2. Bandwidth-Reconfigurable Antennas

The operational bandwidth can also be changed from narrow-band to wide-band, which is usually achieved by employing a matching-network or reconfigurable-filter structure in the feed-line of the antenna [21–23]. The antenna must have good performance over the whole band of interest to cover both the narrow-band and wideband operating spectrums. Another method is to use a modified ground plane, especially for patch or monopole antennas. In [24], a bandwidth reconfigurable antenna was introduced for wireless local area network (WLAN) and worldwide interoperability for microwave access (WiMAX) applications.

The proposed antenna configuration is illustrated in Figure 2a. It has a rectangular radiation patch with a rotated F-shaped slot connected to a microstrip feed-line and a truncated ground plane in the back-side of an FR-4 substrate. A pair of PIN diodes in the middle of the slot on the radiation patch controls the current flow. By switching the operating states of the diodes, the frequency-bandwidth characteristic of the antenna can be reconfigured from narrow-band to wide-band, as illustrated in Figure 2b. The impedance bandwidth ranges from 22% to 78% which makes the antenna suitable for use in various wireless communication systems. In addition, the antenna can also have a reconfigurable characteristic to switch between a multi-resonance and wideband/UWB characteristic. This could be used to avoid interference between the UWB systems and other wireless systems such as fourth generation (4G), WLAN, WiMAX, and satellite communication systems [25]. This characteristic increases the flexibility of the antenna, combining different services in a single unit, suitable for multi-mode wireless communications.

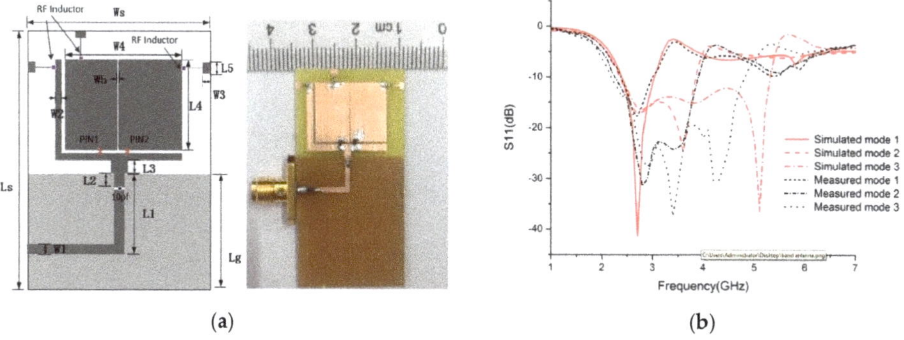

(a)   (b)

Figure 2. (**a**) Bandwidth-reconfigurable antenna configuration and (**b**) its performance for different states of the PIN diodes [24]. Reproduced with permission from [24], Copyright IEEE, 2016.

2.3. Pattern-Reconfigurable Antennas

Pattern reconfigurable antennas are attractive in applications of surveillance and tracking because they produce radiation patterns with different directivities at the same operating frequency [26,27]. Commonly, either structures are designed with the potential to produce pattern diversity or the feeding configurations are reconfigurable, an application of array theory [28,29]. We shall discuss a typical example [30]. In Figure 3a, four identical arc dipoles along with reconfigurable feeding network are shown on different sides of a Rogers 4350B substrate. By changing the states of the diodes, the antenna end-fire radiation pattern can be steered with a 90-degree difference in the azimuthal plane to cover all directions, as illustrated in Figure 3b. The reflection coefficient (S_{11}) characteristic of the antenna for the different conditions of the diodes is constant and covers the frequency range from 2.3–3.2 GHz of 4G and 5G operation bands.

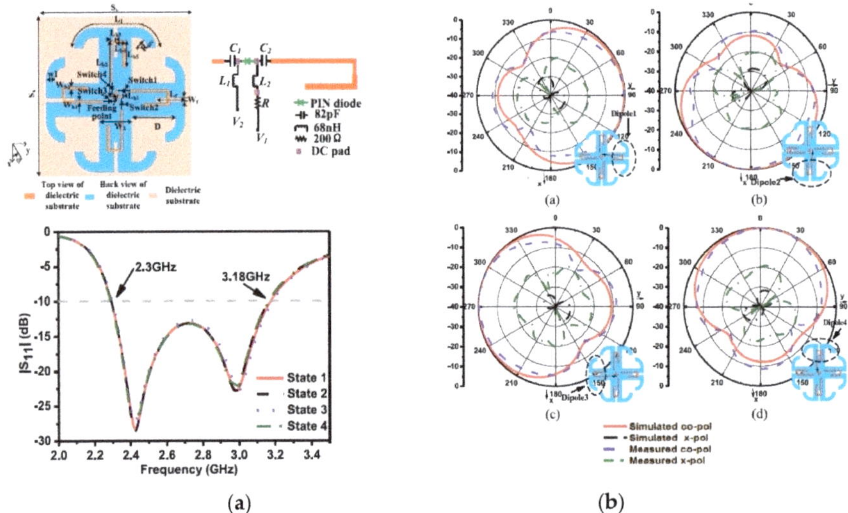

Figure 3. (a) Antenna configuration S_{11} results and (b) the radiation patterns for different diode states [30]. Reproduced with permission from [30], Copyright IEEE, 2018.

Shorting pins, capacitor or reconfigurable ground structure to change the electric field distribution in the antenna substrate are used to change the direction of antenna radiation patterns [31–33]. In [34], a pattern reconfigurable phased array design was proposed for transmission angle sensing in 28 GHz 5G mobile communication. The pattern reconfigurable antenna has been also used in multiple-input multiple-output (MIMO) wireless communications to reduce noise and improve system performance; a suitable example has been studied and its characteristics reported in [35].

2.4. Polarization-Reconfigurable Antennas

Antennas with polarization reconfigurability can offer exceptional multipath fading reduction. With this type, an antenna can be switched to different modes [36]. Most investigations in the polarization-reconfigurable antennas are concerned with switching between right-hand circular polarization (RHCP) and left-hand circular polarization (LHCP) at a desired frequency [37], although linear polarization (vertical or horizontal modes) is also studied [38,39]. Employing modified structures such as slot, slit, parasitic structures, and truncated corners of the main radiator along with embedded active elements are the most popular methods to obtain polarization-reconfigurability [40,41]. Adopting active structures such as meta-surfaces and feeding networks can also provide polarization reconfigurability [42]. Other techniques, such as using reconfigurable external polarizers or phase shifters, can also be useful [43].

In [44], a simple and new design of a polarization reconfigurable antenna with a C-shaped slot was introduced for 2.2–2.8 GHz 4G applications. Its configuration, shown in Figure 4a, consists of a circular radiation patch with a C-shaped slot and two diodes across the concentric circular slot on the radiation patch. As can be observed from Figure 4b, the antenna can switch between vertical and horizontal linear polarizations (VP/HP) modes as well as between LHCP and RHCP circularly polarized modes, and is applicable for use in 5G wireless communications. In [45,46], two designs of polarization reconfigurable patch antennas of compact size were reported for 5G wireless systems and their characteristics were investigated.

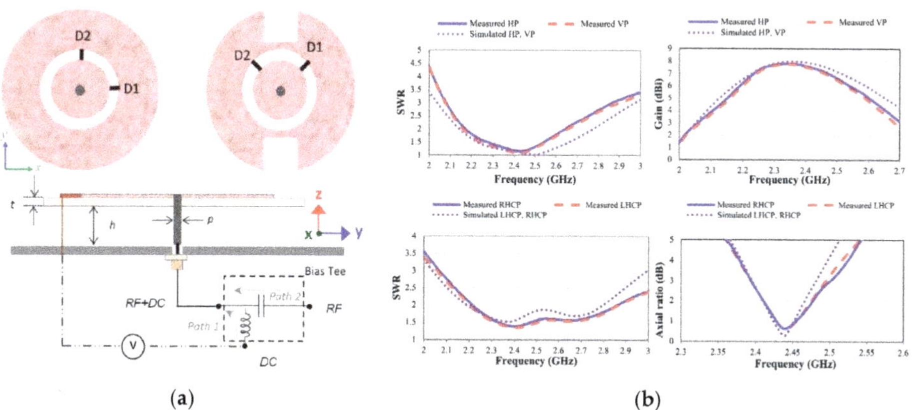

Figure 4. (**a**) Schematic and (**b**) standing wave ratio (SWR), gain, and axial ratio characteristics of the antenna [44]. Reproduced with permission from [44], Copyright IEEE, 2017.

3. Multiple Reconfigurable Features in Antenna

Multi-reconfigurable antennas can have two or more characteristics modified independently:

- frequency and bandwidth
- frequency and radiation pattern
- frequency and polarization
- radiation pattern and polarization
- frequency, radiation pattern, and polarization.

3.1. Frequency and Bandwidth Reconfigurable Antennas

There is an increasing demand in multi-service radios conforming to different spectrum standards for antennas switchable to single-band, multi-band, or wideband operation. This multi-function characteristic has become an important research area [47–55]. There are different antenna design methods which achieve this multimode function:

- to use a reconfigurable band-pass filter on the antenna feed-line to switch from narrow-band to wide-band operation [47–50].
- to use a modified ground plane with embedded parasitic and slit structures and also slots along with active elements such as PIN diodes [51–54].
- to combine different antennas which can give narrowband or wideband operation and could be used in cognitive-ratio (CR) or MIMO mobile communications [55].

Some examples are described below.

In [47–50], the main idea proposed was to integrate a switchable filtering element on the feed-line of a broadband antenna. In [47], a circular disc microstrip-fed monopole antenna with wideband to narrowband frequency reconfiguration was introduced. As shown in Figure 5a, the antenna was integrated with a reconfigurable band-pass filter in the feed-line. The impedance bandwidth could be reconfigured from wideband to narrowband using an active element, as illustrated in Figure 5b. For the narrowband state, a pair of varactor diodes has been used to tune the antenna response from 3.9 to 4.7 GHz with constant isolation. In [48], an elliptical monopole antenna with a rectangular slot in the partial ground plane was developed. The configuration of the antenna is composed of four resonators coupled to the split transmission line at the center along with the integrated reconfigurable band-pass filter (BPF) element onto the 50 Ω feed-line. The frequency performance of the antenna can be switched to operate at 1.8 GHz, 2.4 GHz, 3.5 GHz (WiMAX/5G), and 5.2 GHz (WLAN), using four PIN diodes to switch between bands. With all diodes in their OFF state, the antenna provides wide bandwidth cover of all the four bands. In [49], a new narrow-band antenna into a large ultra-wideband antenna was proposed. Its configurationis the integration of an UWB monopole antenna printed on the top layer and a reconfigurable feeding structure printed on the bottom layer of the substrate. The antenna not only exhibits UWB impedance bandwidth but also could be tuned between six narrow bands, thus the slot antenna with its reconfigurable feeding structure makes the antenna suitable for CR applications.

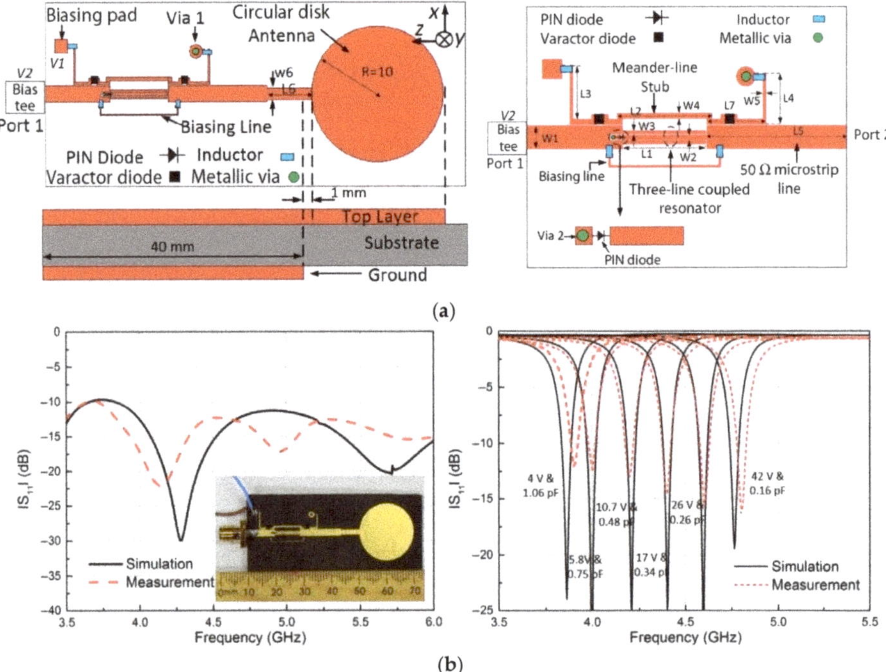

Figure 5. (**a**) Configuration and (**b**) performance of the frequency and bandwidth reconfigurable antenna in [47]. Reproduced with permission from [47], Copyright IEEE, 2015.

Another similar design was proposed in [50]. It is a microstrip-fed antenna including two resonators coupled to the feed-line. The basic antenna has UWB performance while each resonator can provide one narrowband, with the bi-bands mode achieved by activating both resonators at the same time. In [51,52], two designs of integrated reconfigurable MIMO antennas for 3G/4G/5G mobile terminals was presented: an UWB sensing antenna are integrated with reconfigurable MIMO antennas on the same substrate. In the antenna cited in [51], a pair of reconfigurable antennas with meander-line configuration, is employed on the top layer of the board. The UWB antenna on the reverse of the substrate acts as a ground plane for the reconfigurable elements. The UWB antenna exhibits a broad bandwidth of 720–3440 MHz, while the reconfigurable MIMO element can switch between different frequencies in the 573–2550 MHz range. The reconfigurable MIMO element in [52] contained modified printed inverted-F antenna (PIFA) antennas each with their digital biasing circuitry. The switching mechanism of the antenna contained four PIN diodes leading to four distinct modes between 755 and 3450 MHz.

In [53,54], single-port wide to narrowband monopole antennas with uni-planar configurations were proposed. Switching between the wideband and narrow communication bands was obtained using two PIN diodes in a ring-slot filter design in the feed-line. The proposed antenna in [53] exhibited wide-to-narrow band reconfiguration from 1.6–6.0 GHz (UWB) to 3.39–3.80 GHz (5G). A pair of varactor diodes tunes the narrowband operation from 2.55–3.2 GHz [54]. The wideband state provides around 5 GHz bandwidth (1.35–6.2 GHz). For a new combined antenna system containing an UWB antenna for spectrum sensing and a dipole antenna narrowband, communication was proposed in [55], using two feeding ports to obtain wideband coverage of 2–5.5 GHz and narrow-band resonance at 2.8 GHz. The system can cover the spectra of IEEE 802.11ac and 802.11n, universal mobile telecommunications service (UMTS) 2000, and 4G/5G mobile terminals and is suitable for use in front-end CR applications.

3.2. Frequency and Radiation-Pattern Reconfigurable Antennas

In this category, operation frequency or radiation pattern characteristics can be reconfigured [56]. The antenna radiation pattern can also be reconfigured between broad-side, end-fire, and omni-directional modes [57]; radiation pattern reconfiguration can enhance system performance, suppress noise, and save energy by improving signal directions. Frequency tuning is very useful to suppress interference from other wireless systems and also to reduce the number of the antennas required [58]. Due to this, a great number of antenna designs of this type have been reported [59–67]. The usual method in most designs is to switch the frequency operation of the antenna by using active elements, namely varactors or PIN diodes [59–63]. Pattern reconfiguration can be achieved using a modified ground plane with slot or slit structures to change the current distribution of the antenna substrate in order to steer the direction of the antenna radiation pattern [64,65], or can use switchable directors at different sides of the main antenna radiator [66,67].

In [58], a patch antenna design with dual-pattern frequency-reconfigurable characteristics was proposed. The antenna provides radiation patterns with monopolar and broadside modes and its frequency band can be switched using four varactors. A single-fed double-element array design with frequency and radiation pattern reconfigurability was introduced in [59]. The frequency operation of the antenna is tunable between 2.15–2.38 GHz. Beam scanning between $\pm23°$ was obtained using a tuning mechanism. Another frequency and pattern reconfigurable antenna with combined monopole and patch antennas was proposed in [60]. A microstrip-fed patch to resonate at a lower frequency is printed on the upper layer and the monopole to operate at a higher frequency is placed in the bottom ground of a Rogers RO4350 substrate. Five PIN diodes, divided into group A (D_1, D_2, and D_3) and group B (D_4 and D_5), were employed. By changing the states of these two groups, the antenna can be switched into different modes. With group A ON, the antenna acts as an omnidirectional monopole radiator operating at 2.4 GHz. When group B is ON, the design performs as a broad-side patch antenna working at 5.5 GHz. Having both groups OFF, the monopole mode and patch mode are active.

In [61], a reconfigurable slot antenna with a modified ground plane was reported. Two rectangular slots with six PIN diodes in the ground plane provide frequency reconfiguration with different radiation modes. The antenna can operate at eleven different frequencies of S and C band in the frequency range of 2.2–6.5 GHz. A wideband reconfigurable slot antenna for long-term evolution (4G-LTE) and C-band applications was presented in [62]. Its design contains a pair of symmetric slots fed by a fork-shaped feed-line. Pairs of PIN diodes (D_1 and D_2 and D_3 and D_4) are loaded in the slots and feed line to produce frequency and radiation pattern reconfiguration, respectively. Another design of a slot antenna with frequency and pattern reconfigurable properties was introduced in [63], where the antenna, fed by a coaxial port, is composed of four slits with three switches for tuning the scanning angle between ±15°. Two switches placed in the main slot radiator tune between three different narrow bands at 1.82 GHz, 1.93 GHz, and 2.10 GHz. BAR50–02V PIN diodes were used as switches. A flexible frequency and radiation pattern antenna design was proposed in [64]. Its schematic contains two symmetric hexagonal split rings with monopole branches and eight PIN diodes. By changing the diode states, the antenna frequency can switch between 1.9–2.4 GHz with pattern steering capability in two directions.

In [65], a back-to-back F (BTBF) semi-circular antenna was presented (see Figure 6a), composed of a patch radiator with four identical elements. Depending on the states of the embedded diodes, the antenna can work at three different frequency bands. Another design of antenna with frequency and radiation pattern selectivity, shown in Figure 6b, was introduced in [66]; it is enclosed in a circular area, with six symmetrical main radiators and twelve parasitic elements. Three circular slots are etched in a circular ground patch placed on the midline of the microstrip feed line. By changing the diode states, the antenna can operate at the frequency bands of wireless broadband (Wibro), WLAN, Zigbee, and satellite-digital multimedia broadcasting (S-DMB) communications with the capability of steering the antenna radiation pattern to six different angles in each frequency band. In addition, since six radiation elements were used in the design, the antenna pattern can be steered to different angles in 45° steps to provide full coverage.

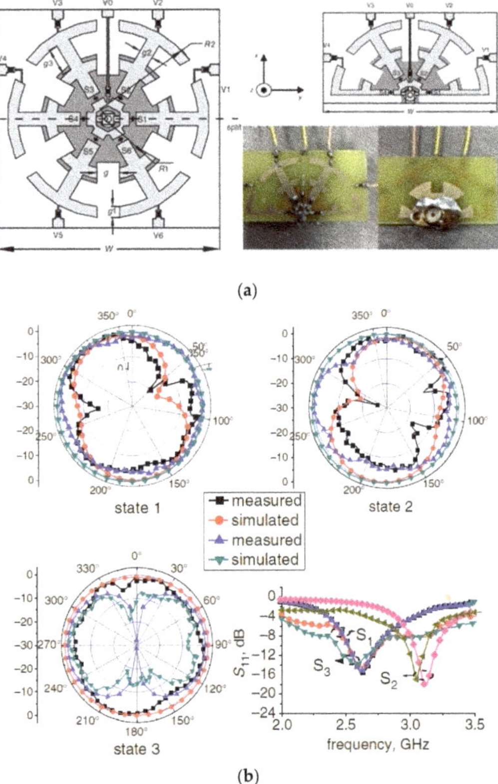

Figure 6. (**a**) Configuration and (**b**) performance of the frequency and pattern reconfigurable antenna in [65]. Reproduced with permission from [65], Copyright IEEE, 2015.

Reference [67] proposed a dipole driven element with calculated pixilation and a pair of parasitic elements to provide frequency and pattern switching. The antenna has modes at 900 MHz, 1800 MHz, or 2.1 GHz, with directional or omnidirectional radiation patterns. In [68], a novel planar antenna containing dual symmetrical radiators with rectangle parasitic elements and an inverted T-shape ground plane was proposed. By loading the parasitic elements and choosing different radiators for the feeding, the antenna can operate at three different frequencies with two different kinds of radiating pattern.

3.3. Frequency and Polarization Reconfigurable Antennas

An antenna of this type is very useful in imaging, sensing, tracking and radar applications [69]. With frequency reconfiguration, the antenna system is able to tune and make use of the available spectrum. Polarization diversity can reduce the effects of multipath and enhance channel capacity. There is a growing interest in using this type of reconfiguration combination and a number of designs have been proposed [70]. One important method makes use of an active electromagnetic band gap (EBG) or metasurface structure along with switches on top of the antenna resonator [69–72]. However, most common is the integration of methods for achieving frequency and polarization reconfigurability using active elements such as varactor diodes and PIN diodes, in addition to some other modifications [73–76]. Different antennas with frequency and polarization reconfigurable functions are reviewed below.

Two such designs were introduced in [69,70]. The configuration of the EBG in [69] composed of rectangular patch arrays on both sides of the substrate. A coplanar waveguide (CPW)-fed monopole antenna with wide bandwidth was placed above the designed EBG surface. By tuning the employed varactors in the EBG structure, the antenna can generate circular polarization at different frequencies from 1.1–1.7 GHz, and can be switched between LHCP and RHCP. For the design proposed in [70], the main radiator is a dipole antenna which can work at four different states, thanks to the reconfigurable EBG structure. By changing the diode states, the antenna frequency can be switched between 4.4–4.6 GHz. The polarization can be dynamically tuned for RHCP, LHCP, and LP.

Another design was introduced in [71]. Its structure consists of a planar slot antenna as the main radiator, and a metallic reflector with a metasurface. By adjusting the positions between the center of the slot radiator and the metasurface, frequencies from 8.0–11.2 GHz and polarization states LP, LHCP, and RHCP could be chosen. A square ring antenna patch antenna with two horizontal gaps [72] provided LP at 2.4 and 4.18 GHz, and LHPC and RHCP characteristics at 3.4 GHz. A square-ring patch radiator cut in the ground plane has a vertical gap across which two PIN diodes bridge. A stub-loaded reconfigurable patch antenna with frequency and polarization agility was proposed in [73]; see Figure 7a. Twelve loading stubs are sited at the four edges of the main radiator. As illustrated in Figure 7b, the frequency can be tuned from 2.4-3.4 GHz with diverse polarization, and so the antenna is suitable for multi-mode 4G/5G communication systems.

A simple monopole antenna with different states of frequency and polarization diversity was introduced in [74]. It includes a monopole radiator, a reflector, and a modified ground plane with two slots. A pair of switches reconfigures the antenna, allowing four states covering 2.02–2.56 GHz, 2.32–2.95 GHz, 1.92–2.70 GHz, and 1.88–2.67 GHz. The radiation is linearly polarized in states 1 and 2, and LHCP and RHCP in states 3 and 4. Another patch antenna design with a corner segment providing frequency/polarization diversity was presented in [75]. A PIN diode between the patch radiator and the embedded corner segment switches the frequency between 5.6–5.8 GHz and the polarization between LHCP and RHCP. In [76], a reconfigurable patch antenna design with dual-probe coupler feed network was presented. The proposed antenna has two layers and an active element (varactor diode) installed between the inner and outer patch on the top layer. Polarization diversity is achieved using the probe feed network. Furthermore, the antenna can tune its operation frequency from 2.05–3.13 GHz by variation of the varactor diode capacitance between 12.33–1.30 pF.

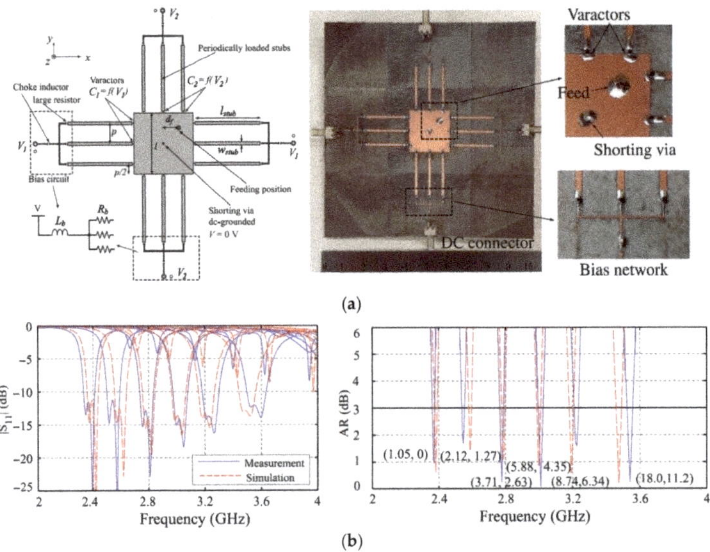

(a)

(b)

Figure 7. (**a**) Configuration and (**b**) performance of the frequency and polarization reconfigurable antenna in [73]. Reproduced with permission from [73], Copyright IEEE, 2015.

3.4. Radiation Pattern and Polarization Reconfigurable Antennas

Radiation pattern and polarization reconfigurable antennas offer benefits such as providing pattern diversity and supporting different polarizations on a single element antenna radiator [77]. They also can improve the capacity of communication systems, increase signal power, improve radiation coverage, and avoid polarization mismatching. Radiation pattern reconfigurability can steer the antenna pattern to satisfy system needs [78], and can improve channel capacity without increasing a radiator's volume [79,80]. Many such antennas have been proposed recently [77–84].

An omnidirectional microstrip-fed patch antenna with a switchable polarization and radiation pattern was introduced in [77]. A pair of back-to-back coupled patches fed at a corner, and a shared ground plane. A phase shifting technique provides a high degree of polarization and radiation pattern reconfigurability. An antenna with switchable radiation patterns and polarization diversity was proposed in [78]. The design concept is to place reconfigurable parasitic elements around the main radiator. Each parasitic element includes a dipole radiator with a PIN diode, whereby the radiation pattern can be switched into different directions by changing the diode states. Furthermore, the antenna can be dynamically reconfigured into three different polarizations. In [79], a PIFA was proposed with radiation pattern and polarization diversity; the main radiator has an inverted F-shaped parasitic structure. Using only one diode, the antenna can switch the antenna pattern and polarization properties in the frequency range of 2.357–2.562 GHz. In [80], a reconfigurable antenna design with switchable radiation pattern and polarization diversity for the band of 2.45–2.65 GHz was introduced. The antenna includes three circular substrate disks for radiator, feeding and switching along with four plastic screws to support the substrates, with an annular slot structure etched in the circular patch radiator. The reconfigurable feed network with one input and four output ports and with L-shaped feeding probes are placed in the middle layer which includes three power dividers. The antenna can provide switchable radiation beams for broadside and conical circular polarized modes.

In [81], a design of a reconfigurable antenna with a metasurface structure was proposed. The overall structure has three layers. The metasurface, to improve the antenna gain, was on the top layer, with the ground plane and the microstrip feed-line on the middle and bottom layers respectively. The antenna is designed to work at 5 GHz, with switchable polarization using the PIN diodes in the ground plane. In addition, the antenna radiation pattern can be steered between $\pm20°$ by controlling the PIN switched in the metasurface. A 0.9 GHz reconfigurable quadrifilar helical antenna with a switchable feeding network was proposed in [82]. Figure 8 shows the cuboid radiator, reconfigurable feeding-network, biasing network, and the fabricated prototype. By switching the diode states, the antenna can provide four different modes with polarization and radiation pattern reconfiguration. It has a low profile and provides wide bandwidth with broad axial ratio characteristics and LHCP/RHCP. Another pattern and polarization reconfigurable antenna was presented in [83]. The radiator is a square patch with four coupled parasitic structures fed by a coaxial cable. Using PIN diodes, the antenna provides three different modes for polarization and pattern diversity at one frequency. A multi-layer wideband reconfigurable circular polarized (CP) antenna with four radiating arms was proposed in [84]. The radiating arms are connected to a reconfigurable feeding network with PIN diodes to provide wideband CP waves in RHCP and LHCP modes. The radiation pattern can be switched from bidirectional to broadside modes and the antenna's placement above a metallic reflector improves its gain. The frequency range is 1.00–2.25 GHz with an axial ratio bandwidth of 1.5–1.9 GHz, thus suitable for global positioning system (GPS), radio-frequency identification (RFID), and mobile communications.

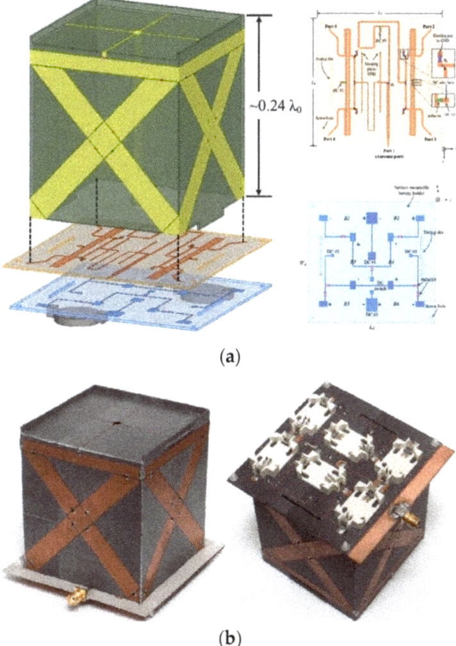

(a)

(b)

Figure 8. The configurations of the (**a**) designed and (**b**) fabricated pattern and polarization reconfigurable antenna [82]. Reproduced with permission from [82], Copyright IEEE, 2018.

3.5. Frequency, Radiation Pattern, and Polarization Reconfigurable Antennas

None of the above designs provide more than one or two reconfigurable features. In the literature there is only one design, proposed in [85], for independently three-parameter reconfiguration. In [86], another design of frequency, radiation pattern, and polarization reconfigurable antenna was proposed, but with limitations in terms of tuning, and, unlike the design in [85], it cannot provide simultaneous, independent reconfiguration. This is a major research gap, albeit a much harder challenge to design an antenna with the flexibility to reconfigure all its fundamental properties in terms of frequency, radiation pattern, and polarization. In the following, the design of [85] is explored.

Figure 9 shows the antenna schematics and the prototype [85], a patch antenna radiator and a parasitic pixel surface with 66 pixels and 60 PIN diodes. It is the switched-grid pixel surface that enables the antenna to be reconfigured simultaneously in frequency, radiation pattern, and polarization. Different configurations of the switches tune the frequency of operation from 2.4–3 GHz. The antenna radiation pattern can be steered between ±30°, and its polarization can be reconfigured between LHCP/RHCP and LP. The design is suitable for different mobile and wireless communication systems.

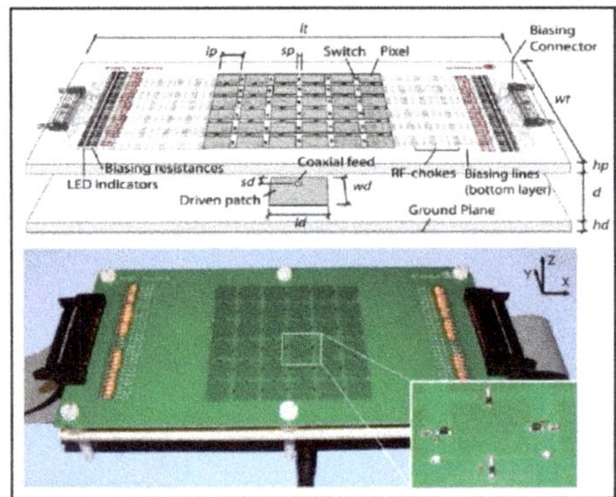

Figure 9. Schematic of the multi-functional reconfigurable antenna [85]. Reproduced with permission from [85], Copyright IEEE, 2014.

Table 1 provides a comparative summary of recently proposed reconfigurable antenna types published in the literature. Fundamental characteristics in terms of antenna type, reconfiguration type and the number of the employed switches, operation frequency, and the overall size of the antennas are provided and have been compared. This is intended to help readers to determine what types of reconfigurable antennas are most suitable for their applications.

Table 1. Comparative summary of reconfigurable antennas described in the literature.

References	Antenna Type	Reconfiguration Type	Reconfiguration Means	Frequency (GHz)	No. of Switches	Antenna Size (mm)
[14]	Monopole	Frequency	PIN-Diode	1.77–2.51	4	50 × 120
[15]	Slot	Frequency	PIN-Diode	1.73–2.28	2	50 × 80
[21]	Patch	Bandwidth	PIN-D/MEMS	2–4	8	12 × 12
[23]	Slot	Bandwidth	PIN-Diode	1.64–2.68	6	10 × 64
[31]	DRA	Rad. Pattern	PIN-Diode	5.8	8	50 × 50
[32]	Patch	Rad. Pattern	PIN-Diode	2.45	8	95.5 × 100
[37]	Bow-Tie	Polarization	PIN-Diode	1.3–1.85	4	58 × 58
[40]	Patch	Polarization	PIN-Diode	2.1–2.6	24	60 × 60
[48]	Monopole	F/B	PIN/varactor	3.9–4.82	3	25 × 75
[52]	Monopole	F/B	PIN/varactor	0.72–3.44	4	65 × 120
[53]	Monopole	F/B	PIN-Diode	3.4–8/4.7–5.4	4	70 × 70
[59]	Patch	F/R	varactor Diode	2.15/2.25/2.38	2	151.5 × 160.9
[62]	Slot	F/R	PIN-Diode	3.6/3.95	2	30 × 40
[63]	Slot	F/R	PIN-Diode	1.8/1.9/2.05	14	130 × 160
[69]	Monopole	F/P	varactor Diode	1–1.6	112	88 × 114
[72]	Patch-Slot	F/P	PIN-Diode	2.4/3.4/4.1	2	60 × 65
[74]	Monopole	F/P	PIN-Diode	1.9–2.7	2	40 × 70
[76]	Patch	F/P	varactor Diode	2–3.1	1	100 × 100
[79]	PIFA	R/P	PIN-Diode	2.5	1	50 × 50
[81]	Slot	R/P	PIN-Diode	5	36	78 × 78
[82]	Helical	R/P	PIN-Diode	0.9	32	81 × 81
[85]	Patch	F/R/P	PIN-Diode	2.4–3	60	31 × 31

4. Conclusions

In this paper, an exhaustive review of reconfigurable antennas with single and multiple reconfiguration features has been presented, classifying the various types of reconfigurable antenna. Several design techniques are discussed in detail with examples, and the fundamental properties of different reconfigurable antenna types are set out. That the complexity of an antenna increases with the number of reconfigurable parameters is clearly stressed in this review. In addition, the presentation highlights a design requirement: the ability to design an antenna with simultaneous and independently flexible reconfigurations of antenna frequency, radiation patterns, and polarization characteristics is not yet achieved, and needs more research and investigation.

Author Contributions: Conceptualization, N.O.P. and H.J.B.; methodology, N.O.P. and H.J.B.; software, N.O.P. and H.J.B.; validation, N.O.P., H.J.B., Y.I.A.A.-Y., R.A.A.-A., and A.M.A.; formal analysis, N.O.P., H.J.B., Y.I.A.A.-Y., R.A.A.-A., and A.M.A.; investigation, N.O.P., H.J.B., and R.A.A.-A.; resources, N.O.P., H.J.B., and R.A.A.-A.; data curation, N.O.P., H.J.B., and R.A.A.-A.; writing—original draft preparation, N.O.P., H.J.B., and J.M.N.; writing—review and editing, N.O.P., H.J.B., R.A.A.-A., and J.M.N.; visualization, N.O.P., R.A.A.-A., and J.M.N.

Funding: This project has received funding from the European Union's Horizon 2020 research and innovation programme under grant agreement H2020-MSCA-ITN-2016 SECRET-722424.

Acknowledgments: Authors wish to express their thanks to the support provided by the innovation programme under grant agreement H2020-MSCA-ITN-2016 SECRET-722424.

Conflicts of Interest: The authors declare no conflict of interest.

References

1. Nadeem, Q.-U.-A.; Kammoun, A.; Debbah, M.; Alouini, M.-S. Design of 5G full dimension massive MIMO systems. *IEEE Trans. Commun.* **2018**, *66*, 726–740. [CrossRef]
2. Bhartia, P.; Bahl, I.J. Frequency agile microstrip antennas. *Microw. J.* **1982**, *25*, 67–70.
3. Bernhard, J.T. *Reconfigurable Antennas*; Morgan & Claypool Publishers: San Rafael, CA, USA, 2007.
4. Peroulis, D.; Sarabandi, K.; Kateh, L.P.B. Design of reconfigurable slot antennas. *IEEE Trans. Antennas Propag.* **2005**, *53*, 645–654. [CrossRef]
5. Christodoulou, C.G.; Tawk, Y.; Lane, S.A.; Erwin, S.R. Reconfigurable antennas for wireless and space applications. *Proc. IEEE.* **2012**, *100*, 2250–2261. [CrossRef]
6. Balanis, C.A. *Antenna Theory Analysis and Design*, 4th ed.; John Wiley & Sons: New York, NY, USA, 1998.
7. Bernhard, J.T. *Reconfigurable Antennas*; Wiley: New York, NY, USA, 2005.
8. Grau, J.R.; Lee, M.J. A dual-linearly-polarized MEMS-reconfigurable antenna for NB MIMO communication systems. *IEEE Trans. Antennas Propag.* **2010**, *58*, 4–17. [CrossRef]
9. Tawk, Y.; Costantine, J.; Christodoulou, C.G. A varactor based reconfigurable filtenna. *IEEE Antennas Wirel. Propag. Lett.* **2012**, *11*, 716–719. [CrossRef]
10. Nikolaou, S.; Bairavasubramanian, R.; Lugo, C.; Carrasquillo, I.; Thompson, D.C. Pattern and frequency reconfigurable annular slot antenna using PIN diodes. *IEEE Trans. Antennas Propag.* **2006**, *54*, 439–448. [CrossRef]
11. Parchin, N.O.; Al-Yasir, Y.; Abdulkhaleq, A.M.; Elfergani, I.; Rayit, A.; Noras, J.M.; Rodriguez, J.; Abd-Alhameed, R.A. Frequency reconfigurable antenna array for mm-Wave 5G mobile handsets. In Proceedings of the 9th International Conference on Broadband Communications, Networks, and Systems, Faro, Portugal, 19–20 September 2018.
12. Ojaroudi, N.; Parchin, N.O.; Ojaroudi, Y.; Ojaroudi, S. Frequency reconfigurable monopole antenna for multimode wireless communications. *Appl. Comput. Electromagn. Soc. J.* **2014**, *8*, 655–660.
13. Ojaroudi, N.; Ghadimi, N.; Ojaroudi, Y.; Ojaroudi, S. A novel design of microstrip antenna with reconfigurable band rejection for cognitive radio applications. *Microw. Opt. Technol. Lett.* **2014**, *56*, 2998–3003. [CrossRef]
14. Hussain, R.; Sharawi, M.S.; Shamim, A. An integrated four-element slot-based MIMO and a UWB sensing antenna system for CR platforms. *IEEE Trans. Antennas Propag.* **2018**, *66*, 978–983. [CrossRef]
15. Hussain, R.; Khan, M.U.; Sharawi, M.S. An integrated dual MIMO antenna system with dual-function GND-plane frequency-agile antenna. *IEEE Antennas Wirel. Propag. Lett.* **2018**. [CrossRef]

16. Hinsz, L.; Braaten, B.D. A frequency reconfigurable transmitter antenna with autonomous switching capabilities. *IEEE Trans. Antennas Propag.* **2014**, *62*, 3809–3813. [CrossRef]

17. Hannula, J.-M.; Saarinen, T.; Holopainen, J.; Viikari, V. Frequency reconfigurable multiband handset antenna based on a multichannel transceiver. *IEEE Trans. Antennas Propag.* **2017**. [CrossRef]

18. Nguyen-Trong, N.; Piotrowski, A.; Fumeaux, C. A frequency reconfigurable dual-band low-profile monopolar antenna. *IEEE Trans. Antennas Propag.* **2017**, *65*, 3336–3343. [CrossRef]

19. Boukarkar, A.; Lin, X.Q.; Jiang, Y.; Chen, Y.J.; Nie, L.Y.; Mei, P. Compact mechanically frequency and pattern reconfigurable patch antenna. *IET Microw. Antennas Propag.* **2018**, *12*, 1864–1869. [CrossRef]

20. Sam, S.; Kang, H.; Lim, S. Frequency reconfigurable and miniaturized substrate integrated waveguide interdigital capacitor (SIW-IDC) antenna. *IEEE Trans. Antennas Propag.* **2014**, *62*, 1039–1045. [CrossRef]

21. Anand, S.; Raj, R.K.; Sinha, S.; Upadhyay, D.; Mishra, G.K. Bandwidth reconfigurable patch antenna for next generation wireless communication system applications. In Proceedings of the International Conference on Emerging Trends in Communication Technologies (ETCT), Dehradun, India, 18–19 November 2016.

22. Anagnostou, E.; Torres, D.; Sepulveda, N. Vanadiom dioxde switch for a reconfigurable bandwdith antenna. In Proceedings of the Loughborough Antennas & Propagation Conference (LAPC 2017), Loughborough, UK, 13–14 November 2017.

23. Meng, L.; Wang, W.; Gao, J.; Liu, Y. Bandwidth reconfigurable antenna with three step-shaped slots. In Proceedings of the Sixth Asia-Pacific Conference on Antennas and Propagation (APCAP), Xi'an, China, 16–19 October 2017.

24. Chen, S.-Y.; Chu, Q.-X.; Shinohara, N. A bandwidth reconfigurable planar antenna for WLAN/WiMAX applications. In Proceedings of the Asia-Pacific Microwave Conference (APMC), New Delhi, India, 5–9 December 2016.

25. Horestani, A.K.; Shaterian, Z.; Naqui, J.; Martín, F.; Fumeaux, C. Reconfigurable and tunable s-shaped split-ring resonators and application in band-notched UWB antennas. *IEEE Trans. Antennas Propag.* **2016**, *64*, 3766–3776. [CrossRef]

26. Abdulraheem, Y.I.; Abdullah, A.S.; Mohammed, H.J.; Mohammed, B.A.; Abd-Alhameed, R.A. Design of radiation pattern-reconfigurable 60-GHz antenna for 5G applications. *J. Telecommun.* **2014**, *52*, 1–5.

27. Zhong-Liang, L.; Xue-Xia, Y.; Guan-Nan, T. A wideband printed tapered-slot antenna with pattern reconfigurability. *IEEE Antennas Wirel. Propag. Lett.* **2014**, *13*, 1613–1616.

28. Chen, S.L.; Qin, P.Y.; Lin, W.; Guo, Y.J. Pattern-reconfigurable antenna with five switchable beams in elevation plane. *IEEE Antennas Wirel. Propag. Lett.* **2018**, *17*, 454–457. [CrossRef]

29. Tang, M.-C.; Zhou, B.Y.; Duan, Y.; Chen, X.; Ziolkowski, R.W. Pattern-reconfigurable, flexible, wideband, directive, electrically small near-field resonant parasitic antenna. *IEEE Trans. Antennas Propag.* **2018**, *66*, 2271–2280. [CrossRef]

30. Jin, G.; Li, M.; Liu, D.; Zeng, G. A simple planar pattern-reconfigurable antenna based on arc dipoles. *IEEE Antennas Wirel. Propag. Lett.* **2018**, *17*, 1664–1668. [CrossRef]

31. Zhong, L.; Hong, J.-S.; Zhou, H.-C. A novel pattern-reconfigurable cylindrical dielectric resonator antenna with enhanced gain. *IEEE Antennas Wirel. Propag. Lett.* **2016**, *15*, 1253–1256. [CrossRef]

32. Lu, Z.-L.; Yang, X.-X.; Tan, G.-N. A multidirectional pattern reconfigurable patch antenna with CSRR on the ground. *IEEE Antennas Wirel. Propag. Lett.* **2017**, *16*, 416–419. [CrossRef]

33. Tran, H.H.; Nguyen-Trong, N.; Le, T.T.; Abbosh, A.M.; Park, H.C. Low-profile wideband high-gain reconfigurable antenna with quad-polarization diversity. *IEEE Trans. Antennas Propag.* **2018**, *66*, 3741–3746. [CrossRef]

34. Zhang, J.; Zhang, S.; Lin, X.; Fan, Y.; Pedersen, G.F. 3D radiation pattern reconfigurable phased array for transmission angle sensing in 5G mobile communication. *Sensors* **2018**, *18*, 4204. [CrossRef]

35. Chamok, N.H.; Yılmaz, M.H.; Arslan, A.; Ali, M. High-gain pattern reconfigurable MIMO antenna array for wireless handheld terminals. *IEEE Trans. Antennas Propag.* **2016**, *64*, 4306–4315. [CrossRef]

36. Wu, F.; Luk, K.M. Widband tri-polarization reconfigurable magnetoelectric dipole antenna. *IEEE Trans. Antennas Propag.* **2017**, *65*, 1633–1641. [CrossRef]

37. Lin, W.; Chen, S.L.; Ziolkowski, R.W.; Guo, Y.J. Reconfigurable, wideband, low-profile, circularly polarized antenna and array enabled by artificial magnetic conductor ground. *IEEE Trans. Antennas Propag.* **2018**, *66*, 1564–1569. [CrossRef]

38. Tong, K.-F.; Huang, J. New proximity coupled feeding method for reconfigurable circularly polarized microstrip ring antennas. *IEEE Trans. Antennas Propag.* **2008**, *56*, 1860–1866. [CrossRef]

39. Wang, K.X.; Wong, H. A reconfigurable CP/LP antenna with cross probe feed. *IEEE Antennas Wirel. Propag. Lett.* **2017**, *16*, 669–672. [CrossRef]

40. Cai, Y.-M.; Gao, S.; Yin, Y.; Li, W.; Luo, Q. Compact-size low-profile wideband circularly polarized omnidirectional patch antenna with reconfigurable polarizations. *IEEE Trans. Antennas Propag.* **2016**, *64*, 2016–2021. [CrossRef]

41. Wu, B.; Okoniewski, M.; Hayden, C. A pneumatically controlled reconfigurable antenna with three states of polarization. *IEEE Trans. Antennas Propag.* **2014**, *62*, 5474–5484. [CrossRef]

42. Zhu, H.L.; Cheung, S.W.; Liu, X.H.; Yuk, T.I. Design of polarization reconfigurable antenna using metasurface. *IEEE Trans. Antennas Propag.* **2014**, *62*, 2891–2898. [CrossRef]

43. Row, J.-S.; Hou, M.-J. Design of polarization diversity patch antenna based on a compact reconfigurable feeding network. *IEEE Trans. Antenna Propag.* **2014**, *62*, 5349–5352. [CrossRef]

44. Mak, K.M.; Lai, H.W.; Luk, K.M.; Ho, K.L. Polarization reconfigurable circular patch antenna with a C-shaped. *IEEE Trans. Antennas Propag.* **2017**, *65*, 1388–1392. [CrossRef]

45. Al-Yasir, Y.; Abdullah, A.; Ojaroudi Parchin, N.; Abd-Alhameed, R.; Noras, J. A new polarization-reconfigurable antenna for 5G wireless communications. In Proceedings of the 9th International Conference on Broadband Communications, Networks, and Systems, Faro, Portugal, 19–20 September 2018.

46. Al-Yasir, Y.I.A.; Abdullah, A.S.; Ojaroudi Parchin, N.; Abd-Alhameed, R.A.; Noras, J.M. A new polarization-reconfigurable antenna for 5G applications. *Electronics* **2018**, *7*, 293. [CrossRef]

47. Qin, P.Y.; Wei, F.; Guo, Y.J. A wideband-to-narrowband tunable antenna using a reconfigurable filter. *IEEE Trans. Antennas Propag.* **2015**, *63*, 2282–2285. [CrossRef]

48. Kingsly, S.; Thangarasu, D.; Kanagasabai, M.; Alsath, M.G.; Thipparaju, R.R.; Palaniswamy, S.K.; Sambandam, P. Multiband reconfigurable filtering monopole antenna for cognitive radio applications. *IEEE Antennas Wirel. Propag. Lett.* **2018**, *17*, 1416–1420. [CrossRef]

49. Hussain, R.; Sharawi, M.S. An integrated frequency reconfigurable antenna for cognitive radio application. *Radioengineering* **2017**, *26*, 746–754.

50. Bitchikh, M.; Rili, W.; Mokhtar, M. An UWB to narrow band and BI-bands reconfigurable octogonal antenna. *Prog. Electromagn. Res. Lett.* **2018**, *74*, 69–75. [CrossRef]

51. Hussain, R.; Sharawi, M.S. A cognitive radio reconfigurable MIMO and sensing antenna system. *IEEE Antennas Wirel. Propag. Lett.* **2015**, *14*, 257–260. [CrossRef]

52. Hussain, R.; Sharawi, M.S. Integrated reconfigurable multiple-input–multiple-output antenna system with an ultra-wideband sensing antenna for cognitive radio platforms. *IET Microw. Antennas Propag.* **2015**, *9*, 940–947. [CrossRef]

53. Chacko, P.; Augustin, G.; Denidni, T.A. Electronically reconfigurable uniplanar antenna for cognitive radio applications. *IET Microw. Antennas Propag.* **2015**, *14*, 213–216.

54. Nachouane, H.; Najid, A.; Tribak, A.; Riouch, F. Reconfigurable and tunable filtenna for cognitive LTE femtocell base stations. *Int. J. Microw. Sci. Technol.* **2016**. [CrossRef]

55. Nachouane, H.; Najid, A.; Tribak, A.; Riouch, F. Dual port antenna combining sensing and communication tasks for cognitive radio. *Int. J. Electron. Telecommun.* **2016**, *62*, 121–127. [CrossRef]

56. Purisima, M.C.L.; Salvador, M.; Augstin, S.G.P.; Cunanon, M.T. Frequency and pattern reconfigurable antennas for community cellular application. In Proceedings of the IEEE Conference TENCON, Singapore, 22–25 November 2016.

57. Nguyen-Trong, N.; Hall, L.; Fumeaux, C. A frequency- and pattern-reconfigurable center-shorted microstrip antenna. *IEEE Antennas Wirel. Propag. Lett.* **2016**, *15*, 1955–1958. [CrossRef]

58. Trong, N.N.; Hall, L.; Fumeaux, C. A dual-band dual-pattern frequency-reconfigurable antenna. *Microw. Opt. Tech. Lett.* **2017**, *59*, 2710–2715. [CrossRef]

59. Zainarry, S.N.M.; Nguyen-Trong, N.; Fumeaux, C. A frequency and pattern-reconfigurable two-element array antenna. *IEEE Antennas Wirel. Propag. Lett.* **2018**, *17*, 617–620. [CrossRef]

60. Li, P.K.; Shao, Z.H.; Wang, Q.; Cheng, Y.J. Frequency- and pattern reconfigurable antenna for multistandard wireless applications. *IEEE Antennas Wirel. Propag. Lett.* **2015**, *14*, 333–336. [CrossRef]

61. Sahu, N.K.; Sharma, A.K. An investigation of pattern and frequency reconfigurable microstrip slot antenna using PIN diodes. In Proceedings of the 2017 Progress in Electromagnetics Research Symposium–Spring (PIERS), St. Petersburg, Russia, 22–25 May 2017; pp. 971–976.

62. Han, L.; Wang, C.; Zhang, W.; Ma, R.; Zeng, Q. Design of frequency- and pattern-reconfigurable wideband slot antenna. *Int. J. Antennas Propag.* **2018**, 1–7. [CrossRef]

63. Majid, H.A.; Rahim, M.K.A.; Hamid, M.R.; Ismail, M.F. Frequency and pattern reconfigurable slot antenna. *IEEE Trans. Antennas Propag.* **2014**, *62*, 5339–5343. [CrossRef]

64. Zhu, Z.; Wang, P.; You, S.; Gao, P. A flexible frequency and pattern reconfigurable antenna for wireless systems. *Prog. Electromagn. Res. Lett.* **2018**, *76*, 63–70.

65. Ye, M.; Gao, P. Back-to-back F semicircular antenna with frequency and pattern reconfigurability. *Electron. Lett.* **2015**, *51*, 2073–2074. [CrossRef]

66. Pan, J.Y.; Ma, Y.; Xiong, J.; Hou, Z.; Zeng, Y. A compact reconfigurable microstrip antenna with frequency and radiation pattern selectivity. *Microw. Opt. Tech. Lett.* **2015**, *57*, 2848–2854. [CrossRef]

67. Li, W.; Bao, L.; Zhai, Z.; Li, Y.; Li, S. An enhanced frequency and radiation pattern reconfigurable antenna for portable device applications. In Proceedings of the 31st International Review of Progress in Applied Computational Electromagnetics (ACES), Williamsburg, VA, USA, 22–26 March 2015.

68. Pan, Y.; Ma, Y.; Xiong, J.; Hou, Z.; Zeng, Y. A compact antenna with frequency and pattern reconfigurable characteristics. *Microw. Opt. Tech. Lett.* **2017**, *59*, 2467–2471.

69. Liang, B.; Sanz-Izquierdo, B.; Parker, E.A.; Batchelor, J.C. A frequency and polarization reconfigurable circularly polarized antenna using active EBG structure for satellite navigation. *IEEE Trans. Antennas Propag.* **2015**, *63*, 33–40. [CrossRef]

70. Huang, Y.C.; Ma, X.; Pan, W.; Luo, X. A low profile polarization reconfigurable dipole antenna using tunable electromagnetic band-gap surface. *Microw. Opt. Tech. Lett.* **2014**, *56*, 1281–1285.

71. Chen, M.S.; Zhang, Z.X.; Wu, X.L. Design of frequency-and polarization-reconfigurable antenna based on the polarization conversion metasurface. *IEEE Antennas Wirel. Propag. Lett.* **2018**, *17*, 78–81.

72. Niture, V.; Govind, P.A.; Mahajan, S.P. Frequency and polarisation reconfigurable square ring antenna for wireless application. In Proceedings of the 2016 IEEE Region, 10 Conference (TENCON), Singapore, 22–25 November 2016.

73. Nguyen-Trong, N.; Hall, L.; Fumeaux, C. A frequency- and polarization-reconfigurable stub-loaded microstrip patch antenna. *IEEE Trans. Antennas Propag.* **2015**, *63*, 5235–5240. [CrossRef]

74. Liu, J.; Li, J.; Xu, R. Design of very simple frequency and polarisation reconfigurable antenna with finite ground structure. *Electron. Lett.* **2018**, *54*, 187–188. [CrossRef]

75. Rahman, M.A.; Nishiyama, E.; Toyoda, I. A frequency diversity reconfigurable antenna with circular polarization switching capability. In Proceedings of the 2017 IEEE International Symposium on Antennas and Propagation & USNC/URSI National Radio Science Meeting, San Diego, CA, USA, 9–14 July 2017; pp. 1367–1368.

76. Kumar, M.S.; Choukiker, Y.K. Frequency and polarization reconfigurable antenna using BLC feed network. In Proceedings of the 2017 IEEE International Conference on Antenna Innovations & Modern Technologies for Ground, Aircraft and Satellite Applications (iAIM), Bangalore, India, 24–26 November 2017.

77. Narbudowicz, A.; Bao, X.L.; Ammann, M.J. Omnidirectional microstrip patch antenna with reconfigurable pattern and polarization. *IET Microw. Antennas Propag.* **2014**, *8*, 872–877. [CrossRef]

78. Gu, C.; Gao, S.; Liu, H.; Luo, Q.; Loh, T.-H.; Sobhy, M.; Li, J.; Wei, G.; Xu, J.; Qin, F.; et al. Compact smart antenna with electronic beam-switching and reconfigurable polarizations. *IEEE Trans. Antennas Propag.* **2015**, *63*, 5325–5333. [CrossRef]

79. Yang, K.; Loutridis, A.; Bao, X. Printed inverted-F antenna with reconfigurable pattern and polarization. In Proceedings of the 10th European Conference on Antennas and Propagation (EuCAP), Davos, Switzerland, 10–15 April 2016.

80. Lin, W.; Wong, H.; Ziolkowski, R.W. Circularly-polarized antenna with reconfigurable broadside and conical beams facilitated by a mode switchable feed network. *IEEE Trans. Antennas Propag.* **2018**, *66*, 996–1001. [CrossRef]

81. Chen, A.; Ning, X.; Wang, L.; Zhang, Z. A design of radiation pattern and polarization reconfigurable antenna using metasurface. In Proceedings of the 2017 IEEE Asia Pacific Microwave Conference (APMC), Kuala Lumpar, Malaysia, 13–16 November 2017.

82. Yi, X.; Huitema, L.; Wong, H. Polarization and pattern reconfigurable cuboid quadrifilar helical antenna. *IEEE Trans. Antennas Propag.* **2018**, *66*, 2707–2715. [CrossRef]

83. Trong, N.N.; Mobashsher, A.T.; Abbosh, A.M. Reconfigurable shorted patch antenna with polarization and pattern diversity. In Proceedings of the 2018 Australian Microwave Symposium (AMS), Brisbane, Australia, 6–7 February 2018.

84. Lin, W.; Wong, H. Polarization reconfigurable wheel-shaped antenna with conical-beam radiation pattern. *IEEE Trans. Antennas Propag.* **2015**, *63*, 491–499. [CrossRef]

85. Rodrigo, D.; Cetiner, B.A.; Jofre, L. Frequency, radiation pattern and polarization reconfigurable antenna using a parasitic pixel layer. *IEEE Trans. Antennas Propag.* **2014**, *62*, 3422–3427. [CrossRef]

86. Selvam, Y.P.; Elumalai, L.; Alsath, G.; Kanagasabai, M.; Kingsly, S.; Subbburaj, S. Novel frequency- and pattern-reconfigurable rhombic patch antenna with switchable polarization. *IEEE Antennas Wirel. Propag. Lett.* **2017**, *16*, 1639–1642. [CrossRef]

Review

Recent Developments of Dual-Band Doherty Power Amplifiers for Upcoming Mobile Communications Systems

Ahmed M. Abdulkhaleq [1,2,*], Maan A. Yahya [3], Neil McEwan [1], Ashwain Rayit [1],
Raed A. Abd-Alhameed [2,*], Naser Ojaroudi Parchin [2], Yasir I. A. Al-Yasir [2] and James Noras [2]

[1] SARAS Technology Limited, Leeds LS12 4NQ, UK; Neil.McEwan@sarastech.co.uk (N.M.);
ashwain.rayit@sarastech.co.uk (A.R.)

[2] School of Electrical Engineering and Computer Science, Faculty of Engineering and Informatics,
University of Bradford, Bradford BD7 1DP, UK; N.OjaroudiParchin@bradford.ac.uk (N.O.P.);
Y.I.A.Al-Yasir@bradford.ac.uk (Y.I.A.A.-Y.); jmnoras@bradford.ac.uk (J.N.)

[3] Computer Systems Department, Ninevah Technical Institute, Northern Technical University, Mosul 41001,
Iraq; dr.maan@ntu.edu.iq

* Correspondence: A.ABD@sarastech.co.uk (A.M.A.); R.A.A.Abd@bradford.ac.uk (R.A.A.-A.)

Received: 15 April 2019; Accepted: 2 June 2019; Published: 6 June 2019

Abstract: Power amplifiers in modern and future communications should be able to handle different modulation standards at different frequency bands, and in addition, to be compatible with the previous generations. This paper reviews the recent design techniques that have been used to operate dual-band amplifiers and in particular the Doherty amplifiers. Special attention is focused on the design methodologies used for power splitters, phase compensation networks, impedance inverter networks and impedance transformer networks of such power amplifier. The most important materials of the dual-band Doherty amplifier are highlighted and surveyed. The main problems and challenges covering dual-band design concepts are presented and discussed. In addition, improvement techniques to enhance such operations are also exploited. The study shows that the transistor parasitic has a great impact in the design of a dual-band amplifier, and reduction of the transforming ratio of the inverter simplifies the dual-band design. The offset line can be functionally replaced by a Π-network in dual-band design rather than T-network.

Keywords: dual-band Doherty power amplifier; LTE-advanced; high-efficiency; phase offset lines; impedance inverter network; phase compensation network

---

1. Introduction

The demands for increasing the amount of data that can be transmitted within a limited bandwidth is continuing to grow rapidly, especially with developments, where users are now being attracted by multimedia data and video streaming, as well as the Internet of Things technology revolution. Hence, the 5G mobile generation will include several technologies that can help to achieve its promised goals. Some of these technologies are: beamforming, carrier aggregation, massive multiple input multiple output (MIMO), and more complex modulation schemes, which produce a high peak to average power ratio (PAPR). The high PAPR requires the power amplifier to be backed off from the most efficient point into a region where the efficiency drops sharply to keep the linearity requirements of any communications standard [1–3]. Working in the back-off region of the power amplifier means that less efficiency will be obtained, where a large amount of supplied power will be converted into heat [1]. With the simple amplifier, there is inevitably a trade-off between high efficiency and linearity. Modern power amplifiers should be designed to produce high efficiency at a large output power back-off (OBO); known efficiency enhancements techniques include: Doherty power amplifier (DPA), envelope elimination and restoration (EER), envelope tracking (ET), and linear amplification using nonlinear components (LINC), and Chireix out-phasing. However, the simplest technique is the Doherty amplifier, where neither signal processing blocks nor additional controlling circuits are required [4–7]. Also, new terms have entered the communications systems which are multi-band and multi-mode, where the term "multi-band" refers to a transmitter which works on two different frequency bands simultaneously; in this case, the number of devices, size, and cost will be reduced. On the other hand, the term "multi-mode" refers to a transmitter who can support and combine different access technologies such as 2G, 3G, 4G and 5G on a single platform [8–13]. To design a dual-band power amplifier, a hybrid configuration can be used but occupies a large area. Ideally, the power amplifier should be designed to handle multi-band and multi-mode systems concurrently.

This paper, following a brief introduction to the Doherty power amplifier mechanism, reviews the techniques that have been used in designing dual-band Doherty power amplifiers, and finally draws conclusions from this.

2. Classical Doherty Power Amplifier Operation

In 1936, W.H. Doherty invented a new combiner designed for broadcasting stations using high power tube amplifiers [5]. A λ/4 transmission lines can be used as a combiner at the output of power amplifiers to achieve a linear output power. The classic DPA consists of two amplifiers known as the carrier (main) amplifier and the auxiliary (peaking) amplifier (Figure 1). A class AB amplifier used for the carrier amplifier whereas a class C amplifier is used for the peaking amplifier. The RF input signal is split between the two amplifiers, where the carrier amplifier is working all the time and should almost reach saturation at the back-off input power due to seeing a high impedance which causes a change in the load-line as shown in Figure 2. At the same power level, the auxiliary amplifier works only in the Doherty region and starts feeding current to the output till it becomes saturated at the peak region, where the two power amplifiers give their maximum designed output power.

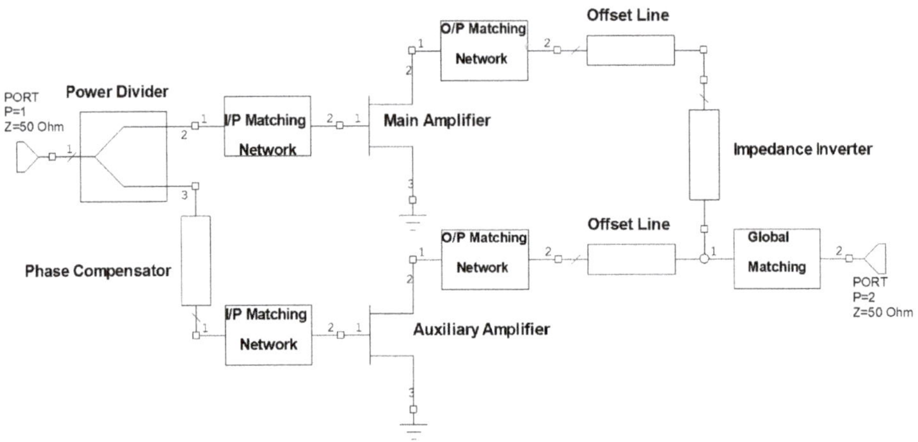

Figure 1. Structure of the Doherty power amplifier.

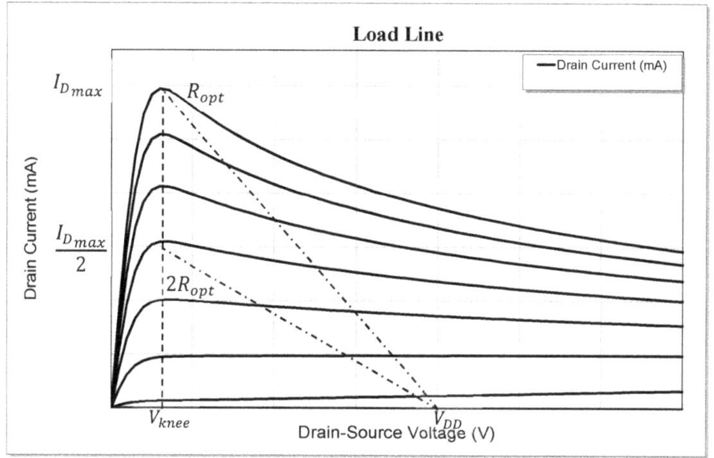

Figure 2. Main amplifier load line.

The idea of the Doherty amplifier depends on the so-called active load-pull technique [1]. Where the DPA operation can be divided into three regions:

The low RF input power region, where the signal level is not sufficient to turn the auxiliary amplifier on, in this case, the auxiliary amplifier can (theoretically) be represented as an open circuit as shown in Figure 3a. At the same time, the main amplifier is amplifying the input signal as an ordinary power amplifier, however, the load is seen by the main amplifier through the impedance inverter (λ/4 transmission line) and is increased because the characteristic impedance of the λ/4 transmission line is higher than the load impedance. In this case, the main amplifier will be almost saturated because its load line has changed, as illustrated in Figure 2. The impedance seen by the main amplifier depends on the following equation:

$$Z_1 = \frac{Z_T^2}{R_L} \tag{1}$$

where:

Z_1: the impedance seen by the main amplifier
Z_T: transmission line characteristic impedance
R_L: the load

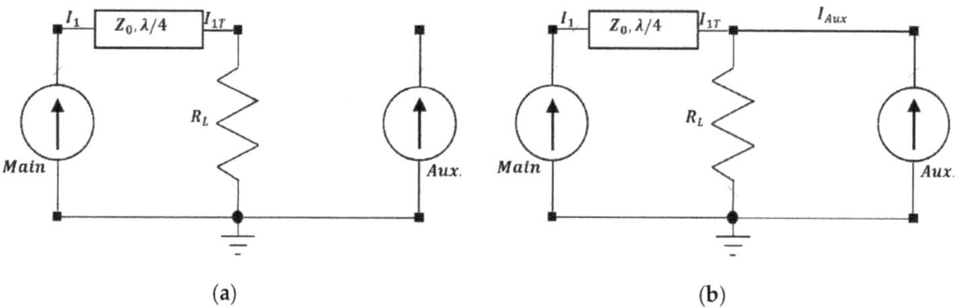

Figure 3. Doherty equivalent circuit diagram (**a**) at low power region and (**b**) at medium and high-power region.

The second region (medium RF input power), where the auxiliary amplifier starts feeding current to the load and acts as an additional current source as illustrated in Figure 3b. As auxiliary amplifier current increases, the apparent load impedance seen by the impedance inverter at the summing node will increase, and hence, the impedance seen by the main amplifier will decrease, and the load line will move as shown in Figure 2. As a result, the output voltage of the main amplifier remains roughly constant, and the total current is increasing which increases the total output power. The following equations show the relationship between the impedance of each amplifier and the amplifiers' currents:

$$Z_2 = R_L\left(1 + \frac{I_{1T}}{I_{Aux}}\right) \tag{2}$$

$$Z_1 = \frac{Z_T^2}{R_L\left(1 + \frac{I_{Aux}}{I_{1T}}\right)} \tag{3}$$

where:

Z_2: the impedance seen by the auxiliary amplifier

I_{1T}: the current after the $\lambda/4$ transmission line

I_{Aux}: current of the auxiliary amplifier

Finally, the high-power region, where both amplifiers work at their maximum output current and the impedance seen by each amplifier is controlled also by Equations (2) and (3). The load modulation occurs in the last two regions, where the benefit of Doherty structure is clear.

The main and the auxiliary current and voltage behavior is shown in Figure 4. The auxiliary amplifier starts contributing its current near the OBO point, whereas the main amplifier voltage remains roughly constant after the OBO point but its current increases.

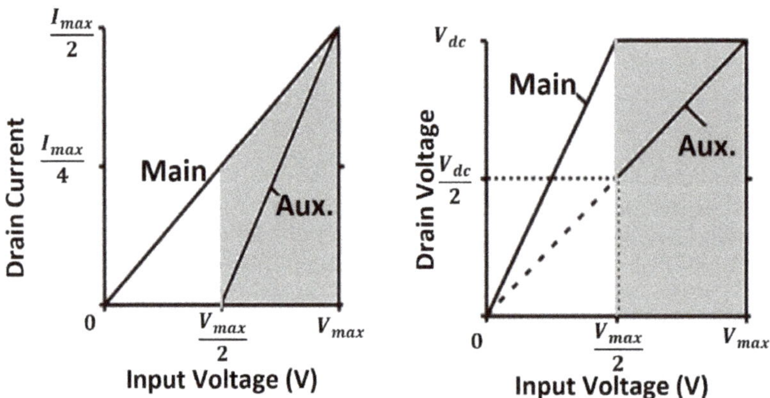

Figure 4. Voltage and current behavior of Doherty amplifier [10].

3. Dual-Band Doherty Power Amplifier

Wireless communications systems have different standards and requirements where each standard can use a certain frequency band. At the same time, mobile communication systems are being developed where the data rate is one of the main targets for each generation; the growth of data rate requires a complex modulation and more bandwidth [11–15]. However, due to the crowded spectrum, the upcoming generation may use more frequency bands in order to fulfil the targeted requirements (e.g., 5G spectrum consists of three bands are 700 MHz, 3.6 GHz and 26 GHz) [16], in this case, the devices of the new generation should be designed to deal with these frequency bands in addition to the compatibility with the previous mobile generations.

Moreover, a new 3GPP protocol which is narrowband Internet-of-Things (NB-IoT) specifies the maximum output power to be at least 23 dBm for long-range communications [17]. This protocol supports the two frequency bands 699–915 MHz and 1710–1980 MHz; the demand for integrating power amplifiers that can support high power and high efficiency on a chip is increasing since it can increase the battery life and reduce the running cost [17].

In order to increase modulation bandwidth efficiency, carrier aggregation technology was introduced, where new wireless communications systems technologies are expected to use multi-band multi-standard applications [18]. To support dual-band simultaneously, two power amplifiers can be used for each band, but this increases the die area and requires more circuitry to combine their outputs. Therefore, dual-band design can provide a reduction in the size and cost of the system, and Doherty can provide good efficiency [19]. The use of dual-band also requires the PA to work in conjunction with software-defined radio (SDR) [20], which can use the same hardware at two different frequencies to perform a similar function.

The term dual-band can be implemented in two ways; the first is called reconfigurable dual-band amplifier, in this case, the power amplifier is designed to work in two frequency bands, one at a time, where the circuit properties and configuration is changed depending on the target band. On the other hand, the concurrent dual-band implies that the amplifier can work on both frequency bands without any modification or changing of the design; intermodulation distortion should be considered especially if the two bands are close to each other because the effect of device nonlinearity will be significant, in this case, some interference will be introduced to the neighboring channels. The next subsections will deal with dual-band limitations, challenges, and the techniques that were used to make the Doherty power amplifier works in two bands.

3.1. Dual-Band DPA Problems and Challenges

The main problem for multi-band/multi-standard features is the RF-front end-stage [21,22], where RF front-end components such as power amplifiers, filters, antennas, and switches should be designed to work properly on the specified frequency bands to fulfil each standard specification.

According to [23], as long as there are dual-band components, it is possible to design a multiband DPA. Hence, theoretically, the DPA can be divided into sub-circuits to be analyzed and designed separately. These sub-circuits are the input power splitter, power amplifier design, output power combiner, and the offset lines.

The input power splitter sub-circuit should be designed to achieve the same performance in both bands; a good solution is to use either a branch line coupler (BLC) or a Wilkinson divider, where both types can divide the input power equally or unequally. The main difference between the two types of splitters is the phase difference at the output; where the phase of the two output at Wilkinson will be same, for the branch line the phase difference between the outputs will be 90° [24]. Unfortunately, both types are narrow band solutions because they are frequency-dependent. To provide a dual-band power divider, a broadband power divider can be used to cover the dual-band and ensure that the matching condition is satisfied.

Power amplifier design sub-circuit includes the device cell, input, and output matching networks; the device cell should be capable of working properly at each designed frequency band, which depends on the device characteristics as well as its designed input and output matching networks. In general, the input matching network is responsible for controlling the gain over the required band, whereas the output matching network is responsible for choosing the output power and efficiency. The simulation in [25] showed that the effect of the second harmonic at the input of the PA is negligible so that the input matching network can be designed only for the fundamental frequency at each band. At the same time, the third harmonic also has little effect on the output matching network; hence, the output matching design can consider only the fundamental frequency and the second harmonic at each band [25]. Nevertheless, modern design for matching networks does not depend on the fundamental frequency matching only; further harmonic effects can be included to achieve a certain performance.

Moreover, by taking the source-pull and load-pull data with the harmonic effects of a transistor, the required impedance for the transistor can be matched so that good performance can be achieved.

The main limitations of the DPA is the $\lambda/4$ quarter wavelength (impedance inverter) and the offset lines which limit the bandwidth due to their natural characteristics [25–30], so that to design a DPA that can work in dual-band, the quarter-wave impedance inverter and the offset line should be replaced with more complex networks which give equivalent functions in split bands.

3.2. Π-And T-Networks for Dual-Band DPA

The classic DPA has two simple $\lambda/4$ transformers: the impedance inverter and global matching, as shown in Figure 1. These two transmission lines are different in their purpose of work. A simple $\lambda/4$ transformer (global matching) can be replaced with a two-section transmission line form which is easily adjusted to give the exact impedance transformation at two frequencies, as shown in Figure 5 (more details of dual-band transformer can be found in [28]). The inherently larger bandwidth of the two sections can be maximized when no matched operation is required between the two operating bands. However, the two-section form cannot be assumed as an impedance inverter, due to the different of work purpose; instead, the impedance inverting section is realized for split band operation as a Π-network by adding two stub sections to a single near-$\lambda/4$ section [25].

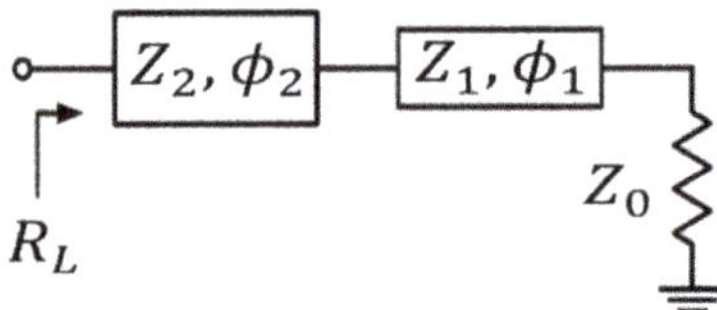

Figure 5. Dual-band global matching network.

In a comprehensive passive design of a dual-band DPA (1.8 GHz and 2.4 GHz) presented by [25], the design of each amplifier depended on the load-pull/source-pull simulation to find optimum load and source impedances. For dual-band harmonic control, two shunt stub lines in addition to (TL$_1$ and TL$_2$) are used as shown in Figure 6. The first is a short circuit terminated with a length of $\lambda/4$ at f2, where it used to cancel the effect of the second harmonic at the upper band at point A, whereas the second with a length of $\lambda/8$ at f1 is open to suppress the second harmonic of the lower band at point B. The designed amplifier of [25] was showing good performance.

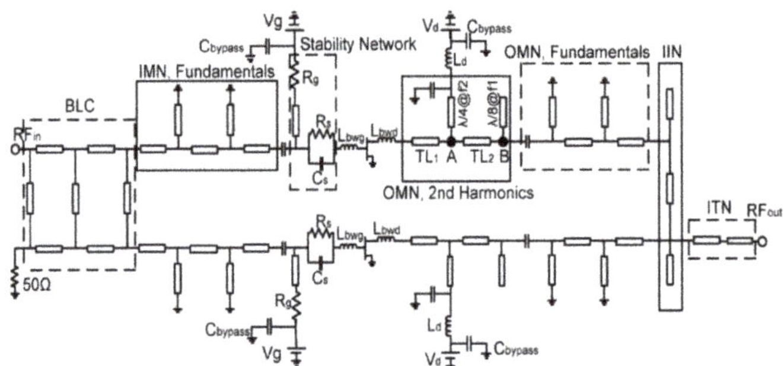

Figure 6. Dual-band Doherty power amplifier (DPA) schematic designed by [25].

Additionally, a T-network was used in a dual-band DPA working at 900 MHz and 2000 MHz in [7] as an alternative component for the $\lambda/4$ inverter, as shown in Figure 7. In the output matching network, a T-network with a shorted shunt line in addition to an open stub were used [7]; the open stub was used to adjust the phase of output, whereas the T-network with a shorted shunt was used as a dual-band impedance transformer and as a dc feeder at the same time.

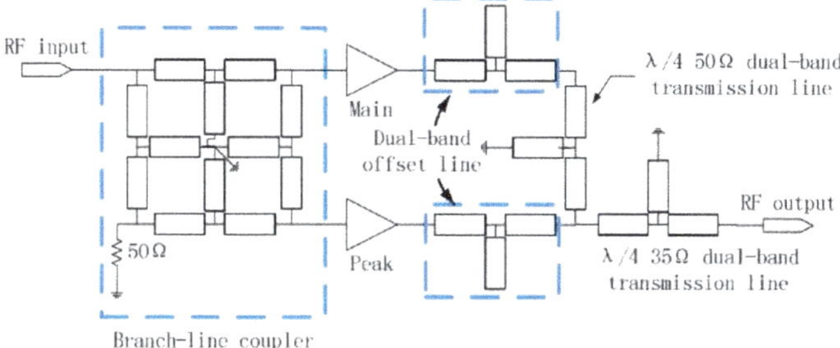

Figure 7. Dual-band DPA schematic [7].

The suggested solutions of using Π- or T-topology in [7,25] were only used for impedance inverters, nevertheless, a modified Π-network is employed as an output combiner [18], which eliminates the need for using an offset line by absorbing the output parasitic of both the main and auxiliary amplifiers; three networks are used in the dual-band design as illustrated in Figure 8, the first is a short-circuited Π-network at the main path, which works as dual-band impedance inverter, and double Π-networks at the auxiliary path work as a −180°/0° impedance transformer. The linearizability of the design in [18] is enhanced by providing small low-frequency impedances for both amplifiers; The shunt shorted-stub of the Π-network helps to reduce the effect of output parasitic capacitor of both amplifiers, so that the bandwidth limitations will be reduced; the shorted stubs can also be used as a dc feeder for both transistors.

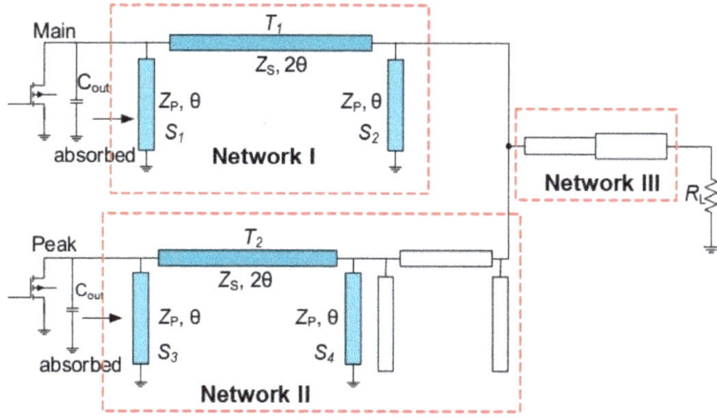

Figure 8. Schematic dual-band circuit proposed by [18].

Besides, Dual-band DPA attempt targeting 2.14 GHz and 3.5 GHz was done by [29]. It explained alternative components that can work as dual-band input power splitter, impedance inverter network, and phase compensation network. In general, a cascade of two transmission line sections is used for the impedance transformer network; however, as noted, this configuration cannot be used as an impedance inverter network because the relationship between the voltage of one side does not depend on the current of the other side, so that, Π-network or T-network is used for input power splitter and impedance inverter network. The attempted designed dual-band DPA in [29] was found to work only in the lower band due to an incorrect estimate of the auxiliary output capacitance C_{ds}, where a low impedance was seen from the common node towards the auxiliary amplifier, and in this case, one of the Doherty operating conditions were not satisfied.

Furthermore, dual-band three stage DPA was designed by [30]. The design was targeting 1.7 GHz and 2.4 GHz using three devices of 10 W Cree transistors so that 9 dB back-off can be achieved; the R_{opt} for the main, auxiliary 1 and auxiliary 2 power amplifier are 100 Ω, 50 Ω, and 50 Ω respectively. The phase offset lines and the quarter wave transformer are designed and realized using Π-type. The designed PA was tested using wideband code division multiple access (WCDMA) and worldwide interoperability for microwave access (WiMAX) signals with 7 dB PAPR, where WCDMA signal had a bandwidth of 3.5 MHz and the WiMAX had a 5 MHz bandwidth. DPD was used to improve the linearity of the system.

Form the discussed approaches; it can be noticed that both Π- and/or T-networks can be used as an equivalent circuit for the impedance inverter, at the same time, the shorted end of either network can be used as a dc feeding path for the designed power amplifier.

3.3. Offset Lines in Dual-Band Doherty Power Amplifier

The offset lines (Figure 1) are one of the major contributors to the bandwidth limitation in a DPA so that, the offset line should be designed to work in both bands, the offset lines circuit in [25] was implemented using the same circuit that was used for the final impedance transforming network (two-section transformer), where the same phase can be achieved at both frequencies.

Another approach of designing a dual-band offset line is using either T- or Π-networks. The most notable issue in designing offset lines using T- and Π-networks is that the phase difference between the input and output of these networks at different frequency bands, where it is generally a ±90°. Table 1 summarizes the differences at the two operating frequencies.

Table 1. Phase delay introduced by T- and Π-networks [29].

	Dual-Band		Single Band
	T-Topology	**Π-Topology**	**Both**
S_{21}Phase@f_1	−90°	−90°	90°
S_{21}Phase@f_2	+90°	−90°	

As can be seen from Table 1, the T-network is difficult to use in a dual-band DPA due to the different phase response in the two bands. Hence only the Π-network can be used as a dual-band line offset. The primary purpose of the offset line is to absorb the parasitic effect of the main and the auxiliary amplifier, so that as the parasitic of the transistor increase, a longer offset line is required, a solution for reducing the length of the offset line is to reduce the effect of the transistor parasitic as was done by [18], where the use of offset lines was eliminated by absorbing the transistors' output parasitic; at the input side, an unequal power divider was realized using a three-section Wilkinson coupler to compensate for the lower transconductance of the auxiliary amplifier. The harmonic termination of both amplifiers was carefully designed to avoid Smith chart sensitive regions. Since the reactive part of the output impedance of the auxiliary amplifier is not the same for the two band frequencies, there is a need to design an offset line that can produce different phases to increase the impedance so that there will be no power leakage toward the auxiliary amplifier when it is off.

3.4. Impedance Modification for Dual-Band DPA

According to [22], the design will be physically large in size if Π- or T-network topology is used for impedance inversion because the ratio between the upper and lower band will affect the size of the network. However, a compact design approach was used in [22], for a dual-band DPA where the ratio of the impedance transformation was 1:2.85 instead of the traditional ratio of 1:4, and in this case the common load impedance will be increased to 1.4 × Ropt/2. Two transistors used in this design were 10 W and 25 W gallium nitride high electron mobility transistors (GaN HEMTs), where the main amplifier output matching network was designed to be matched to 100 Ω, whereas the gate of the main, the gate, and drain of the auxiliary amplifier are designed to match 50 Ω, to reduce the phase difference between the main and the auxiliary amplifier due to the difference in the biasing voltage, the offset line was used at the input of the main amplifier. A small offset line was used at the output of the auxiliary amplifier to reduce the leakage from the main amplifier towards the auxiliary amplifier. Moreover, a higher drain supply voltage of 36 V was used for the auxiliary amplifier in order to increase its contributed power and to improve the load modulation [22]. It can be noticed that reducing the impedance inversion ratio of the quarter-wavelength will reduce the bandwidth limitations.

Moreover, a novel dual-band DPA was designed in [31], as shown in Figure 9, where the main concept is to use two quarter-wavelength transmission lines at the output of the auxiliary amplifier and only one at the output of the main amplifier, where the impedances are 70.7 Ω and 50 Ω in the auxiliary output path and 70.7 Ω in the main output path. The main purpose of these additional lines is to improve the broadband performance compared to the traditional DPA, where only one quarter-wavelength line is used at the output of the main amplifier. The improvement was made by reducing the impedance transformation ratio, as shown in the following equations, so that the loaded Q factor will be reduced to increase the band covered. In the classical DPA, the impedance transformation ratio is 1:4 to transfer 25 Ω to 100 Ω so that the loaded Q factor will be:

$$Q_L = \sqrt{\frac{100\ \Omega}{25\ \Omega} - 1} = 1.73 \tag{4}$$

whereas, in the design of [31], the impedance ratio is 1:2 making the loaded Q factor equal to 1:

$$Q_L = \sqrt{\frac{100\ \Omega}{50\ \Omega} - 1} = 1 \tag{5}$$

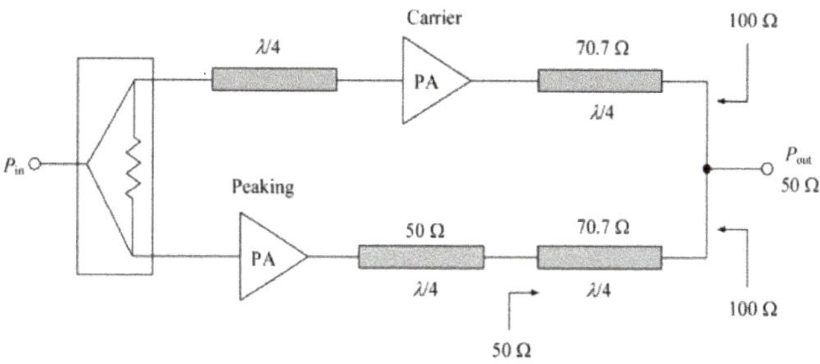

Figure 9. Modified Doherty amplifier proposed by [31].

Hence, there is an increment of 1.73 times in terms of frequency bandwidth. At the back-off region, the main amplifier will see 100 Ω due to using the 70.7 Ω quarter wavelength line, whereas both amplifiers would see a 50 Ω at the saturation point.

Besides, a dual-band Doherty amplifier design using class-J amplifier was studied in [32], where the imperfect high impedance seen towards the peaking amplifier at the back-off region was utilized to be part of the output matching of the main amplifier at the fundamental frequency matching. Whereas the control of harmonics was carried out using a post matching network after examining the load pull data of the main amplifier to determine the location of the second harmonic.

Shao et al. [19] modified the structure of a DPA so that they eliminated the effect of the impedance inverter at the output of classical DPA, and in addition used four transmission line sections (two in series and two in shunt), as shown in Figure 10, in order to make the proposed DPA works in dual-band according to the equations proposed by Chuang [33]. They suggested matching the main amplifier to 50 Ω and the auxiliary amplifier to 100 Ω. In this case, at low input power, the main and auxiliary amplifiers will see 50 Ω and ∞ Ω respectively, whereas at the high-power region, they will see 25 Ω and

100 Ω respectively. Two simplified designs for dual-Band DPA were done by [34], where the designs were tested, the main advantage of these new structures is the impedance inverter at the output power of DPA is eliminated as shown in Figure 11 [34]. This was done by changing matching the auxiliary output impedance to 100 Ω instead of 50 Ω. In this case, only one quarter-wavelength impedance inverter will be needed at the output.

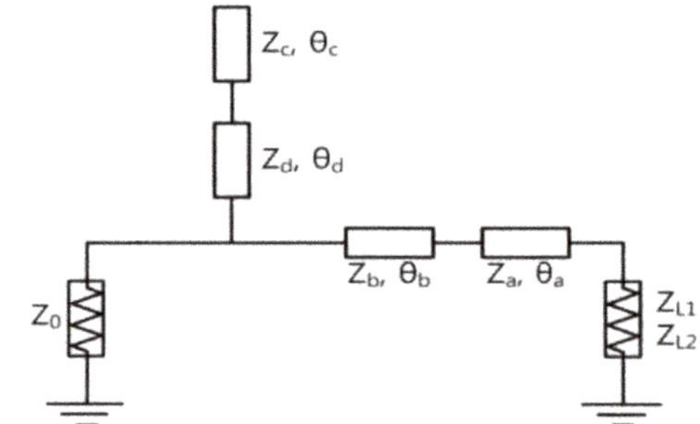

Figure 10. Dual-band matching structure used in [19].

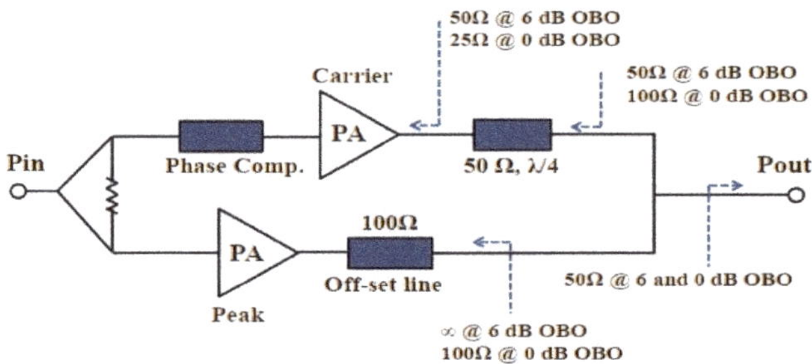

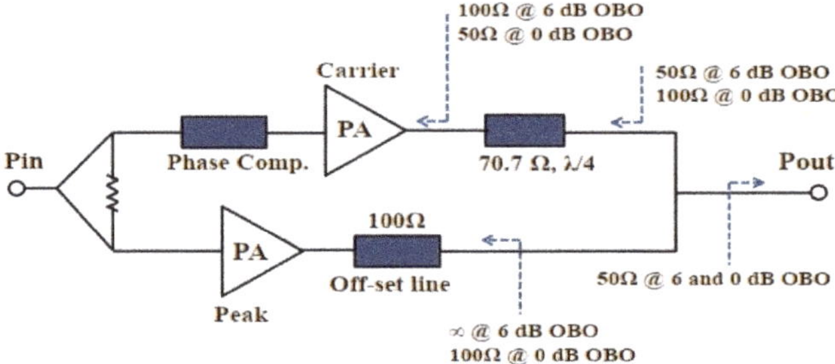

Figure 11. Simplified structure of DPA [34].

Monolithic microwave integrated circuit (MMIC) was used to design dual-band Doherty amplifier [14,35–37], because the power ratio between the main and the auxiliary amplifier which determines the back-off of the DPA depends on the availability of devices in the market; however, this can be ignored when MMIC technology is used. The linearity of the DPA was improved in [35] by cancelling the effects of the third-order intermodulation (IM3) by using a tunable capacitor at the input to make sure that the phase difference between IM3 of the main amplifier and IM3 of the auxiliary amplifier is equal to 180°, and their IM3 magnitudes are same.

In addition, MMIC dual-band multi-mode power amplifier was designed by [36], where at the lower band (3.5 GHz), the amplifiers work as Doherty amplifier; whereas at the upper band (5.8 GHz) both amplifiers biased as class AB. It should be noted that the load seen by both amplifiers are modified to be Z_0 instead of $Z_0/2$ for the purposing of reducing the transformation ratio. At the same time, high and low pass filters were used at the input side of both the main and peaking amplifier respectively to compensate the phase difference between the amplifiers due to the frequency band change.

3.5. Dual-Band Design Improvements

The linearity of the DPA can be improved by employing the ET technique at the gate of each transistor [38], where the effect of the third order modulation can be cancelled. In addition, varactor diodes have also been used adaptively for improving the bandwidth [39,40]; these techniques have very complex circuity. Moreover, dual-band reconfigurable Doherty amplifier was introduced by [21] where a PIN diode was used adaptively in to remove the reactance components of the load each frequency band. There was an enhancement about 5% of power added efficiency (PAE) compared with a traditional DPA, at the same time, the diode was used to switch between the two bands, where the state of the diode determines the selected band. Moreover, a large back-off compared to previous works been achieved in [23,41,42], which describe a dual-band multiway DPA.

A tri-way DPA was designed to cover the two bands 0.9 GHz and 2.31 GHz using an unequal power divider as shown in Figure 12, where three Wilkinson power dividers are used to produce a quad-way power divider, the first power divider was split the input power unequally, whereas the next two power splitter split the power equally in order to avoid adjusting the auxiliary gate bias by showing the importance of power splitter to manipulate the gain of the auxiliary amplifier that varies depending on the frequency [43]. The saturated designed power was 46 dBm, whereas 61% and 44% efficiency at 9 dB back-off power were achieved at both bands respectively.

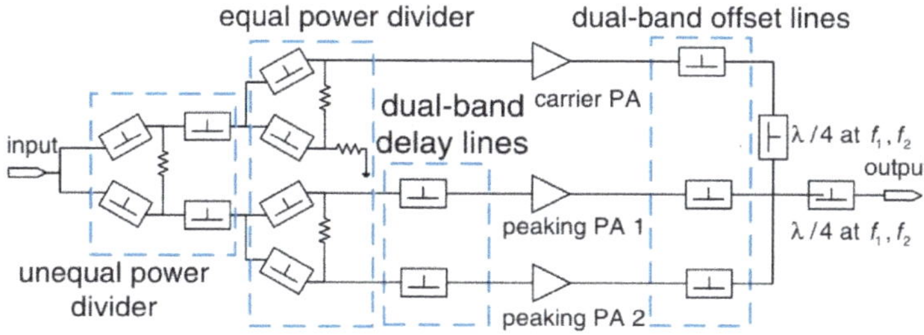

Figure 12. Tri-way Dual-band DPA with frequency dependent power division [43].

It can be noticed that the amount of output current of the auxiliary amplifier is less than for the main amplifier due to the lower gain of the auxiliary amplifier which is deeply biased; in this case, less efficient load-modulation will occur at a high-power level. For this reason, an appropriate selection of the power ratio should be chosen to maximize the load-modulation effects. Depending on the probability density function of a signal with a non-constant envelope, i.e., it has a substantial value of PAPR, the output signal spends most of the time near to the average output power. Hence, the power amplifier should be designed to produce its highest efficiency near the average power.

Adaptive input power control for a dual-band DPA was used in [44], which relied on the adaptive power splitter invented by Nick et al. [45]. The adaptive input can be used to deliver more power to the main amplifier than the auxiliary amplifier when the DPA is working in the low power region and vice versa. The main idea of [45] is to make the auxiliary amplifier turn on later so that, at the OBO, early load modulation can be avoided and after that, more power will be delivered to the auxiliary amplifier to compensate the low gain due to the biasing condition and improving the load modulation.

Extending the bandwidth of a dual-band of DPA using a novel output combiner was realized in [46] where a resonant shunt LC tank was added at the combining node to maintain the main amplifier back-off impedance and reactance compensation by reducing the reactance variation with frequency, however, this technique is limited by the LC band. In the work presented by [46], the 20 W peak power

DPA used two GaN HEMT devices with different capabilities, where a 10 W device was used for the main amplifier and operated as class F, whereas a 25 W device was used for the auxiliary amplifier and biased as class C. This amplifier showed a PAE of 42% when the dual-band input contained long term evolution (LTE) and WCDM signals concurrently with 15 MHz bandwidth for each.

In [47,48], the main and auxiliary power amplifiers operating classes was changed to class ABJ, where harmonics control was employed. The main thing to note is the linearity measurements in terms of adjacent channel power ratio (ACPR) were less than 25 dBc for both bands. New technology elements were used in the designs of [49,50], where micro-electro-mechanical switches (MEMS) were used to select the operating dual-band of quarter wave transmission line, where the switches move the operation center frequency from one band to another depending on the state of the switches. This technique depends on the reliability of switches and their load power limitations. At the same time, this technology suffers from slow switching speed and high actuation voltage in addition to temperature sensitivity and dielectric charging problems [51].

In [52], the combiner section was simplified by matching both amplifiers to 70 Ω as illustrated in Figure 13, at the same time, the load impedance was still 50 Ω, in this case, however, a mismatch between the load and the matching networks will be introduced (−15 dB return loss) in the cost of simplifying the combiner. Further work of reconfigurable as well as concurrent dual-band DPAs was described by [53] where the reconfigurability had been achieved by using switches.

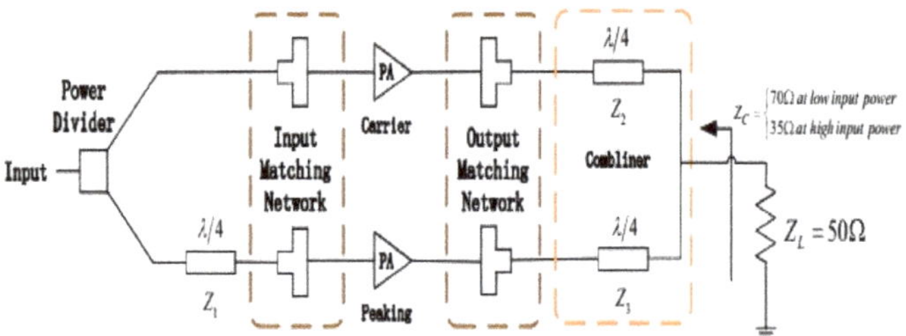

Figure 13. Load modulation proposed by [52].

The proposed design of digital DPA in [17] targeting 850 MHz and 1.7 GHz bands depends on a 3-coil current-mode combiner which is used as a power combiner of 4-way. The transformer is maintained to reduce the ground parasitic and minimize the mismatch between the two PAs. At the peaking power, both PAs are working, and the currents at the two primary coils are in-phase so that the impedance seen by each amplifier is 50 Ω. However, the load impedance seen by each PA stays the same when the output power is backed off if the two PAs are controlled synchronously, so that, in [17], both amplifiers are controlled independently in order to perform load modulation. At the same time, a switched capacitor is used because of its good performance in terms of linearity and efficiency. Another attempt targeting mm-wave was done by [54]. A simulation and detailed passive design are discussed in this paper for mm-wave: the lower band was 24 GHz and the upper bad was 28 GHz with a bandwidth of 1 GHz.

A recent work using digital dual-input power was done by [55] targeting a dual-band Doherty power amplifier, where the difference of the reactance impedance that the transistor needs to see at the input side for both bands is eliminated by using a short-circuited shunt stub. Moreover, the load modulation of the designed circuit was improved by using different drain voltages, where 22 V and 36 V were used for both the main and the auxiliary amplifier, the difference in drain voltages helped the main amplifier to be more efficient at the back-off and more current to be got from the auxiliary amplifier, where the load modulation can be improved.

3.6. Tri-Band and Quad-Band DPA

In addition to the dual-band DPA, there are researches which focus on tri- and quad-band DPA [56–60]. A design approach was used in [58] to cover three frequency bands. In the design, it was taking into considerations that all sections of the DPA should be able to give good performance at each band. However, the implemented design showed a relative shift in the band center frequencies.

A tri-band DPA targeting three types of mobile applications—LTE, universal mobile telecommunications service (UMTS), and WiMAX—was studied in [60], where two-stage DPA was designed for an average output power of 35 dBm. The results in [60] showed a good scattering parameters response. At the input side of the designed power amplifier, a broadband divider was used to cover the selected frequency bands, on the other hand, the impedance inverter network and the impedance transforming network were designed to fulfil the requirements of the tri-band. The designed PA was fabricated and tested using 10 MHz LTE for the lower frequency band, and two carriers with a WCDMA signal of 10 MHz for the upper bands. Each testing signal is used individually for the design evaluation.

The work of [58] has been extended in [56] to cover quad-band instead of tri-band operation. The chosen bands were 0.9 GHz, 1.5 GHz, 2.1 GHz and 2.6 GHz. Two transistors were used to implement asymmetrical DPA: these transistors are 10 W and 25 W with different power supplies for the main and the carrier. Although asymmetrical DPA was used, the authors in [56] were looking for only 6 dB OBO. The most complicated part in multiband DPA is the multiband impedance inverter network which should give good impedance inversion over the selected bands so that, in [56], at the input side, a wideband Wilkinson power divider was used to cover the four bands.

Proof of multi-band Doherty was presented by [61], where three boards were designed targeting three bands; the boards were identical in terms of the artwork, however for each band different capacitors and resistors were used. The work of [61] can be improved by using digitally controlled switches so that one reconfigurable board can be used for the three bands. In [62], a 3-way Doherty power amplifier targeting 1.8–2.2 GHz frequency bands suitable for mobile base-stations was achieved depending on LDMOS RFIC technology, a 47% efficiency accomplished at 12 dB back-off.

3.7. Linearization of Dual-Band DPA

Finally, power amplifiers in general often need to be linearized, especially when they operates at the nonlinear region to avoid interfering with the next adjacent channel. This can be achieved, for example, by using digital pre-distortion (DPD) which is widely renowned in its ability to improve the linearity performance of the designed dual-band due to its reconfigurable capability, accuracy, and its implementation in the digital domain. Using DPD will improve the signal linearity; however, the system complexity will be increased, as there has to be an additional circuit added to the system [63]. However, there is a deterioration in the performance of concurrent dual-band power amplifiers since it is not easy for the DPD to correct the signal at two different frequency bands at the same time. This is obvious because the nonlinearity behavior of the power amplifier is different from one frequency band to another. A possible solution is to use a new technique called two-dimensional digital predistortion (2D-DPD), which, though more complex than ordinary DPD, can deal will dual-band PA [30]. At the same time, the crest factor also leads to performance degradation in dual-band PA. A two-dimensional crest factor reduction was studied and used in [30] to improve the linearity. Table 2 shows a comparison of performance achieved in dual-band DPA in addition to the signal test type that was used in each work; it can be noticed that most of the works were done using GaN HEMT.

Table 2. Samples of achieved dual-band DPA.

Reference	Lower Band					Upper Band					Test	
	Frequency (GHz)	PPeak (dBm)	Efficiency @ PPeak %	Back-Off dB	Efficiency @OBO %	Frequency (GHz)	PPeak (dBm)	Efficiency @ PPeak %	Back-Off dB	Efficiency @OBO %	Signal Type	Signal Bandwidth
[17]	0.85	28.9	36.8 *	6	29.9	1.7	27	25.4 *	6	16.8 *	12-subcarrier NB-IoT/64-QAM/256-QAM WLAN	12 * 180k / 20 MHz
[18]	2.15	47.3	NA	6	52	3.4	47	NA	6	51	WCDMA/LTE	20 MHz
[25]	1.8	43	64 *	6	60 *	2.4	43	54 *	6	44	CW/WCDMA/LTE	− / 10 MHz
[7]	0.9	41.9	NA	5	39.8	2	41.2	NA	5	41.4	CW	-
[29]	2.14	39	50	6	45	3.5	NA	NA	NA	NA	CW	-
[64]	0.88	41	40 *	6	33 *	1.96	40	38 *	6	30 *	CW	-
[22]	2	42	NA	6	49	2.72	42	NA	6	44	CW/WCDMA/LTE	− / 5 MHz
[21]	2.1	50	NA	6	47 *	2.2	50	NA	6	47 *	CW/WCDMA	− / 5 MHz
[43]	0.9	46	66	9	65	2.15	46	56	9	46	CW	-
[23]	0.92	41.5	NA	7	33 *	1.99	41	NA	8	29 *	CW	-
[31]	2.14	45.5	NA	6.5	45	2.655	45.5	NA	6.5	40	WCDMA	5 MHz
[44]	0.85	43	47.5 *	6	45 *	2.33	42.5	32 *	6	41 *	CW/LTE1/LTE2	− / 10 MHz / 15 MHz
[46]	0.75	43	NA	9.4	42 *	0.9	43	NA	9.4	72 *	CW/WCDMA/LTE	− / 15 MHz
[19]	0.6	38.5	65	6	50	1	40.5	57	6	37	CW	-
[41]	1.7	45	NA	7	51	2.4	45	NA	9	45.5	CW/WCDMA/WiMAX	−/ 3.84 MHz / 5 MHz
[30]	1.9	36.8	NA	6.8	50.1	2.6	36.8	NA	6.8	50.1	LTE1/LTE2	10 MHz / 15 MHz
[35]	2.3	42.5	NA	7.2	30*	2.7	42.5	NA	7.2	44.2 *	CW/LTE	− / 10 MHz
[34]	0.6	40	76.3	6	54.9	1	40.2	54	6	38	CW	-
[65]	1.8	44	72	6	63	2.6	44	60	6	51	CW/LTE	− /20 MHz
[52]	2.02	43.9	62.1	6	51.1	2.63	44.6	71.4	6	53.3	CW	-
[55]	0.85	44.6	58	9	58	2.1	44.2	68	9	68	CW	-

NA: Not available. CW: Continuous wave. *: Power added efficiency PAE.

4. Conclusions

The techniques for designing dual-band Doherty amplifier have been reviewed and presented. The main limitation of each technique has been also addressed. The dual-band DPA was found as a promising technique that can be used in modern and future communications systems (e.g., satellite communications and mobile communications) where carrier aggregation and SDR can be applied. It was concluded that the offset line length mainly depends on the parasitic effect of the amplifier, in addition to the matching network characteristics. Using modern RF switches, which may include MEMS, can help to make the design of dual-band power amplifiers easier. The linearization process depends on characteristics of the PA devices, which vary depending on the operational frequency band, so that, in multi-band power amplifiers, the linearization process will be more complex and may be extremely so when the frequency bands are used concurrently. A solution for dual-band linearization is to use 2D-DPD, which will be able to achieve the required linearity at the cost of complexity. The DPA back-off depends mainly on the load modulation between the main and auxiliary amplifier, and at the same time, can be increased as the power ratio between the peaking power and the main power increases; however, selecting device sizes for the main and the auxiliary amplifiers is not always easy because it depends on the market availability of the devices. The use of MMIC technology gives flexibility in selecting the device peak power ratio; nevertheless, the output power of this technology is still low. For low frequency up to 2.2 GHz, LDMOSFET technology is often preferable; however, the GaN HEMT has smaller input and output capacitance due to its small periphery, and its transition frequency is high. Currently, the GaN HEMT has a good broadband performance, high efficiency, and large breakdown voltage, due to its features. It has given the designers a better ability to design an efficient PA.

The market for mobile communications covering more than one frequency band is growing. In this frame, the dual-band or multi-band Doherty amplifier can support the operation of the PA to cover two or more frequency bands with good efficiency at the back-off power, for which, the cost and the size can be reduced at the same time to preserve energy efficient operation.

Author Contributions: Conceptualization, A.M.A. and M.A.Y.; Methodology, A.M.A., N.M. and M.A.Y.; Software, A.M.A., R.A.A.-A. and M.A.Y.; Validation, A.M.A., R.A.A.-A., N.M. and A.R.; Formal analysis, A.M.A., R.A.A.-A., J.N., N.M., N.O.P., Y.I.A.A.-Y. and A.R.; Investigation, A.M.A., M.A.Y., R.A.A.-A., J.N., N.M., N.O.P., Y.I.A.A.-Y. and A.R.; Resources, A.M.A., M.A.Y., R.A.A.-A., J.N. and Y.I.A.A.-Y.; Data curation, A.M.A., M.A.Y. and R.A.A.-A.; Writing—original draft preparation, A.M.A., M.A.Y. and R.A.A.-A.; Writing—review and editing, A.M.A., M.A.Y., J.N. and R.A.A.-A.; Visualization, A.M.A., M.A.Y., R.A.A.-A. and J.N.

Funding: This project has received funding from the European Union's Horizon 2020 research and innovation programme under grant agreement H2020-MSCA-ITN-2016 SECRET-722424.

Acknowledgments: Authors wish to express their thanks for the support provided by the innovation programme under grant agreement H2020-MSCA-ITN-2016 SECRET-722424.

Conflicts of Interest: The authors declare no conflict of interest.

References

1. Cripps, S.C. *RF Power Amplifiers for Wireless Communications*; Artech House: Norwood, MA, USA, 2006.
2. Kervhé, E.; Belot, D. *Linearization and Efficiency Enhancement Techniques for Silicon Power Amplifiers: From RF to mmW*; Elsevier Science: New York, NY, USA, 2015.
3. Colantonio, P.; Giannini, F.; Limiti, E. *High Efficiency RF and Microwave Solid State Power Amplifiers*; John Wiley & Sons Ltd.: West Suss, UK, 2009.
4. Camarchia, V.; Pirola, M.; Quaglia, R.; Jee, S.; Cho, Y.; Kim, B. The Doherty Power Amplifier: Review of Recent Solutions and Trends. *IEEE Trans. Microw. Theory Tech.* **2015**, *63*, 559–571. [CrossRef]
5. Doherty, W.H. A new high-efficiency power amplifier for modulated waves. *Bell Syst. Tech. J.* **1936**, *15*, 469–475. [CrossRef]

6. Kim, H.; Seo, C. Improvement of Power Added Efficiency and Linearity in Doherty Amplifier using Dual Bias Control and Photonic Band-Gap Structure. In Proceedings of the 2007 Asia-Pacific Microwave Conference, Bangkok, Thailand, 11–14 December 2007; pp. 1–4.

7. Li, X.; Chen, W.; Zhang, Z.; Feng, Z.; Tang, X.; Mouthaan, K. A concurrent dual-band doherty power amplifier. In Proceedings of the 2010 Asia-Pacific Microwave Conference, Yokohama, Japan, 7–10 December 2010; pp. 654–657.

8. Saad, P.; Piazzon, L.; Colantonio, P.; Moon, J.; Giannini, F.; Andersson, K.; Kim, B.; Fager, C. Multi-band/multi-mode and efficient transmitter based on a Doherty Power Amplifier. In Proceedings of the 2012 42nd European Microwave Conference, Amsterdam, The Netherlands, 29 October–1 November 2012; pp. 1031–1034.

9. Gustafsson, D.; Andersson, C.M.; Fager, C. A Modified Doherty Power Amplifier With Extended Bandwidth and Reconfigurable Efficiency. *IEEE Trans. Microw. Theory Tech.* **2013**, *61*, 533–542. [CrossRef]

10. Barakat, A.; Thian, M.; Fusco, V.; Bulja, S.; Guan, L. Toward a More Generalized Doherty Power Amplifier Design for Broadband Operation. *IEEE Trans. Microw. Theory Tech.* **2017**, *65*, 846–859. [CrossRef]

11. Lin, J.; Xie, Z. Analysis and Design of a High Efficient Dual-Frequency Doherty Power Amplifier. In Proceedings of the 2018 International Conference on Microwave and Millimeter Wave Technology (ICMMT), Chengdu, China, 7–11 May 2018; pp. 1–3.

12. Chen, X.; Chen, W.; Huang, F.; Ghannouchi, F.M.; Feng, Z.; Liu, Y. Systematic Crest Factor Reduction and Efficiency Enhancement of Dual-Band Power Amplifier Based Transmitters. *IEEE Trans. Broadcast.* **2017**, *63*, 111–122. [CrossRef]

13. Kelly, N.; Cao, W.; Zhu, A. Preparing Linearity and Efficiency for 5G: Digital Predistortion for Dual-Band Doherty Power Amplifiers with Mixed-Mode Carrier Aggregation. *IEEE Microw. Mag.* **2017**, *18*, 76–84. [CrossRef]

14. Chen, W.; Chen, X.; Zhang, S.; Feng, Z. Energy-efficient concurrent dual-band transmitter for multistandard wireless communications. In Proceedings of the 2014 Asia-Pacific Microwave Conference, Sendai, Japan, 4–7 November 2014; pp. 558–560.

15. Chen, X.; Chen, W.; Gongzhe, S.; Ghannouchi, F.M.; Feng, Z. A concurrent dual-band 1.9–2.6-GHz Doherty power amplifier with Intermodulation impedance tuning. In Proceedings of the 2014 IEEE MTT-S International Microwave Symposium (IMS2014), Tampa, FL, USA, 1–6 June 2014; pp. 1–4.

16. GSMA. 5G Spectrum GSMA Public Policy Position. Available online: https://www.gsma.com/latinamerica/5g-spectrum-gsma-public-policy-position/ (accessed on 15 February 2019).

17. Yin, Y.; Xiong, L.; Zhu, Y.; Chen, B.; Min, H.; Xu, H. A compact dual-band digital doherty power amplifier using parallel-combining transformer for cellular NB-IoT applications. In Proceedings of the 2018 IEEE International Solid-State Circuits Conference (ISSCC), San Francisco, CA, USA, 11–15 February 2018; pp. 408–410.

18. Liu, M.; Golestaneh, H.; Boumaiza, S. A concurrent 2.15/3.4 GHz dual-band Doherty power amplifier with extended fractional bandwidth. In Proceedings of the 2016 IEEE MTT-S International Microwave Symposium (IMS), San Francisco, CA, USA, 22–27 May 2016; pp. 1–3.

19. Shao, J.; Poe, D.; Ren, H.; Arigong, B.; Zhou, M.; Ding, J.; Zhou, R.; Kim, H.S.; Zhang, H. Dual-band microwave power amplifier design using GaN transistors. In Proceedings of the 2014 IEEE 57th International Midwest Symposium on Circuits and Systems (MWSCAS), College Station, TX, USA, 3–6 August 2014; pp. 559–562.

20. Ghannouchi, F.M.; Rawat, K. Doherty power amplifiers in software radio systems. In Proceedings of the 2011 XXXth URSI General Assembly and Scientific Symposium, Istanbul, Turkey, 13–20 August 2011; pp. 1–4.

21. Park, J.; Yook, J.; Kim, Y.; Lee, C.H. Dual-band switching Doherty power amplifier using phase shifter composed of PIN diode. In Proceedings of the 2011 6th European Microwave Integrated Circuit Conference, Manchester, UK, 10–11 October 2011; pp. 300–303.

22. Bathich, K.; Gruner, D.; Boeck, G. Analysis and design of dual-band GaN HEMT based Doherty amplifier. In Proceedings of the 2011 6th European Microwave Integrated Circuit Conference, Manchester, UK, 10–11 October 2011; pp. 248–251.

23. Xiang, L.; Wenhua, C.; Zisheng, L.; Zhenghe, F.; Yaqin, C.; Ghannouchi, F.M. Design of dual-band multi-way Doherty power amplifiers. In Proceedings of the 2012 IEEE/MTT-S International Microwave Symposium Digest, Montreal, QC, Canada, 17–22 June 2012; pp. 1–3.

24. Vizmuller, P. *RF Design Guide: Systems, Circuits, and Equations*; Artech House: Norwood, MA, USA, 1995.
25. Saad, P.; Colantonio, P.; Piazzon, L.; Giannini, F.; Andersson, K.; Fager, C. Design of a Concurrent Dual-Band 1.8–2.4-GHz GaN-HEMT Doherty Power Amplifier. *IEEE Trans. Microw. Theory Tech.* **2012**, *60*, 1840–1949. [CrossRef]
26. Wang, G.; Zhao, L.; Szymanowski, M. A Doherty amplifier for TD-SCDMA base station applications based on a single packaged dual-path integrated LDMOS power transistor. In Proceedings of the 2010 IEEE MTT-S International Microwave Symposium, Anaheim, CA, USA, 23–28 May 2010; pp. 1512–1515.
27. Abdulkhaleq, A.M.; Al-Yasir, Y.; Ojaroudi Parchin, N.; Brunning, J.; McEwan, N.; Rayit, A.; Abd-Alhameed, R.A.; Noras, J.; Abduljabbar, N. A 70-W Asymmetrical Doherty Power Amplifier for 5G Base Stations. In Proceedings of the Broadband Communications, Networks, and Systems, Faro, Portugal, 19–20 September 2018; pp. 446–454.
28. Monzon, C. A small dual-frequency transformer in two sections. *IEEE Trans. Microw. Theory Tech.* **2003**, *51*, 1157–1161. [CrossRef]
29. Colantonio, P.; Feudo, F.; Giannini, F.; Giofrè, R.; Piazzon, L. Design of a dual-band GaN Doherty amplifier. In Proceedings of the 18th International Conference on Microwaves, Radar and Wireless Communications, Vilnius, Lithuania, 14–16 June 2010; pp. 1–4.
30. Chen, W.; Chen, X.; Zhang, S.; Feng, Z. Energy-efficient power amplifier techniques for TD-SCDMA and TD-LTE multi-standard wireless communications. In Proceedings of the 2014 XXXIth URSI General Assembly and Scientific Symposium (URSI GASS), Beijing, China, 16–23 August 2014; pp. 1–4.
31. Grebennikov, A.; Wong, J. A Dual-Band Parallel Doherty Power Amplifier for Wireless Applications. *IEEE Trans. Microw. Theory Tech.* **2012**, *60*, 3214–3222. [CrossRef]
32. Yang, Z.; Li, M.; Yao, Y.; Dai, Z.; Li, T.; Jin, Y. Design of Concurrent Dual-Band Continuous Class-J Mode Doherty Power Amplifier With Precise Impedance Terminations. *IEEE Microw. Wirel. Compon. Lett.* **2019**, *29*, 348–350. [CrossRef]
33. Chuang, M. Dual-Band Impedance Transformer Using Two-Section Shunt Stubs. *IEEE Trans. Microw. Theory Tech.* **2010**, *58*, 1257–1263. [CrossRef]
34. Ren, H.; Shao, J.; Arigong, B.; Zhou, M.; Fu, S.; Ding, J.; Kim, H.; Zhang, H. Simplified Doherty power amplifier structures. In Proceedings of the 2015 Texas Symposium on Wireless and Microwave Circuits and Systems (WMCS), Waco, TX, USA, 23–24 April 2015; pp. 1–3.
35. Jee, S.; Yunsik, P.; Cho, Y.; Lee, J.; Seokhyeon, K.; Bumman, K. A highly linear dual-band Doherty power amplifier for femto-cell base stations. In Proceedings of the 2015 IEEE MTT-S International Microwave Symposium, Phoenix, AZ, USA, 17–22 May 2015; pp. 1–4.
36. Lv, G.; Chen, W.; Liu, X.; Feng, Z. A Dual-Band GaN MMIC Power Amplifier With Hybrid Operating Modes for 5G Application. *IEEE Microw. Wirel. Compon. Lett.* **2019**, *29*, 228–230. [CrossRef]
37. Lv, G.; Chen, W.; Chen, X.; Ghannouchi, F.M.; Feng, Z. A Compact Ka/Q Dual-Band GaAs MMIC Doherty Power Amplifier With Simplified Offset Lines for 5G Applications. *IEEE Trans. Microw. Theory Tech.* **2019**, 1–12. [CrossRef]
38. Moon, J.; Kim, J.; Kim, I.; Kim, B. A Wideband Envelope Tracking Doherty Amplifier for WiMAX Systems. *IEEE Microw. Wirel. Compon. Lett.* **2008**, *18*, 49–51. [CrossRef]
39. Qureshi, J.H.; Nan, L.; Neo, E.; Rijs, F.V.; Blednov, I.; de Vreede, L. A wide-band 20W LDMOS Doherty power amplifier. In Proceedings of the 2010 IEEE MTT-S International Microwave Symposium, Anaheim, CA, USA, 23–28 May 2010; p. 1.
40. Sarkeshi, M.; Leong, O.B.; van Roermund, A. A novel Doherty amplifier for enhanced load modulation and higher bandwidth. In Proceedings of the 2008 IEEE MTT-S International Microwave Symposium Digest, Atlanta, GA, USA, 15–20 June 2008; pp. 763–766.
41. Barthwal, A.; Ajmera, G.; Rawat, K.; Basu, A.; Koul, S.K. Design scheme for dual-band three stage Doherty Power Amplifiers. In Proceedings of the 2014 IEEE International Microwave and RF Conference (IMaRC), Bangalore, India, 15–17 December 2014; pp. 80–83.
42. Liu, M.; Fang, X.; Huang, H.; Boumaiza, S. Dual-band 3-way Doherty Power Amplifier with Extended Back-off Power and Bandwidth. *IEEE Trans. Circuits Syst. II Express Briefs* **2019**, 1. [CrossRef]
43. Li, X.; Chen, W.; Feng, Z.; Ghannouchi, F.M. Design of dual-band tri-way GaN doherty power amplifier with frequency dependent power division. *Electron. Lett.* **2012**, *48*, 797–798. [CrossRef]

44. Chen, W.; Zhang, S.; Liu, Y.; Ghannouchi, F.M. A Concurrent Dual-Band Uneven Doherty Power Amplifier with Frequency-Dependent Input Power Division. *IEEE Trans. Circuits Syst. I Regul. Pap.* **2014**, *61*, 552–561. [CrossRef]

45. Nick, M.; Mortazawi, A. Adaptive Input-Power Distribution in Doherty Power Amplifiers for Linearity and Efficiency Enhancement. *IEEE Trans. Microw. Theory Tech.* **2010**, *58*, 2764–2771. [CrossRef]

46. Abadi, M.N.A.; Golestaneh, H.; Sarbishaei, H.; Boumaiza, S. An extended bandwidth Doherty power amplifier using a novel output combiner. In Proceedings of the 2014 IEEE MTT-S International Microwave Symposium (IMS2014), Tampa, FL, USA, 1–6 June 2014; pp. 1–4.

47. Carrubba, V.; Ture, E.; Maroldt, S.; Mußer, M.; Raay, F.V.; Quay, R.; Ambacher, O. A dual-band UMTS/LTE highly power-efficient Class-ABJ Doherty GaN PA. In Proceedings of the 2015 European Microwave Conference (EuMC), Paris, France, 7–10 September 2015; pp. 1164–1167.

48. Carrubba, V.; Ture, E.; Maroldt, S.; Mußer, M.; Raay, F.V.; Quay, R.; Ambacher, O. A dual-band UMTS/LTE highly power-efficient class-ABJ Doherty GaN PA. In Proceedings of the 2015 10th European Microwave Integrated Circuits Conference (EuMIC), Paris, France, 7–8 September 2015; pp. 313–316.

49. Kalyan, R.; Rawat, K.; Koul, S.K. Design of reconfigurable concurrent dual-band quarter-wave transformer with application of power combiner/divider. In Proceedings of the 2015 IEEE MTT-S International Microwave and RF Conference (IMaRC), Hyderabad, India, 10–12 December 2015; pp. 169–172.

50. Kalyan, R.; Rawat, K.; Koul, S.K. Design of reconfigurable concurrent dual-band power amplifiers using reconfigurable concurrent dual-band matching network. In Proceedings of the 2016 IEEE MTT-S International Wireless Symposium (IWS), Shanghai, China, 14–16 March 2016; pp. 1–4.

51. Lahiri, S.K.; Saha, H.; Kundu, A. RF MEMS SWITCH: An overview at- a-glance. In Proceedings of the 2009 4th International Conference on Computers and Devices for Communication (CODEC), Kolkata, India, 14–16 December 2009; pp. 1–5.

52. Fan, M.; Yu, C.; Yu, Q.; Liu, Y. Design of a dual-band Doherty power amplifier utilizing improved combiner. In Proceedings of the 2016 IEEE International Conference on Computational Electromagnetics (ICCEM), Guangzhou, China, 23–25 February 2016; pp. 313–315.

53. Kalyan, R.; Rawat, K.; Koul, S.K. Reconfigurable and Concurrent Dual-Band Doherty Power Amplifier for Multiband and Multistandard Applications. *IEEE Trans. Microw. Theory Tech.* **2017**, *65*, 198–208. [CrossRef]

54. Taghian, F.; Abdipour, A.; Mohammadi, A.; Roodaki, P.M. Design and nonlinear analysis of a dual-band Doherty power amplifier for ISM and LMDS applications. In Proceedings of the 2011 IEEE Applied Electromagnetics Conference (AEMC), Kolkata, India, 18–22 December 2011; pp. 1–4.

55. Kalyan, R.; Rawat, K.; Koul, S.K. A Digitally Assisted Dual-Input Dual-Band Doherty Power Amplifier With Enhanced Efficiency and Linearity. *IEEE Trans. Circuits Syst. II Express Briefs* **2019**, *66*, 297–301. [CrossRef]

56. Nghiem, X.A.; Negra, R. Design of a concurrent quad-band GaN-HEMT Doherty power amplifier for wireless applications. In Proceedings of the 2013 IEEE MTT-S International Microwave Symposium Digest (MTT), Seattle, WA, USA, 2–7 June 2013; pp. 1–4.

57. Li, X.; Helaoui, M.; Ghannouchi, F.; Chen, W. A Quad-Band Doherty Power Amplifier Based on T-Section Coupled Lines. *IEEE Microw. Wirel. Compon. Lett.* **2016**, *26*, 437–439. [CrossRef]

58. Nghiem, X.A.; Negra, R. Novel design of a concurrent tri-band GaN-HEMT Doherty power amplifier. In Proceedings of the 2012 Asia Pacific Microwave Conference Proceedings, Kaohsiung, Taiwan, 4–7 December 2012; pp. 364–366.

59. Rawat, K.; Gowrish, B.; Ajmera, G.; Kalyan, R.; Basu, A.; Koul, S.; Ghannouchi, F.M. Design strategy for tri-band Doherty power amplifier. In Proceedings of the WAMICON 2014, Tampa, FL, USA, 6 June 2014; pp. 1–3.

60. Giofré, R.; Costanzo, F.; Vikraman, D.N.M.; Colantonio, P.; Giannini, F. Designing a tri-band concurrent Doherty power amplifier. In Proceedings of the 2015 Integrated Nonlinear Microwave and Millimetre-wave Circuits Workshop (INMMiC), Taormina, Italy, 1–2 October 2015; pp. 1–3.

61. Yanduru, N.K.; Jeckeln, E.; Monroe, R.; Brobston, M. 50W GaN based multi-band Doherty amplifier proof of concept for LTE with 48% efficiency. In Proceedings of the WAMICON 2014, Tampa, FL, USA, 6 June 2014; pp. 1–3.

62. Blednov, I. Wideband 3 way Doherty RFIC with 12 dB back-off power range. In Proceedings of the 2016 11th European Microwave Integrated Circuits Conference (EuMIC), London, UK, 3–4 October 2016; pp. 17–20.

63. Rawat, K.; Ghannouchi, F.M. Design Methodology for Dual-Band Doherty Power Amplifier with Performance Enhancement Using Dual-Band Offset Lines. *IEEE Trans. Ind. Electron.* **2012**, *59*, 4831–4842. [CrossRef]

64. Chen, W.; Bassam, S.A.; Li, X.; Liu, Y.; Rawat, K.; Helaoui, M.; Ghannouchi, F.M.; Feng, Z. Design and Linearization of Concurrent Dual-Band Doherty Power Amplifier With Frequency-Dependent Power Ranges. *IEEE Trans. Microw. Theory Tech.* **2011**, *59*, 2537–2546. [CrossRef]

65. Pang, J.; He, S.; Dai, Z.; Huang, C.; Peng, J.; You, F. Novel design of highly-efficient concurrent dual-band GaN Doherty power amplifier using direct-matching impedance transformers. In Proceedings of the 2016 IEEE MTT-S International Microwave Symposium (IMS), San Francisco, CA, USA, 22–27 May 2016; pp. 1–4.

 *electronics*

Review

A Survey on RF and Microwave Doherty Power Amplifier for Mobile Handset Applications

Maryam Sajedin [1,2,*], **I.T.E. Elfergani** [1,*], **Jonathan Rodriguez** [1,2], **Raed Abd-Alhameed** [3] and **Monica Fernandez Barciela** [4]

[1]   Instituto de Telecomunicações, Campus Universitário de Santiago, 3810-193 Aveiro, Portugal; jonathan@av.it.pt

[2]   Departamento de Electrónica, Telecomunicações e Informática, Universidade de Aveiro, 3810-193 Aveiro, Portugal

[3]   Faculty of Engineering and Informatics, Bradford University, Bradford BD7 1DP, UK; r.a.a.abd@bradford.ac.uk

[4]   Department of Signal Theory and Communications, University of Vigo, 36310 Vigo, Spain; monica.barciela@uvigo.es

*   Correspondence: Maryam.sajedin@av.it.pt (M.S.); i.t.e.elfergani@av.it.pt (I.T.E.E.); Tel.: +351 218-418-454 (I.T.E.E.)

Received: 2 May 2019; Accepted: 19 June 2019; Published: 25 June 2019

Abstract: This survey addresses the cutting-edge load modulation microwave and radio frequency power amplifiers for next-generation wireless communication standards. The basic operational principle of the Doherty amplifier and its defective behavior that has been originated by transistor characteristics will be presented. Moreover, advance design architectures for enhancing the Doherty power amplifier's performance in terms of higher efficiency and wider bandwidth characteristics, as well as the compact design techniques of Doherty amplifier that meets the requirements of legacy 5G handset applications, will be discussed.

Keywords: High power amplifiers; high efficiency; Doherty power amplifier; 4G; 5G; GaN-HEMT

---

1. Introduction

5G communications is an international initiative that aims to deliver next generation services that are power hungry and data intensive. To achieve the targeted 5G performance indicators will rely on phased-array MIMO (multiple-input and multiple output) antennas, new spectrum availability, and small cell technology, which in synergy aims to provide a communication platform to provide not only enhanced broadband connectivity, but to enable the Internet of things service coverage for smart manufacturing applications, and provide Ultra reliable and Low Latency services. This paradigm will put stringent design requirements on the system architecture in place, and beyond that on the RF front-end. A Radio Frequency Front-End (RFFE), as shown in Figure 1, as one of the key components of a mobile terminal, is powered by a low-voltage source, or even batteries, which has to cover a vast number of frequency bands in order to provide a high-level of integration [1]. In the design architecture of Figure 1, power amplifier modules combine multiplexers, filters, and RF switches blocks to provide highly integrated transmitter and receiver, which helps to reduce the manufacturer's time-to-launch. A key design requirement for power amplifiers (PAs) is energy efficiency at the required output power levels and the targeted operating frequency. This requirement is even more pronounced in 5G cellular networks to not only minimize operational expenditure, but also to reduce the carbon footprint that is associated with the PA lifecycle [2]. Moreover, in conventional RF front-end configuration, power amplifiers are optimized for a specific frequency band, which results in a narrow-band matching scheme; therefore, the PA's operating range at higher frequencies is limited. On the other hand, spectrally efficient multi-carrier signal exhibits time-varying amplitude and phase characteristics due to wide and rapid variation of the instantaneous transmit power, resulting in a high peak-to-average ratio (PAPR) signal and wider occupied bandwidth [3]. The adoption of high PAPR modulated signals forces the power amplifier to operate at a large output back-off (OBO) to satisfy the stringent linearity requirements that are imposed by the wireless communication standards. This provides an amplifier device with 8 to 15 percent efficiency, and the implementation might be acceptable if the RF power requirements are very low [4]. As RF output power increases, power wastage can take significant cost, which translates into various forms, such as higher temperatures, more expensive heat transfer solutions and higher operating costs. Therefore, power amplifiers with higher back-off efficiency and linearity are required to enhance the overall transmitter performance.

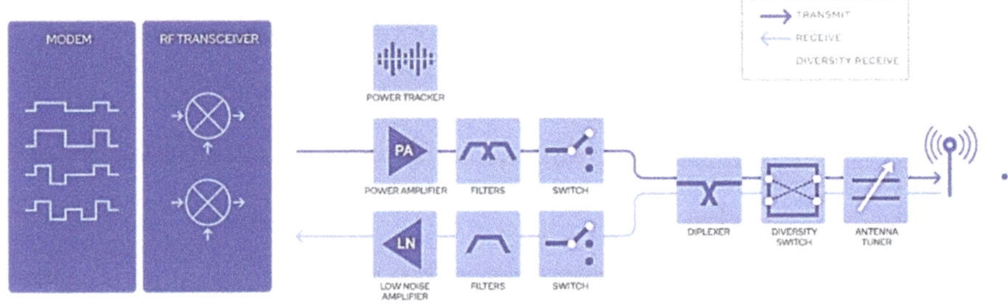

Figure 1. Qualcomm Radio Frequency Front-End (RFFE) solution [2].

Wide varieties of two-way power amplifier architectures have been introduced for efficiency enhancing without distorting linearity. In Envelope Tracking (ET) [5] and Envelope Elimination and Restoration (EER) techniques [6], based on the bias adaptation principle, the collector/drain supply of the RF power dynamically changes with the output envelope, thus the transistor operates with higher efficiency over a wide dynamic range of output power [7]. The ET PA can support modulation bandwidth of up to 80 MHz by utilizing digital predistortion [8]; however, the tracking bandwidth strongly relies on the supply modulator that needs the bandwidth enhancement for complex modulation signals. Several techniques have been applied to increase the bandwidth of modulator with penalty of complexity reflecting the additional circuit [9]. The EER technique generates constant envelope signal by changing the characteristics of high PAPR signal. One drawback of the EER/polar and ET amplifiers is the dependency of supply modulator performance on the amplifier efficiency, bandwidth, peak power, and dynamic range, which, in practice, severely restricts the instantaneous modulation bandwidth [10]. The operation of the Doherty Power Amplifier (DPA) that was originally proposed by W. H. Doherty in 1936 [11] and Outphasing by H. Chireix [12] are based on an active load modulation mechanism. The Outphasing architecture performs linear amplification by nonlinear components and it provides efficiency levels of 20–60%, and bandwidths of up to 40 MHz. In fact, the wideband Outphasing causes serious baseband overhead [13]. DPA operates in an optimal load impedance trajectory, which varies according to the amplitude of the input signal, which results in increasing the average efficiency of the Doherty PA without compromising its linearity. The DPA architecture provides an efficiency of 20–45% and bandwidths of up to 500 MHz [14].

The survey contribution is expressed in Section 2, we briefly address two main design challenges and potential strategies for 5G cm-wave/mm-wave DPA design, namely, the efficiency and bandwidth enhancement techniques. In this context, the most important characteristics of mobile handset power amplifiers, such as output power and power added efficiency, are targeted, which not only determine battery life, but also address the linearity/efficiency compromise in the handset amplifiers. An overview of device technologies that are nowadays highly considered as promising technologies for the design of high efficiency, high power, and high linearity power amplifiers will be discussed in Section 3. Various design challenges of the RF and the microwave DPAs and effective solutions to overcome these issues have been introduced in Sections 4 and 5 respectively, along with the review on the recent research in the design and fabrication of the DPAs. Additionally, Section 6 discusses the bandwidth limiting factors and wideband design approaches of DPA. Next, the design methodologies for Multi-Band Doherty PAs will be introduced in Section 7. Then, the elaborated compact DPA circuit for handset applications will be discussed in Section 8. Finally, some conclusion will be given in Section 9.

2. The Survey Contribution

As a first contribution, this survey goes on to propose key developments of the Doherty topology to allow for exploiting complex modulation standards, when considering the available device technologies. Second, it suggests that practical solutions serve to overcome manufacturing limitations, which address the DPA-bandwidth degradation due to its contributing electrical components. A general significance of work comes up with looking to prior art for a deeper understanding of the design features challenges and applicable solutions for realizing energy-efficient, low cost, small size, and low complexity 5G mobile handset applications. These organization is important milestone towards building a superposition Doherty technique that can be deployed in supporting simultaneous transmission of multiple carriers that are formed by the carrier aggregation of non-contiguous spectrum.

3. The Choice of Device Technologies for RF-Front End Power Amplifiers

5G modern handset power amplifiers require lower output power than those that are currently used in 4G LTE due to utilizing higher Cm-Wave/Mm-Wave carrier frequencies and massive MIMO [15] technology. Moreover, mobile devices will need to support a wider set of RF bands, enable reliable connectivity, and require longer battery lifetime and the efficient use of electrical energy. The operation frequency band and the output power are the determining factors in choosing the semiconductor technology for power amplifiers design. GaN inherently shows high efficiencies, resulting in a reduction in system power consumption and presenting fewer thermal management challenges, which could ultimately lead to improved battery life [16]. Moreover, GaN devices can be downsized in fabrication, leading to much higher impedances that are more convenient for broadband matching. GaAs HBT is widely used in low power mobile devices, since it requires a single supply voltage, which is deemed to be a positive feature in any application, where a battery supplies the circuit. GaAs pHEMTs delivers excellent bandwidth, linearity, and efficiency, as shown in Table 1, for devices under one watt with low battery voltage, and thus serves as a strong candidate to develop millimeter wave PAs above 20 GHz [17]. SiGe RF PAs for handsets have become used in billions of RF FEMs for 4G handsets and WLAN products [18]. The LDMOS and GaN HEMT device technologies are widely used in base-stations due to their strong linearity attributes, besides being low-cost. DPAs based on CMOS technologies are also investigated due to their capability for co-integration and flexibility.

Table 1. Performance comparison of common semiconductors [15–18].

Tech.	Frequency (GHz)	Power (W)	Gain (dB)	PAE%
LDMOS	<3	300	<15	Up to 70
CMOS	2.4	0.2	18	45
GaAs MESFET	12	0.08	5	65
HBT	2–8	2	9	20
	24–26	2.2	5	42
	3.5–10	1.3	8	25
P-HEMT	8–14	3.5	8.4–14	40
	28	1.6	16	35

4. Design Challenges of Doherty Power Amplifier

The two-way Doherty power amplifier implements by Carrier and Peaking active device stages. As depicted in Figure 2, it consists of a power splitter to properly divide the input signal to the device gates, and power combining network, including an impedance inverter to sum in phase the signals that arise from the two active devices; and, an impedance transformer that was connected to the output load. Finally, the phase variation that was introduced by the impedance inverter is compensated at the input of peaking amplifier.

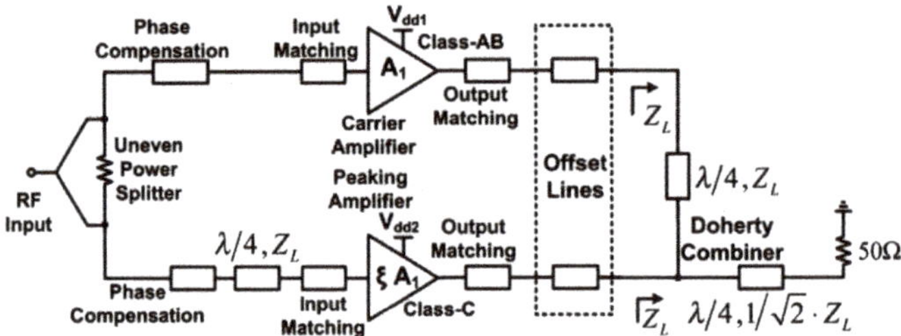

Figure 2. Two-way Doherty power amplifier scheme [19].

If the carrier and peaking transistors in the conventional architecture are represented by equivalent voltage controlled current sources, which are linearly proportional to the input signal voltage, the constant voltage at the carrier PA will be transformed into a constant current at the peaking PA, regardless of the load value. However, the practical challenges of a non-constant transconductance, non-ideal harmonics, knee-voltage, and effects of peaking amplifier's Class-C bias condition need to be addressed. However, the implementation of DPA presents other issues, such as nonlinearity at operating frequency, due to the device non-ideality. The Back-Off efficiency challenges of the Doherty PA are:

- Gain degradation: The peaking amplifier modulates the carrier one to deliver the maximum efficiency at the maximum power, this impedance variation and deviation from optimal load, results in a reduction of gain, Therefore, the individual transistors' gain response will be nonlinear.
- Phase distortion: the different conduction angles of the carrier and peaking devices result in non-similar output current profiles, which impose phase offset and gain imbalanced between two amplifiers stages [20]. Moreover, the parasites of real transistors will cause phase distortion and leads to a back-off efficiency drop and poor linearity.
- Poor inter-modulation distortion (IMD) performance: the peaking amplifier may cause a large distortion due to a low biasing condition (class C). One method of solving this issue is to deploy the intermodulation products of carrier amplifier to add up with that of the peaking transistor at the load destructively to eliminate the IMD [21].

It should be noted that the instantaneous efficiency of power amplifier is a function of output power. In different classes of amplification, the instantaneous signal envelope can adapt the quiescent current. High quiescent current of Class A amplification causes the low IMD and low harmonic levels, which enable the amplifier to operate close to the maximum capability of transistor; however, the saturation voltage of the transistor deteriorates the efficiency [22]. Thus, Class A is typically used in applications with high gain and high linearity requirements. The quiescent current of Class B is fixed to minimize the crossover distortion at low output power, which enables linear and efficient amplification. In fact, the linear amplification refers to the short-circuited of all voltage harmonics of sinusoidal output signal. Increasing the load impedance, which provides larger voltage swing, can enhance the efficiency of this amplifier [23]. Class B is typically used in battery-operated, mobile radios, and base station amplifiers. By decreasing the conduction angle, the efficiency can be enhanced in the Class C mode. However, drive signal tends to increased, with the output power reduction, which results in low gain. Moreover, Class C mode is not often utilized in solid state amplification at high and microwave frequencies, because the reverse breakdown condition of transistor [24].

- Narrow bandwidth: The inherently narrow bandwidth behavior of DPA has originated from the quarter wave impedance inverter, which is usually applied for load modulation [25]. Moreover, the Doherty architecture integration into a single chip is a nightmare task due to the large size of quarter wavelength impedance network.
- Parallel parasitic losses: In the low power levels, the peaking device is in an open-circuit condition to avoid the current leakage to the carrier device. The traditionally adapted quarter-wavelength Impedance Inverter Network (IIN) can correctly perform load modulation only for real impedances in an ideal DPA [26]. However, the output parasitic reactances of real devices involve an imaginary part to the load, which must be eliminated. Furthermore, the output matching network can compensate one specific load impedance parasitic at saturation, which means that, for all other impedance introduced by load modulation, the reactive parasitics are not properly eliminated, and an unwanted phase rotation influences the load modulation [27], which results in lower back-off efficiency and DPA nonlinearity.

The common solution for restoring the optimum load-modulation is the insertion of two offset lines in the carrier and peaking output matching networks with characteristic impedance that is equal to the load impedance at saturation. An alternative method to compensate for parasitic effects is to integrate the device output matching networks and output combiner, which can reduce the device size [28]. Recently, the co-design method has been further improved while using the black-box technique at the output combiner, [29]. In this approach, the phase difference between two devices add more degree of freedom to achieve higher efficiency and extended the bandwidth.

5. Advanced Doherty Amplifier Architectures

The DPA high back-off efficiency can be achieved when considering some factors, including lower peak power level of carrier PA, than that of DPA, efficient operation of carrier PA at back-off, and small impedance matching loss. In this respect, the varieties of techniques have been proposed in order to restore the optimum active load modulation behavior. This section will present some of them.

5.1. Asymmetrical Doherty Power Amplifier (ADPA)

If the DPA implements by the same size of carrier and peaking amplifiers, the maximum output current of the peaking amplifier is smaller than that of the carrier amplifier due to the small conduction angle, which results in a reduction of the maximum output power. In fact, for an ideal behavior of the DPA, either the peaking device should be roughly double the size of the carrier device or the input power splitting should be asymmetrical [30]. However, increasing the device size asymmetry causes severe gain degradation due to the stronger influence of the inherently lower gain Class-C peaking stage, which reduces the power that is delivered to the carrier stage and it affects the overall performance.

One of the most common techniques is using uneven power divider in favor of the peaking PA. Delivering more RF input power to the peaking PA, rather than to the carrier device, allows for the generation of sufficient current for the peaking PA to achieve a proper load modulation that increases the drain efficiency ranges from 10 to 13% [31]. Bias adapted DPA is depicted in Figure 3a, and it consists of two separate circuits, a fully matched Doherty amplifier, and an envelope shaped voltage generator for the carrier amplifier. Adaptive gate-bias control circuit continuously modifies the gate voltage of the carrier PA, which results in a reduction of the power consumption in the low power level when only the carrier device is operating. However, the main drawback of this solution is increased bias circuit complexity which requires additional cost to implement. Figure 3b shows a model of fabricated gate bias controlled DPA [32].

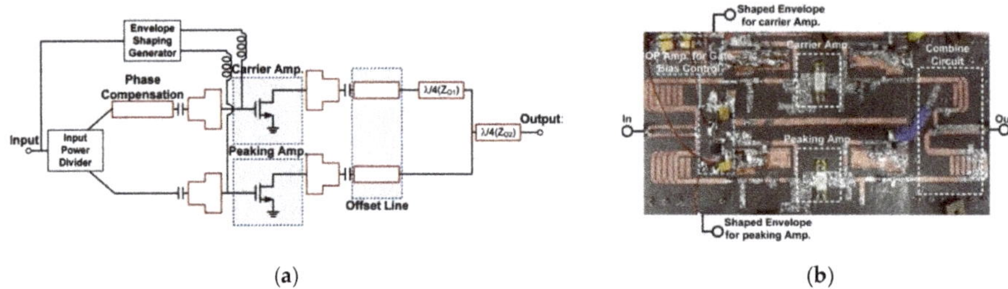

Figure 3. (**a**) Schematic of the Doherty amplifier with gate-bias adaptation, (**b**) A model of fabricated Doherty Power Amplifier (DPA) and presented in [32].

Table 2 compares the performance of some ADPAs and Bias adapted DPAs in the literature. The results show that both the bias adapted-DPA and ADPA structures enhance the efficiency and output power characteristics with respect to the conventional DPA. The gain reduction of the ADPA in the high power levels results in a poor power-added efficiency. On the other hand, bias adapted-carrier PA ensures the full load modulation and the maximum output power.

Table 2. Measured performance summary of similar topology DPAs.

Ref.	DPA Type	Frequency	Output Power	Efficiency (PAE)	Gain
[30]	Gate bias Adaptation	3.4–3.6 GHz	49.3 dBm	45%	12.3 dB
[31]	Gate bias Adaptation	1.94 GHz	44.35 dBm	60.5%	12.75 dB
[32]	ADPA	2.5 GHz	47 dBm	52%	15 dB
[33]	ADPA	2.14 GHz	42 dBm	48%	15 dB

5.2. Multiway and Multistage DPA

The conventional DPA is considered to be one of the promising approaches for improving efficiency over the 6 dB output power back-off. However, the power usage profile shows that, most of the time, the power amplifier operates at average transmitted output power in the range of 9–12 dB level below the maximum power; therefore, the 6 dB back-off efficiency improvement of the conventional DPA is insufficient and it results in the poor system efficiency [34]. It is possible to use more than two amplifiers, the so-called "Multistage DPA" configuration, to maintain the efficiency throughout the back-off region. The combination of the output power of several power amplifiers increases the linearity (larger OBO). The Multistage DPA constitutes multiple Doherty PAs working in a parallel configuration, overcoming the issues that are related to the adoption of single very large device. In three-stage DPA that is shown in Figure 4a, the first peaking PA modulates the load of the carrier PA, and the second peaking PA modulates the load of the previous Doherty PA [35]. It can be observed that, in Figure 4b, the three-stage DPA provides additional peak in efficiency curve, which enables DPA for further enhancement on average efficiency.

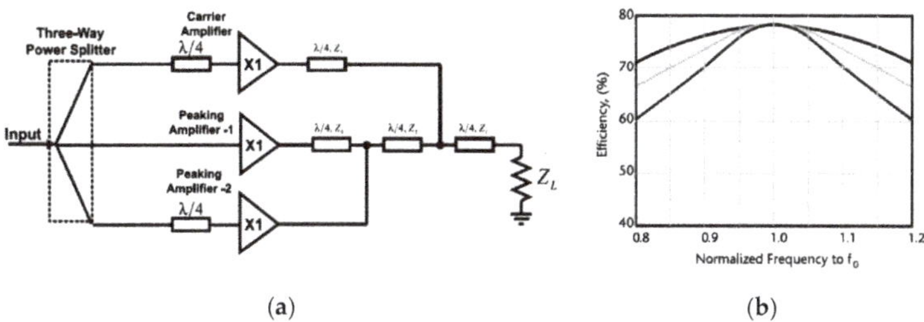

Figure 4. (**a**) Three-stage DPA architecture, (**b**) Efficiency versus frequency of improved three-stage DPA in [35].

The multi-stage DPAs can deliver highly efficient amplification of a modulated signal. However, these amplifiers are not very popular due to their complicated circuit structure, which poses problems in implementation. It is assumed that all of the carrier and peaking amplifiers reach their maximum current levels at the maximum output power, even though their biases are different. However, due to the lower supply voltage of peaking amplifiers, they cannot generate their respective full powers and the reduced output powers directly affect the efficiency, as well as linearity. Additionally, at the same current level, the smaller fundamental current levels of the peaking devices cause the modulated load impedances to become higher than those of the ideal operation, subsequently non-optimum load modulation. Table 3 compares the performance of some recent works on three-way and three-stage DPAs, which achieve reasonable compromise between the output power and efficiency without any further linearization method.

Table 3. Measured performance summary of similar topology DPAs.

Ref.	Frequency	Output Power	Efficiency (PAE)	Gain
[34]	2.14 GHz	40 dBm	35.2%	9 dB
[35]	3.5 GHz	40 dBm	37.3%	11.1 dB
[36]	2.5 GHz	42 dBm	30.85%	23.84 dB
[37]	2.14 GHz	35 dBm	39%	10 dB

5.3. DPA Combined with the Envelope Tracking Technique

Employing the input signal envelope tracking technique in the DPA is applicable for highly efficient handset PAs for operation in the low power region. In this amplifier, the gate bias of the peaking amplifier is controlled according to the magnitude of the envelope [38]. This technique can solve the common problem with the DPA, which is the fixed bias condition of the peaking PA. If both the peaking and carrier amplifiers adjust the gate bias, the overall system efficiency will be obtained by multiplying the efficiency of DPA with that of the supply modulator [39]. Envelope tracking can also be deployed on the carrier amplifier, leading to provide efficiency enhancement in low power level, then, while it goes to saturation in higher power and would bring about the modest efficiency. Figure 5a shows the block diagram of a DPA employing envelope tracking. In this architecture, the DC current for each PA decreases to a half, leading to larger impedance. First, the device is biased for operation in low power levels, and then the bias is adjusted as the input power level increases making the peaking amplifier conduct more current to the load, while the carrier PA maintains the peak voltage. At the maximum power level, the peaking amplifier will have the same bias as the carrier amplifier, and its current contribution to the load will be equal to the carrier amplifier. Thus, the bias adaptation can help the peaking amplifier to compensate for the extra margin that it requires for synchronizing with the carrier amplifier. Figure 5b indicates The function of envelope shaping for the Doherty amplifier when the supply voltage is clipped to 3 V at back-off; thus, the extreme peak of envelope signal is reduced, and it results in an extension of dynamic range as well as high efficiency of the supply modulator [40].

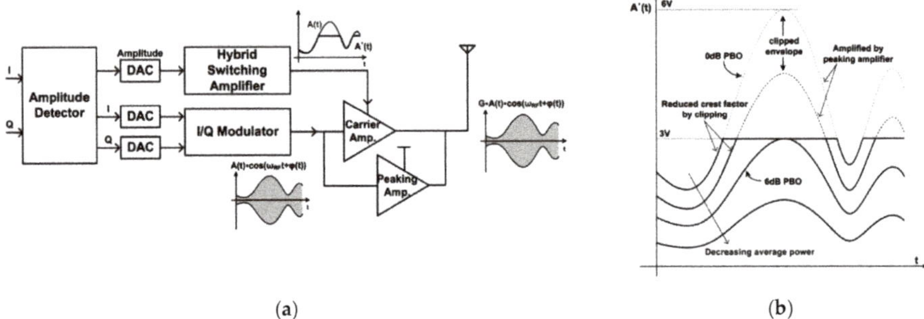

(a)   (b)

Figure 5. (a) Block diagram of a Doherty PA with Envelope Tracking. (b) Carrier PA supply voltage at different output power levels [41].

This technique is attractive, because it can exactly synthesize the peaking amplifier current profile. However, providing tuning for ET-Doherty PA with three-port and additional circuit elements may increase the complexity and cost. Table 4 lists the published articles, which applied ET technology for the integration with DPAs for higher efficiency. The results show that using the supply modulation method reduces the gain reduction and provides a good PAE.

Table 4. Performance comparison of envelop tracking combined with Doherty amplifiers.

Ref.	Signal	Frequency	Output Power	Efficiency (PAE)	ACLR (dBc)
[42]	WCDMA	2.14 GHz	38 dBm	39.4%	−30
[43]	WiMAX	1 GHz	37 dBm	55.5%	−33.15
[44]	WCDMA	2.14 GHz	42 dBm	50.9%	−26.5
[45]	LTE	2.14 GHz	38.4 dBm	27.8%	−45

5.4. Class-F Doherty PA

Doherty can employ the switching mode for tuning two main harmonic components that enable amplifier to provide higher efficiency and power in order to restore the proper load modulation. Among the available techniques, the so-called saturated Class-F Doherty scheme has drawn the most attention because of its capability of outputting high power and providing high PAE performance [46]. The Class-F amplifier terminates the device output with the short-circuited of even harmonic frequencies of the fundamental component and open-circuited of odd harmonics to synthesize a perfect square waveform of drain voltage combination with half-sinusoid drain current [47]. Therefore, the overlap between the current and voltage waveform is near zero leading to reduce the dissipated power, the size, and weight of the power-amplifier. The second and third harmonic voltage components have the main contribution on Class-F design and higher order imply impractical suggest, because of circuit complexity. In fact, the harmonic generating mechanism of Class-F PA enables the Doherty PA to operate at quite high frequency (up to the X-band) for smart manufacturing applications [48]. A number of practical aspects that are related to the finite number of harmonics, which can be efficiently controlled in actual devices, are addressed in [49]. In [50], the authors propose a new asymmetric Class-F^{-1}/F GaN Doherty while using Fourier transforms to compensate the low output current of peaking device, which mitigates the improper load modulation associated in conventional Doherty. The implemented device in this reference is depicted in Figure 6b. In other work [51], a blended Class-EF mode and load-pull technique for fundamental-frequency are proposed. The fabricated

DPA delivers acceptable peak output power of 40.4 dBm at 84.4% drain efficiency. In addition, many linearization techniques have been adapted to variable envelope systems to overcome the nonlinearity issue associated with Class-F DPA, due to operating in saturation mode [52]. Figure 6a presents the schematic of Class-F DPA, where the wave-shaping network includes the harmonic control parts and the fundamental matching network.

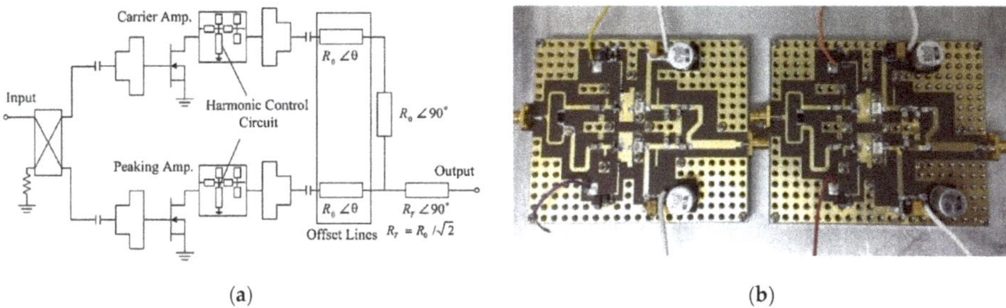

<table>
<tr><td align="center">(a)</td><td align="center">(b)</td></tr>
</table>

Figure 6. (**a**) Schematic of Class-F based DPA [49], (**b**) Realized Class F^{-1}/F^{-1} left, and Class F^{-1}/F right DPA in [50].

Table 5 indicates the employed technology, center frequency, output power, and efficiency of key works on DPAs employing the class F mode. It shows that the theoretical efficiency improvement can be achieved by providing the proper output harmonic loading conditions.

Table 5. Performance comparison of Class-F^{-1}/F Doherty Amplifiers.

Ref.	Tech.	Frequency	Output Power	Efficiency (PAE)	Signal	Gain
[47]	GaN HEMT	3.5 GHz	40 dBm	37.7%	WiMAX	8.8 dB
[48]	GaN HEMT	2.4 GHz	44 dBm	70%	LTE	15 dB
[49]	GaN HEMT	2.4 GHz	37.3 dBm	68%	LTE-Advanced	15 dB
[51]	GaN HEMT	2.14 GHz	43 dBm	49%	WCDMA	6.7 dB

5.5. Class E Doherty PA

Class E PA has similar non-linearity to the class C PA, but higher PAE performance. The transistor in Class-E mode operates as a switch to summing up the DC and RF currents and charge the drain capacitance. In the optimal condition, the drain voltage tends to become zero, when the transistor turns on, leading to the elimination of the losses for charging the capacitor. In fact, the shunt capacitor minimizes the overlap between the current and the voltage waveforms in the circuit of Class-E power amplifier. Figure 7a displays the schematic of a Digital Doherty PA while using the Class-E topology. Assuming that both on-resistance switch in the carrier and peaking path vary from 1000 to 0.1 [52]. At the beginning, the input power is low, and the on-resistance of the peaking amplifier switch is infinity. With the increase in input power, the on-resistance of the carrier amplifier switch ranges from 1000 to 0.1, yield an increase of output power and efficiency [53]. When the on-resistor of the carrier PA reaches its minimum value of 0.1, the obtained efficiency is 96.9% and is close to the ideal value of 100% [54]; then, the load of the peaking PA decreases from infinity to 0.1 with a further increase of the input current. Another efficiency peak is reached when the on-resistor of the peak amplifier is also 0.1 and it achieves 98% efficiency. Unlike the conventional DPA, it is obvious that the digitally controlled Class-E based DPA can potentially deliver high efficiency at the maximum and back-off power levels. Figure 7b shows the photograph model of a realized Class-E DPA.

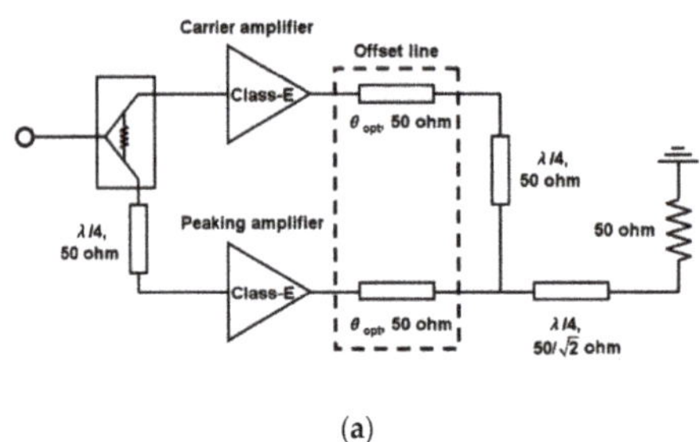
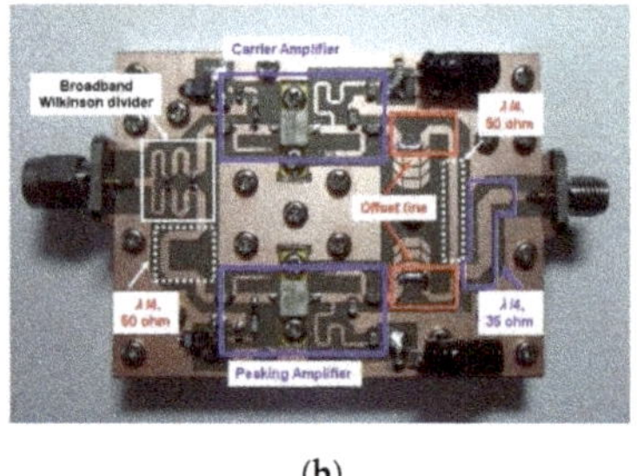

(a)   (b)

Figure 7. (**a**) Basic schematic of Class E based DPA. (**b**) Picture of Class-E DPA using GaN HEMT presented in [54].

Table 6 summarizes the performance of the Doherty power amplifier that is based on harmonic tuning; taking advantage of the soft-switching operation offered by Class E leads to high efficiency operation.

Table 6. Performance summary of Class-E Doherty Amplifiers operation

Ref.	Tech.	Frequency	Output Power	Efficiency (PAE)	Signal	ACRL
[51]	GaN HEMT	2.14 GHz	43.1 dBm	56.1%	WCDMA	−27.2 dBc
[52]	GaN HEMT	2.85 GHz	40 dBm	42.9%	WCDMA	−26 dBc
[53]	GaN HEMT	2.655 GHz	42 dBm	49.3%	WiMAX	−23.1 dB
[54]	GaN HEMT	2.4 GHz	45 dBm	68%	CW	−26.6 dB

The power matching circuits of amplifiers should be appropriately designed to cancel the IMD over power ranges across the wide bandwidth [55]. The theoretical load of 100 Ω (2*Ropt*) for the main amplifier is not actually the optimum load value, because of the knee effect of the transistor. The efficiency of traditional symmetrical DPAs will be enhanced by adopting an output impedance for the peaking device different from an open circuit and modifying the phase delay of the input matching network.

5.6. Inverted Doherty PA

In the inverted DPA structure, the inverter transmission line is connected to the drain of peaking PA instead of the carrier PA. A 25 ohm load is observed by the main PA to deliver maximum efficiency within the low power region by reversing the Doherty combining point. The impedance inversion in the conventional Doherty is accomplished with the 50 ohm λ/4 line, which is incorporated into the peaking output matching network, and the output is taken from the combining node. However, in inverted Doherty topology, the inversion λ/4 line is connected to the carrier amplifier and the offset line provides the off-state condition for the peaking PA, as depicted in Figure 8a, [56,57]. In fact, the measured output reflection coefficient that determines the lengths of the offset lines is rotated to a high impedance in the conventional DPA, while it is rotated to a low impedance for the inverted DPA. The inverted DPA provides the high efficiency at the low drive level; however, the key challenge of the inverted DPA is designing the output matching networks of the carrier and peaking amplifiers. Figure 8b shows an example of fabricated 190W IDPA with the final stage in 14 × 8 cm^2 case [58].

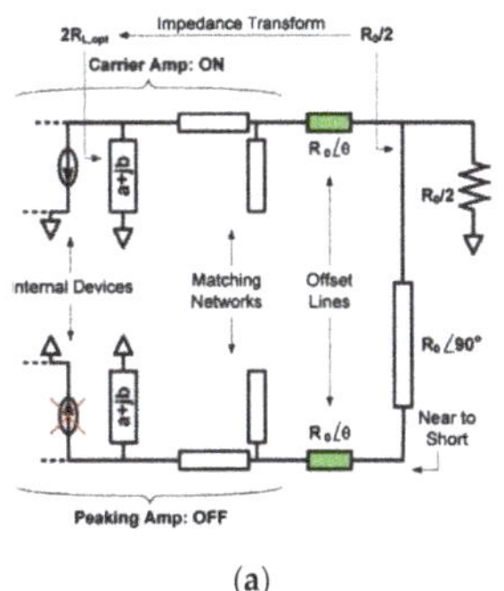

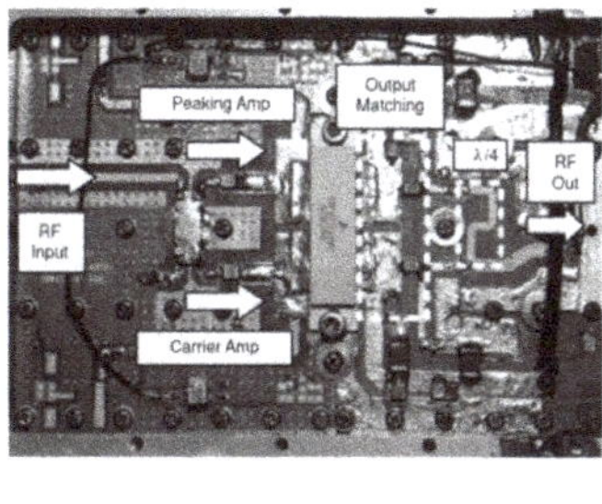

(a)   (b)

Figure 8. (**a**) Load network of inverted DPA operation [57], (**b**) Picture of the 190W IDPA described in [58].

5.7. Digital Doherty PA

An optimal load modulation behavior over a wide frequency band is achievable with good phase synchronization between the carrier and the peaking PAs. This can be obtained through reengineering the basic DPA structure, which leads to the so-called "digital DPA" [59,60]. This solution relies on a dual-input architecture for digitally assisted control over the input signals of amplifiers to mitigate the hardware impairments of the analogue circuit [61]. According to this strategy, the carrier and peaking PAs are individually driven from the baseband. Such a digital DPA eliminates the signal splitting device and minimizes the wasted energy of the driving power into the peaking path.

The Dual-Input DPA transmitter in Figure 9 consists of a DSP and a signal conversion block [62]. The DSP performs the modulation, interpolation, and digital pre-distortion (DPD) on an input signal. In this architecture, the DSP is applied to provide a main signal component along the carrier PA path, and a peaking signal component along the peaking PA path from a baseband input signal. The carrier and peaking PAs amplify the signals that are individually up-converted to an RF frequency. Precise alignment between the two paths at the baseband and isolation between the carrier and the peaking signals is crucial, which makes the implementation of the dual-input Doherty architecture challenging. Moreover, several limitations of digital DPA, like high oversampling, extensive pre-distortion requirements, and timing synchronization, restrict its application in multi-Gb b/s modulations [63,64] and limit the modulation speed.

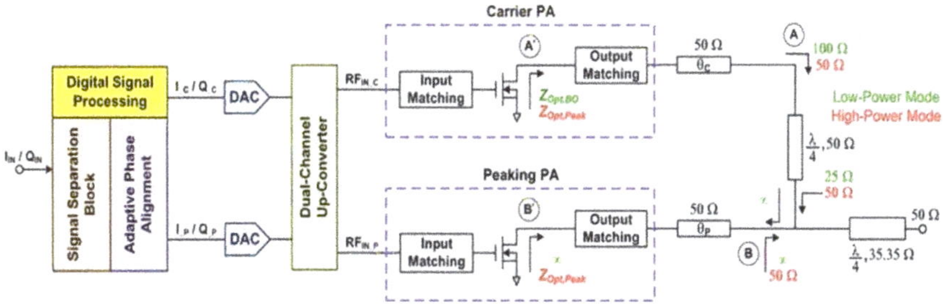

Figure 9. Block diagram of the dual-input digital Doherty PA architecture [62].

On the other hand, the analog PAs mostly show a significant distortion and gain compression. The paper work of [65] proposes an advanced hybrid use of mixed-signal Doherty-PA (MSDPA) in order to overcome these limitations. In this work, the MSDPA combines analog carrier PA with a digital peaking one, which is driven by a complex modulated signal. Additionally, in this architecture, several peaking PAs are applied to extend the DPA compression point. The MSDPA extends the dynamic range by acting as an analog mm-wave PA in low power regime, in contrast to the digital DPAs with limited dynamic range due to the time-sampled interpolations of envelop signal. Besides, in high power levels, it works in mixed-mode to enhance the back-off efficiency by controlling the turning on instance of peaking PA. Finally, taking advantage of envelop-varying complex modulation at input, the MSDPA reduces the bandwidth expansion in mm-wave DPA.

Table 7 summarizes the recent research on RF and microwave DPAs. This table reports the employed technology, frequency target, output power, target OBO, and OBO efficiency. It is important to notice that all of the proposed techniques aim to enhance efficiency, gain, and linearity.

Table 7. Recent research publications for RF and microwave Doherty power amplifiers (PA).

Ref.	Year	Technology	f_o (GHz)	Power Gain	POUT (dBm)	PAE at 6 dB OBO	Signal
[66]	2013	SiGe HBT	2.4	11 dB	21	10%	LTE
[67]	2017	GaN HEMT	2.14	10 dB	35.5	39%	LTE
[68]	2015	GaN HEMT	3.6	11 dB	41	48%	LTE
[69]	2017	GaN HEMT	1.8–2.2	12 dB	50	50%	LTE
[70]	2017	GaN on SiC	15	7 dB	38	28%	WiMAX
[71]	2014	GaAs	23–25	12.5 dB	30.79	20%	WiMAX
[72]	2015	GaN HEMT	10	9.2 dB	36.02	47%	LTE

5.8. Linearization Techniques on Doherty PA

An extensive research on the linearization techniques of Doherty PA has been performed in order to suppress its poor intermodulation distortion (IMD) performance. The most applied techniques include adaptive digital predistortion (DPD), post-distortion, feedback, and feedforward, as well as their combination. The regulatory agencies require the spectral emissions masked at close offsets (1.5 MHz) to be at least −45 dB, while, the nonlinear power amplifier exceeds this level by 15 to 20 dB, resulting in a significant distortion [73]. The linearizing techniques are focused on the harmonic cancellation of the power amplifier by optimizing the circuitry for modern applications. Various dedicated linearization techniques for DPA generate the second and forth order of the nonlinear signal at the output of peaking transistor through the band-pass filter, and feed into the carrier device with the fundamental signal, to suppress the third and fifth order intermodulation products [74]. An analog or digital splitter converts the input signal to dual-input independent branch signals in a typical predistortion approach, which can practically limit the efficiency and linearity performance due to a static function in Figure 10. Moreover, the conventional pre-distortion methods are restricted to low IMD cancellation, because the memory effect induces lower and upper spurious emissions [75].

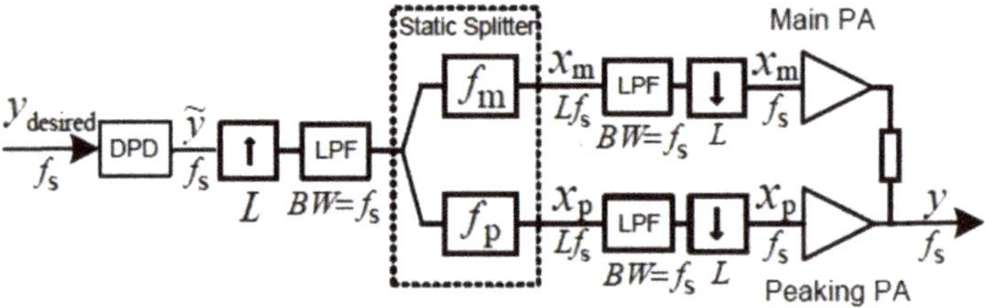

Figure 10. Linearization scheme in dual-input Doherty PA. f_s is the initial sampling rate, L denotes the up sampling factor, x_m, x_p are input signals of carrier and peaking PAs [74].

The dual-input Doherty PA, referred to as a digital DPA, generates two independent input RF signals to control the carrier and peaking PAs, leading to an extra degree of freedom that can be used to optimize its linearity and efficiency [76]. The varactor-based linearization architecture that was proposed in [77] adapted for dual-band DPA by utilizing additional input signal. In this method, if a proper envelope signal is injected, the transfer function of the DPA can be obtained by mapping function between the input and output signal and it can be inversed as a pre-distortion function. This scheme provides better performance in comparison with the single input linearization; however, it is not able to eliminate all of the intermodulation products of dual-input DPA. The linearized DPA commonly uses high sampling rates and fast digital to analog converters result in a spectral regrowth. In this respect, a digital predistortion method for a dual-input Doherty PA is introduced in [78], which performs the third-order Volterra-based function to drive the additional RF input signal to the peaking PA. Although, this dual-input DPD outperforms the single-input model in the reduction of the adjacent channel leakage power ratio, its outweigh complexity provides inadequate performance in high order nonlinearity. Figure 10 shows the block diagram of a conventional pre-distortion technique on Doherty PA in the digital domain.

The linearity of the high power Doherty amplifier can be improved by post-distortion-compensation, as proposed in [79], where the optimization is achieved by the cancellation of upper and lower symmetric sideband third order IMD products of devices by applying the two-tone input signal. In this work, a linearized transfer function of PA is modeled by the Taylor series function, when the gain expansion curve of the peaking device compensates the gain compression curve of the carrier PA, as can be seen in Figure 11. In another effort, a highly efficient feed-forward DPA is investigated [80] for base station application. This linear DPA optimizes the length of peaking compensation line and its gate bias by finding the output impedance of this device; however, it delivers a low PAE of 6–10%.

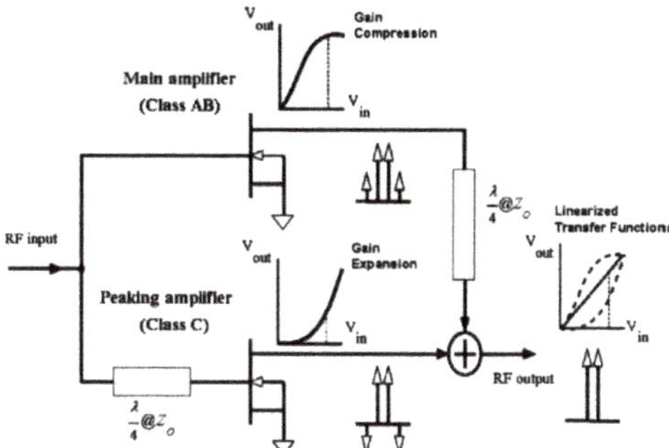

Figure 11. Linearization scheme of Doherty PA using post-distortion [79].

6. Doherty Bandwidth Extension

The efficiency enhancement of DPA was only maintained over narrow frequency bands and it offers a fractional bandwidth of smaller than 10% [81]. Although the conventional DPA is able to satisfy the modern handset requirements for efficiency, linearity, and output power, it is unable to meet the bandwidth requirement of the modern handset amplifier. Therefore, at present, the main challenge in DPA design is to extend its bandwidth. In conventional PAs, the output power or the gain define the bandwidth and the bandwidth range is achieved by the actual load power divided by the maximum power that could be delivered by the generator (available power). Beside, since DPAs are used to enhance the back-off efficiency, in this case a proper definition of bandwidth, is the frequencies range for which the back-off efficiency peak remains close to the maximum value achieved at the center frequency [82]. The DPA bandwidth increases with the increase in the input signal amplitude. While, the input power increases to provide the full voltage swing at the output, the bandwidth is not limited at all and the DPA delivers 78.5% efficiency over the whole frequency range. Therefore, more attention has to be paid to the bandwidth behavior at the back-off power.

The degradation of the efficiency and output power of DPA from their maximum levels, over the entire range of frequencies can be observed in Figure 12a,b, respectively [83]. It can be noticed that the maximum output power is reduced by 25% with deviation from the center frequency. Therefore, the first efficiency peak does not actually represent an exact 6-dB back-off from the maximum output power, but actually higher than the 6-dB level. The main sources of frequency limitation that were typically observed in various implementations of the classical two-way DPA can be divided into two categories:

- Theoretical limitations: are directly related to the selected DPA topology and realize in the impedance inverter frequency dispersion, as well as the carrier and peaking transistors' current profiles.
- Practical limitations: attributed to the imperfections in the building blocks of the DPA (e.g., frequency dependence of the phase compensation network and offset lines, device non-idealities, and output/input matching networks).

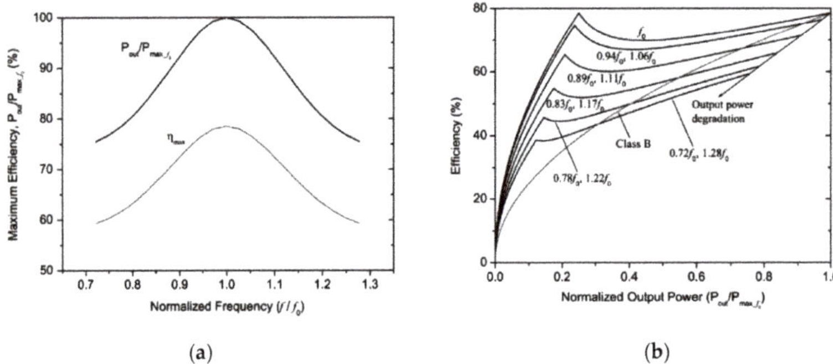

(a)   (b)

Figure 12. (**a**) Maximum efficiency and output power of DPA versus normalized frequency. (**b**) Efficiency of the DPA versus normalized output power at various frequency deviations from the center frequency [83].

The following sub-sections introduces the theoretical and practical bandwidth limiting components that contribute to constructing the DPA and present bandwidth extension techniques that have been proposed in recent research. Multiband capability in the conventional two-stage Doherty amplifier can be achieved when all of the Doherty constituent components are designed to deliver the corresponding performance over the required bandwidth of operation.

6.1. Impedance Inverters

When considering the carrier and peaking transistors as equivalent current sources in the two-way and three stages Doherty PAs in Figure 13a,b, respectively, the impedance that is seen by the carrier amplifier can be changed whilst the voltage swing across it has to remain constant, leading to maximized efficiency. Once the carrier amplifier saturates, the peaking amplifiers reduce the impedance that is seen at the output of the carrier PA by delivering current, which results in directing more current to the load, even when the carrier amplifier is saturated. Therefore, it is necessary to impose an impedance inverting network between the load and the carrier source, whereby the load impedance of each amplifier can be derived by the active load-pulling principle.

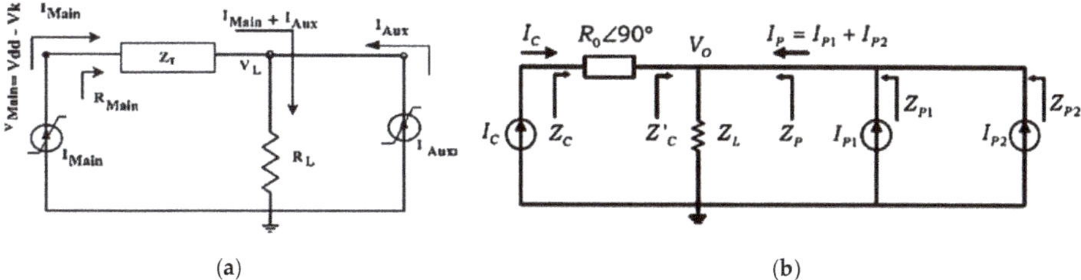

Figure 13. Structure of equivalent circuit of (**a**) Two-way, and (**b**) Three-stage DPA [84].

Impedance inverters, which are usually originated by means of a quarter-wave length transmission lines, provide perfect impedance inversion at the center frequency. By deviating from the center frequency, the length of impedance inverter has changed, and it has no longer $\lambda/4$ length, which shifts the purely resistive load that is seen by the carrier device (2*Ropt*) into a complex load disturbing the load modulation; that gets worse as the frequency deviation increases [85]. The frequency response analysis of the conventional DPA, in which the output combiner is realized by means of a $\lambda/4$ transformer, has been discussed in [86]. The optimum wideband operation of the DPA would require the transmission line (TL) to be a perfect impedance inverter over the whole desired frequency band, where not changing the physical properties of the TL is not possible; since the electrical length of the TL linearity increases with frequency. There have been various research works on the DPA to overcome the size constraint for handset applications, in which alternative impedance inverters, as surveyed in the following, replace the transmission line:

Three different impedance inverters, namely the coupled-line impedance inverter, short-circuit $\lambda/4$ TL compensated impedance inverter, and the open-circuit $\lambda/2$ TL compensated impedance inverter are proposed in [87], where the higher average efficiency could be obtained over a significant frequency range. The coupled-line impedance inverter consists of a coupled line section (CLS) and a single transmission line, where the electrical length of the coupled line and the feedback TL are both 90 degrees. This configuration provides the wideband impedance inversion properties that can be utilized to extend the bandwidth [88]. A short-circuit transmission line is the much simpler structure for a wideband impedance inverter by assuming that the coupling coefficient for the coupled lines is zero. However, the improvement is obtained by the expense of the bandwidth degradation at full power level for the coupled line impedance inverter and the open-circuit $\lambda/4$ TL compensated impedance inverter. Moreover, an additional $\lambda/2$ TL compensation line is connected to the quarter wave transmission line (QWTL) in the open-circuit $\lambda/2$ compensation transmission line, which can provide a wideband DPA, if the impedance of the compensation line is chosen to be equal to the optimum load impedance of the peaking device, and a proper input phase compensation network is applied.

Moreover, the effective load impedances of the Doherty amplifier can be expressed as a $[ABCD]$ parameters of the impedance transformer transistor, when the transistors are considered to be ideal current sources [89], as can be seen in Figure 14a. The $[ABCD]$ is the transmission matrix of the impedance transformer at a given frequency. In the low-power region, the effective resistance of the carrier PA has a maximum value of 2*Ropt* only at the center frequency, and it experiences a reduction as the frequency deviates. This leads to a back-off efficiency degradation as the frequency deviation increases, particularly when the DPA is operated over a wide bandwidth. The effective resistance of the carrier amplifier load impedance in the low-power region should be increased to extend the operation frequency band of the DPA. The resonance inductor-capacitor network (or *LC* tank) introduced in [90] is one of the possible solutions for reactance compensation. Additionally, a $\lambda0/4$ short-circuited stub with a characteristic impedance of Z0 can extend the operation frequency, as illustrated in Figure 14b. However, the shunted *LC* tank or the short-circuited stub operations may affect the load impedances of both the carrier and peaking amplifiers at saturation, which can degrade the DPA performance.

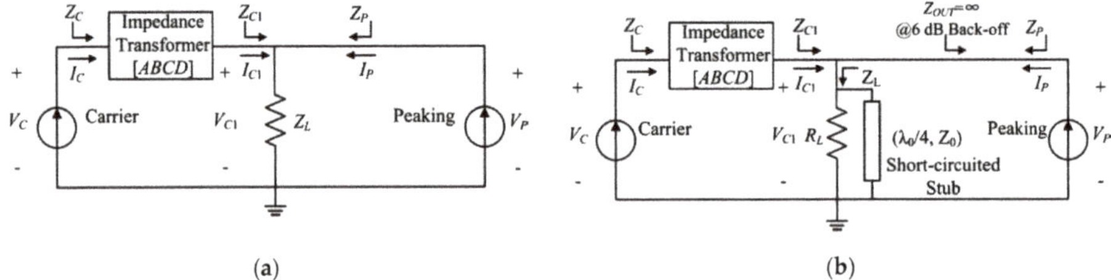

(a)   (b)

Figure 14. (**a**) Equivalent-circuit diagram of a DPA with the carrier and peaking amplifiers represented by current sources. (**b**) DPA using a shunted short-circuited stub [89].

6.2. The Output Power Combiner Network

The Doherty output combining circuit in Figure 15a consists of a quarter-wave transmission line with a characteristic impedance of 50 Ω to provide the required phase delay, and a quarter-wave transmission line of 35.35 Ω impedance that transforms the common load impedance to the final load impedance of 50 Ω. When the peaking amplifier is off, the output power combiner network acts as a 1:2 impedance transformer, which results in a loaded quality factor of 1.73 at 3 dB output-power that limits the broadband operation. In the high power levels, particularly at the maximum power point when the carrier and peaking amplifiers deliver equal amounts of output power, the combining circuit functions as a 1:1 combiner. With the maximum reflection coefficient Γ_m, the fractional bandwidth $\Delta f / f_0$ of the quarter-wave transmission line can be expressed by:

$$\Delta f / f_0 = 2 - 4/\pi \cos^{-1}\left[\left(\frac{\Gamma_m}{\sqrt{1-\Gamma_m^2}}\right).\left(2\sqrt{Z_0 Z_L}/|Z_L - Z_0|\right)\right] \tag{1}$$

where Z_L and Z_0 is the impedance pair for inversion [91]. Reducing the transformation ratio of the quartet-wave impedance inverter, making Z_L and Z_0 closer together, can enhance the bandwidth of a quarter-wave TL. If the common load impedance is increased to a higher value, and then the transformation ratio of interconnection transmission line between peaking and carrier amplifiers will be reduced, which extends the bandwidth. Recent research proved that better bandwidth enlargement of the DPAs is obtained when the adopted output combiner circuit scheme is different from the conventional one. In [92], the authors proposed a DPA architecture for DPA center frequency, four sections of quarter-wave transmission lines are adopted instead of two, with different characteristic impedances, where two of the four characteristic impedances can be assumed as the free parameters; thus being usable to define the bandwidth of the DPA. This approach can optimize both the performance and frequency response of the DPA.

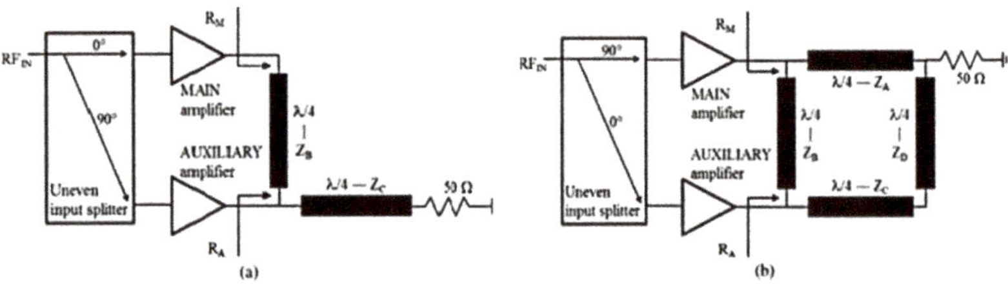

(a)   (b)

Figure 15. (**a**) Standard output power combiner and (**b**) four-section transmission line topology [92].

Quarter-wave TL stub can be replaced by an LC tank to reduce the size and broaden the bandwidth, as can be seen in Figure 16a. The resonator circuit provides open-circuit at the fundamental frequency and the short-circuit harmonics by large capacitor(C_t), while a short-circuit quarter-wave stub behaves differently at harmonics. It only provides short-circuit at even harmonics; therefore, the impedance of the carrier and peaking transistors are no longer the same at the maximum output power level [93]. In contrast, the Doherty combiner with a resonant tank holds the prime active load modulation of the carrier amplifier over the broadband. Figure 16b illustrates that the tank inductance L_t acts as the biasing feed for both the carrier and peaking amplifiers to provide small baseband impedance at the drain.

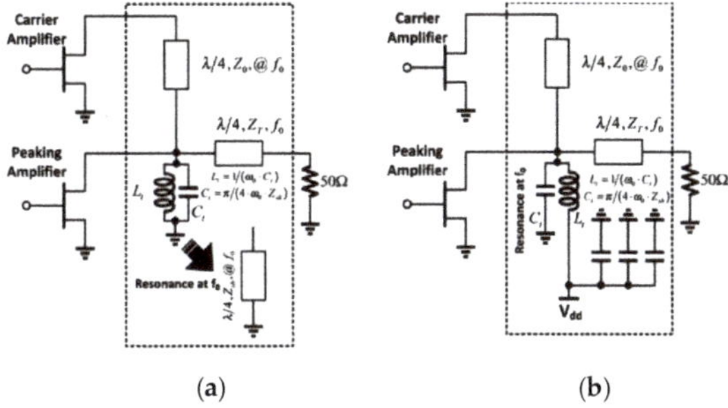

Figure 16. The broadband output combiner employs a resonance *LC* tank at the output of the peaking PA (**a**) The short-circuited, quarter-wave stub as a resonator can be approximated by a parallel LC tank [93]. (**b**) The tank inductor Lt can act as the biasing feed for both the carrier and peaking amplifiers [94].

The digital assist Doherty mentioned in previous sections can be adopted, in which a pre-compensation mechanism acts on the input power distribution and the phase variation between the carrier and peaking amplifiers to compensate the frequency-selective behavior of the DPA to overcome the problem of bandwidth restrictions imposed by the quarter-wavelength transformers of the Doherty combining network [94]. The pre-compensation function will be dynamically adjusted based on the modification of current factors, which is used to model the variation of the output current of the carrier amplifier and that of the peaking device, respectively, in response to the changes in the injected input power at the center frequency.

6.3. Offset Lines

The offset lines are the essential components following the carrier and peaking amplifiers for optimum load modulation. The load impedance that was observed by the carrier PA at low power levels is twice $(2Z_0)$, and the output matching network is designed to match the optimum impedance determined in load pull and $2Z_0$, subsequently the efficiency and gain are maximized; whilst, at the output of the peaking amplifier, the offset line is adjusted to a high impedance. Thus, the offset line that was commonly used to compensate the output parasitic effects of the transistors and to modulate the output load impedances of both amplifiers to keep them close to the ideal values. Although this approach can be effective for single ended PAs, depending on the length of the offset line, as the operating frequency deviates from the center frequency, the output impedance of the peaking PA decreases. Thus, power leaks into this amplifier.

The offset lines and the impedance inverter in the DPA topology should be eliminated in order to enhance the bandwidth [95,96]. If the output capacitance of the PA device can be compensated in a more wide-band manner, the overall bandwidth of the DPA will be significantly improved. The linear output capacitance of a transistor can be absorbed by the quasi-lumped equivalent circuit of a quarter-wave line [97]. By doing so, the efficiency versus frequency improves, since, in this situation, the DPA is only limited by the bandwidth of the quasi-lumped transmission line. In [98], the authors eliminate the capacitive reactance by adding a parallel inductor at the device output capacitance, resulting in an *LC* resonant circuit that has a small bandwidth and it does not address the practical implementation of the DPA. Another attractive technique is to absorb the output capacitance along-with the connecting bond-wires in the TL forming of the impedance inverter [99]. Figure 17a shows a wideband DPA with in-phase combination of both devices' output powers and Figure 17b shows the related equivalent schematic, which includes the output capacitance and connecting bond-wires at the active devices. Note that the transmission line along with bond wires and the output capacitance of the PA devices result in a quasi-lumped TL. Consequently, if the length and the characteristic impedance of this TL are properly adapted, it will act as an impedance inverter at the center frequency.

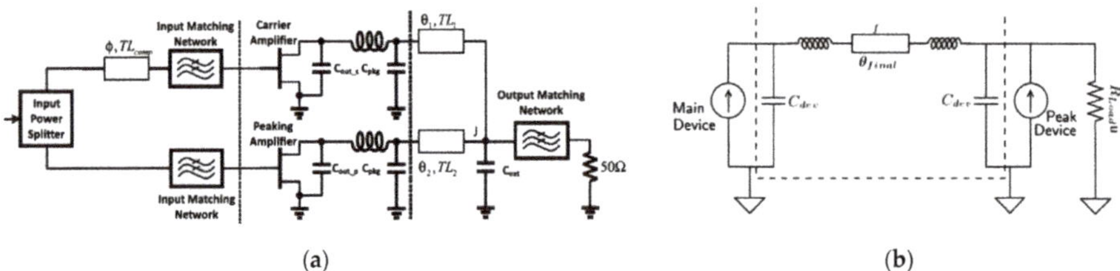

(a) (b)

Figure 17. (**a**) Simplified scheme of broadband Doherty PA [95]. (**b**) Schematic of a DPA, which absorbs the device output capacitances and bond-wires into the quasi lumped transmission line impedance inverter [99].

6.4. Input Dividing Network

The DPA behavior, as a function of the input power, is usually divided in two regions: saturation and back-off. The Wilkinson power divider in Figure 18a is utilized in most of the Doherty designs to achieve power division and isolation. It consists of quarter-wavelength micro strip lines for output ports and resistors for isolation. Moreover, it is better than the direct power divider in terms of bandwidth, because the power division ratio is determined by the input reflection coefficient of each amplifier, resulting in higher efficiency. However, it is desirable to drive more power to the carrier at low power to prevent the peaking amplifier from turning on early, and it improves the gain and efficiency. Moreover, more power allocated to the peaking path at high power level leads to proper load modulation, desired power generation, and good linearity by IMDs cancellation.

(a) (b)

Figure 18. (**a**) Topology of unequal Wilkinson power splitter which uses a resistor that is normally open circuited (even mode) and does not generate loss (**b**) Compact micro-strip three-way Wilkinson power divider [100].

For a multiband operation with the center frequency ratio at each of the frequency bands of two or greater, the input divider can be configured as a multi-section Wilkinson power divider [100], which consists of stepped transmission-line sections with different characteristic impedances and electrical lengths. Figure 18b shows the compact micro-strip three-way Wilkinson power divider designed to operate over a frequency range of 1.7 to 2.1 GHz, with minimum combining efficiency of 93.8%, maximum amplitude imbalance of 0.35 dB, and isolation better than 15 dB in [101].

6.5. Phase Compensation Networks

The phase compensation networks are required to ensure the in-phase combination of the carrier and peaking amplifiers' output currents at saturation. The input and output matching networks may be not identical because of using different transistors based on bias condition and output power. Even with the same transistors, the matching networks may be different, since the two branches are working under different bias conditions, and they may have different input and optimum load impedances. Due to their different bias and matching conditions, the phase behaviors of the main and peaking amplifiers may not be identical over the bandwidth, consequently the phase compensation is difficult to achieve across the frequencies band while using a fixed-length offset line. A possible initiative to combat this challenge could be applying the high pass filter (HPF) with the same impedance transformation ratio in the input matching network of the amplifiers. This approach compensates the phase variation of the quarter-wavelength transformer and delay [102]. An alternative solution is to individually inject the input power to the amplifiers, so that the phase of the input signal can be adapted for each amplifier at each frequency [103]. However, this solution significantly increases the circuit complexity of the wideband DPA.

6.6. Input Matching Network

The appropriate impedance transformations at the transistor drain levels are realized while using impedance-matching networks. In the peaking PA, the input impedance increases as the input power increases leads to the gain expansion, whereas the input impedance is nearly constant and the gain compresses at high power for the carrier PA biased in class AB/B. Hence, the input powers delivered to the PAs should be properly adjusted to cancel gain expansion and compression. Proper input impedance matching is essential for highly efficient and linear operation since the input impedance of the peaking PA varies with the input power level using various matching circuits.

It is possible to use a multi-section matching transformer that consists of stepped transmission-line sections with different characteristic impedances and electrical lengths to provide an input broadband matching network [104]. Figure 19 shows the multi-section impedance transformer with N-1 sections matches in which the characteristic impedance of the transmission line are set to Z_0, to avoid impedance deviation in the matched bands. Such a structure is convenient in practical implementation, since there is no need to use any tuning capacitors.

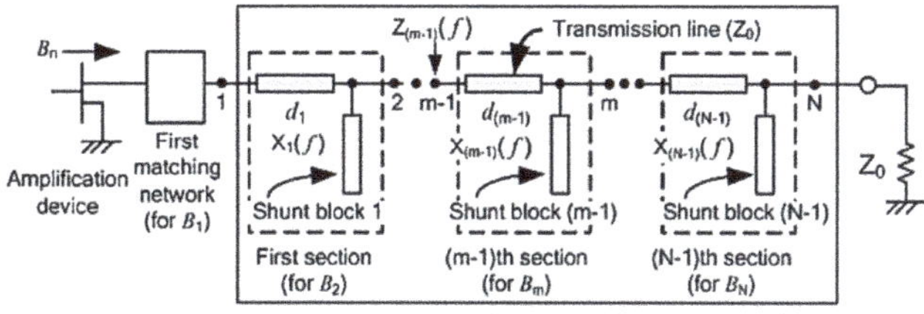

Figure 19. Basic topology of multi-section multi-band matching network [104].

6.7. Output Matching Networks

The bandwidth limitation arises when the load that is seen by the quarter wave transformer is unlike its characteristic impedance in the conventional DPA, where the load value matches at saturation, which is R_{opt}. This clearly means that in back-off, where the output load is $2R_{opt}$, impedance mismatch occurs. In a broadband PA design, the output matching network normally employs high-order topologies with filter structures to realize a certain impedance ratio in the operation band, which is necessary for high power amplifiers, because of the high impedance ratio [105]. Meanwhile, these high-order topologies provide an appropriate pass band with impedance ratio, and sufficient out-of-band rejection to ensure the transmission of the power. Table 8 outlines the comparison of various reported wideband DPAs in the literature in terms of output power, fractional bandwidth, power added efficiency at maximum output power, and at 6 dB back-off output power in different frequency ranges.

Table 8. Performance summary of broadband Doherty Amplifiers.

Ref.	Year	Technology	Frequency Range	Fractional Bandwidth	Output Power	Efficiency at Peak	Efficiency 6 dB OBO	Signal
[106]	2014	GaN HEMT	2 GHz	36%	37 dBm	70%	41%	WiMAX
[107]	2019	GaN HEMT	480 MHz	21%	43 dBm	63%	50%	LTE
[108]	2017	InGaP-HBT	800 MHz	30%	27.5 dBm	45.3%	41.2%	LTE
[109]	2013	GaN HEMT	300 MHz	14%	42 dBm	65%	50%	WCDMA
[110]	2016	GaN HEMT	1.1 GHz	49%	41 dBm	57%	50%	LTE
[111]	2018	GaN HEMT	2.3 GHz	87%	43.01 dBm	55%	33%	WiMAX
[112]	2018	GaN HEMT	1.1 GHz	51%	43.8–45.2 dBm	56%–75.3%	46.5%–63.5%	WCDMA

7. Multiband Doherty Integration

The multi-band operation of a DPA is obtained through passive structures, such as impedance matching networks, phase compensation network, and impedance transformer network, due to their frequency dependency behavior. In fact, in the two close frequencies, these components introduce different phase relation between input and output signals [113]. In this respect, the multi-band Doherty constructing transmission lines are designed to work at the independent frequencies. The performance of dual-band is similar to single-band DPAs while taking advantage of choosing the operating bands, since the passive elements can be optimized for two operating bands at the same time. In [114], the authors propose a fully integrated 28/37/39 GHz multiband Doherty for massive MIMIO applications. The proposed power-award prototype is implemented in 0.13 μm SiGe BicMOS and it achieves high efficiency and extended carrier bandwidth. In this scheme, a compactness output network on-chip transformer to reduce the loaded quality factor and the impedance transmission ratio, which results in broadening the Doherty bandwidth, replaces the conventional output power combiner. Moreover, this adaptive feeding scheme dynamically modulates the peaking PA load impedance to increase its output current, leading to an enhancement of power gain. Some design approaches of dual-band passive components are discussed, as follows.

Dual-band impedance inverter functions at two uncorrelated frequency bands and it introduces phase shifting of $\mp 90°$ between its input and output ports. This condition is obtained when the real load resistance is transformed to the real resistance at the input port of impedance network. The equivalent impedance inverters at two arbitrary frequencies that are commonly realized by T or π-network in Figure 20a, which are formed by two transmission lines with characteristic impedances of Z_1 and Z_2. The design equations for the T-network are given in [115]. In Figure 20a, the shunt stubs are 90° short-circuited and the line in the middle has electrical length of 180° at $f_0 = (f_1 + f_2)/2$. This inverter provides $\mp 90°$ phase shift at f_1 and f_2, where the output capacitance of the carrier and the peaking PAs can be integrated into the shunt stubs. Moreover, the shunt stubs improve the linearity of the circuit by serving as biasing feed. On the other hand, the quarter-wavelength line cannot provide multiband real to real impedance transformation operation due to its frequency dependency behavior. Therefore, cascaded transmission lines with different characteristic impedance that is introduced in [116], can provide a multipole response with different electrical lengths and different characteristic impedance ratio. The dual band two-section impedance transformation proposed in [117] is shown in Figure 20b.

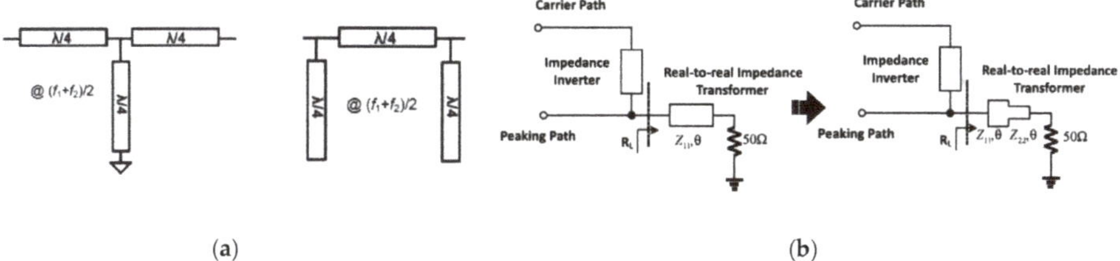

(a)   (b)

Figure 20. (**a**) Dual-band quarter-wave impedance inverter, T-network and π-network [115]. (**b**) A real-to-real impedance transformer is used to transform the output load to the required resistance at the Doherty common node: single-band and to dual-band [117].

The dual-band divider can provide equal power division for carrier and peaking PAs or send more power to the peaking PA and it can be configured as an in-phase multi-section Wilkinson divider or directional coupler. A suitable phase compensation network is required at the input to introduce the same phase shifting of the impedance inverter network. The branch-line coupler that is composed of four dual-band quarter wavelength transmission-line impedance-inverter can be based on π-section impedance transformers, or made up by four *T*-section quarter-wave transmission lines [118,119]. The phase compensation network can be directly absorbed into this component, providing appropriate output port connections.

The dual-band impedance matching network constructed at dual fundamental frequencies and second harmonics to ensure an appropriate gain and deliver maximum power. The dual-band phase offset lines are tuned for proper load modulation at the two frequencies with two arbitrary electric lengths, which leads to performance improvement. In Figure 21a, the Doherty load modulation circuit deploys the π-section network for dual-band implementations and Figure 21b shows the significant component on a sample of dual band DPA operating at 1.96 GHz and 3.5 GHz. The active load modulation varies with frequency due to different bias condition [120]. In fact, designing multi-band Doherty is critical, where all of the components require simultaneously addressing multiple band.

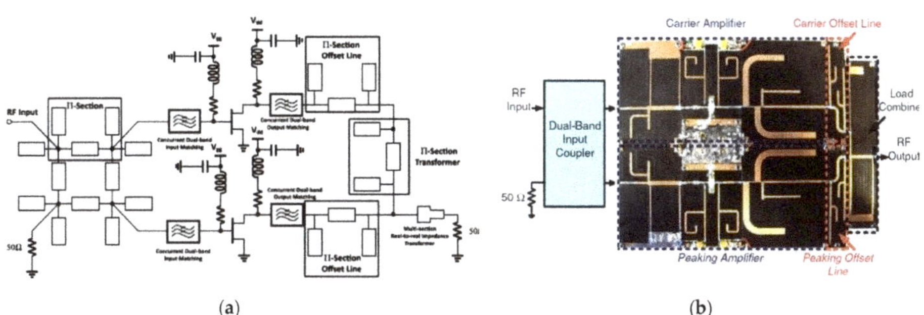

(a)   (b)

Figure 21. (**a**) Dual-band DPA architecture with Π-section structures [119]. (**b**) Dual-band DPA presented in [120].

The paper [121] presents the theoretical analysis and design technique of a load modulation balanced amplifier that is based on a specified output power and back-off level. In this architecture, the active matching network is performed by an adaptive output coupler, which controllers the reflection coefficient that is seen by amplifiers, Therefore, it does not need the bandwidth limited conventional output matching network. In this work, the design equations of output combiner are implemented based on an arbitrary black box network. The schematic of load modulation balanced amplifier is depicted in Figure 22, in which the Class-C control PA injects a signal to the output of the balanced amplifier to perform the load modulations, and the RF-input signal asymmetrically divided between the balanced PA and controlled PA. Moreover, this paper compares the performance of single-frequency load modulation balanced amplifier with different load modulation architectures in terms of device periphery scaling between the carrier and control devices at the back-off power range for efficiency enhancement, and concluded that the conventional DPA and balanced PA show similar analysis of dynamic range, gain compression, and power division factor of splitter.

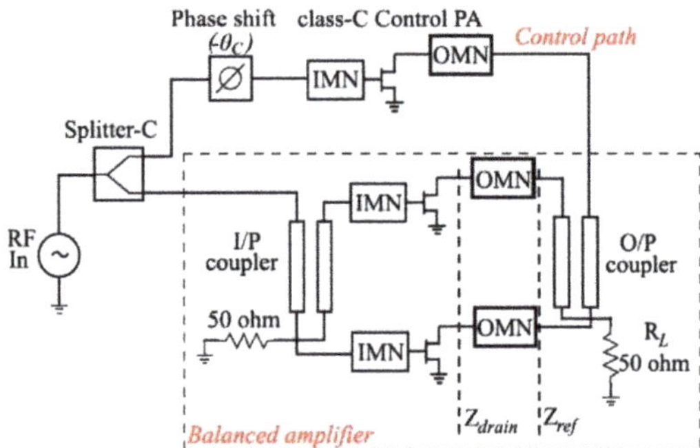

Figure 22. Simplified schematic of the RF-input load modulation balanced amplifier [121].

In [122], the authors develop the combination of Doherty and Chireix techniques in 2 GHz dual-input hybrid PA, in which the Outphasing angle is dynamically adapted with the incident power for proper load modulation. Recently, the Doherty-Chireix architecture has been investigated while using an output combiner with variable transmission lines in [123], and the realistic combiner when considering parasitics for wideband performance is discussed in [124]. The recent studies have shown that not only this continuum structure benefits from both Doherty and Chireix characteristics, but also it has the advantage of more design space to enhance the tradeoff between efficiency and linearity in comparison with single –input Chireix Outphasing PA mentioned in [125]. The Hybrid Chireix-Doherty (HCD) enhances the average efficiency of the modulated signal by reducing the efficiency drop between back-off and peak power observed in the conventional Doherty PAs. However, HCD complicated design increases the complexity and cost. In the mentioned work of [122], four fundamental current and voltage ratio factors of carrier and peaking devices generalize the HCD continuum theory, and the drain voltage of carrier device is set as equal to supply voltage at back-off and full power for higher efficiency. The implementation evaluation of HCD PA drain efficiency and back-off power range, proves that the HCD has a superiority performance in comparison with the Doherty PA for both the CW and modulated signals. Figure 23 shows the proposed HCD prototype in [122], where the carrier PA operates in Class-F mode of operation and peaking PA is biased in Class-C at the package reference plane. In this design, the two port output combiner is synthesized and implemented while using the Z-parameters, which is connected to the carrier and peaking PAs at two individually ports. Moreover, a stepped-impedance matching technique is applied for conjugate matching at fundamental frequency and the elimination of higher order harmonics. In the low power levels, the peaking PA is turned off in the Chireix-Doherty PA, similar to the Doherty PA.

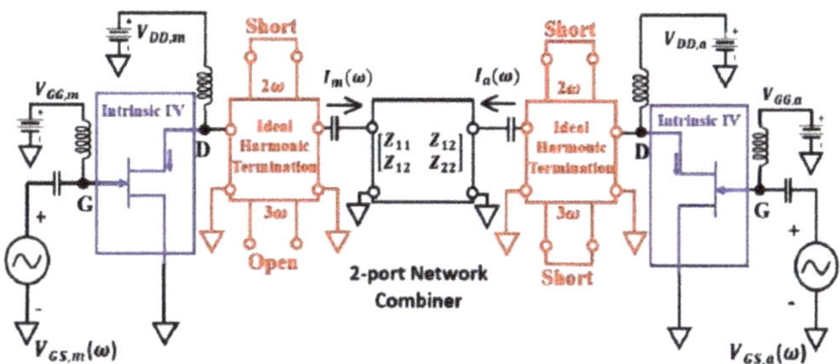

Figure 23. Hybrid Chireix-Doherty (HCD) PA prototype [122].

8. Compact Design of DPA for Handset Applications

The Doherty amplifier is less popular for handset application because of the size, bandwidth, and complex circuit topology. However, those problems can be solved while using compact design of the Doherty to integrate into a single chip [126]. By merging conventional components, such as power divider, the quartet-wavelength line for load modulation, and offset-lines for imaginary impedance modulation, the compact size and stable operation will be achieved. For mobile applications, the conventional bulky input power divider of Wilkinson splitter can be substituted by a direct power splitter to deliver more power to the gate of carrier amplifier in low power region due to its higher admittance, leading to compensate low gain, and peaking amplifier receives more power in high power region to increase the gain and output power [127]. The carrier amplifier could be matched to 50ohm for all power levels, because its impedance variation is small. Smaller size of the output circuit implementation can be achieved using inductance of the bias lines to resonate out the output capacitances of carrier and peaking amplifiers. Therefore, the output matching networks do not need to transfer the imaginary part of impedance. Moreover, the bulky quarter-wave inverter can be replaced by an equivalent lumped LC network for a Monolithic Microwave Integrated Circuit (MIMIC) handset application. Harmonic load condition and a compact design are the determining factors to choose appropriate lumped network type [128]. The lumped inverter of π-type shown in Figure 24a, reduces the size of the network since the two large capacitors provide short second harmonic load conditions at the output of carrier and peaking amplifiers; therefore, the second harmonic has a minor effect on linearity. Figure 24b illustrates the DPA implemented chip photograph with 1.1×1.2 mm^2 area size.

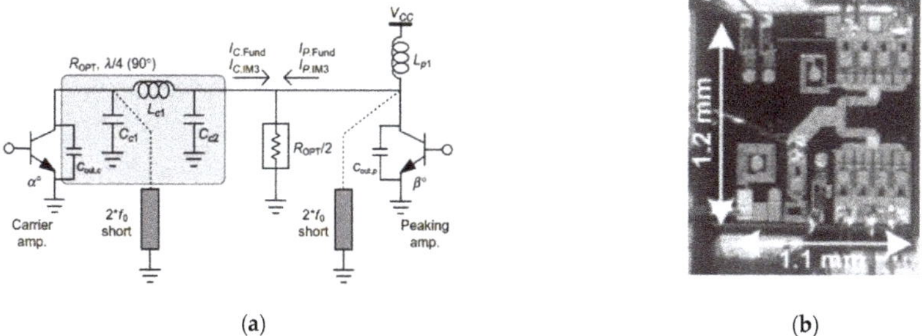

(a) (b)

Figure 24. (a) Doherty amplifier using a lumped low-pass π-type quarter-wave inverter. (b) Picture of fabricated DPA using an InGaP/GaAs HBT described in [128].

The compact design of DPA employ lumped components with on-chip integration has been presented in literature [66,129] while applying different bias control techniques, including fixed, step, dynamic biasing, and logical. These techniques adjust the gate and drain biases as a function of the average output power to reduce the current and voltage at the low power level, resulting in a reduction of power consumption. Figure 25a shows a block diagram of a MIMIC DPA while using an integrated bias adaptation circuit on chip with the same size of two-stages active devices and Figure 25b shows a sample of a module of printed circuit board as small as 1 mm × 1 mm [130]. At a low power level, the gate bias of carrier PA is fixed and tuning the drain bias according to the power level for high efficiency performs the load modulation. Once the same optimum impedance for both amplifiers are achieved at the higher power levels, the dynamic base bias control circuit increases the bias point of the peaking device to satisfy the linearity.

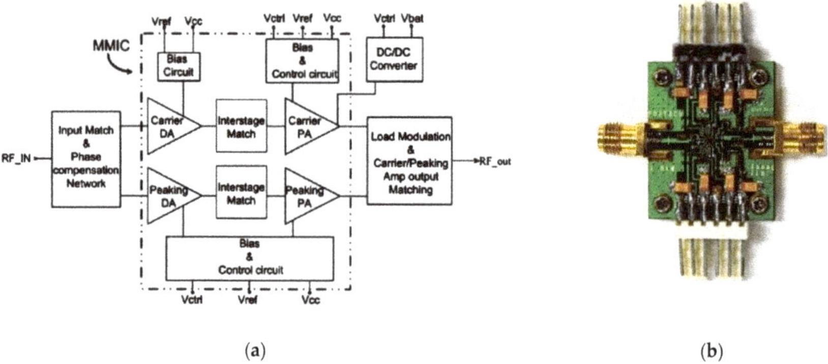

(a)   (b)

Figure 25. (a) MIMIC DPA using dynamic bias control. (b) A module of RF power test presented in [130].

A transformer-based DPA architecture is developed in standard 90 nm CMOS, adapting asymmetrical series combining transformer (SCT), allowing for chip implementation [131]. The SCT reduces the impedance matching loss in the silicon substrate by combination of multiple low voltage amplifiers. In this fully integrated un-even DPA architecture, by applying the proper number of Class-AB/C amplifiers according to the modulated envelope, a smooth transition for the whole power range and wide modulation bandwidth are achieved. It should be noted that the SCT is modeled using coupled lossy inductors, where the transformer ratio is the ratio of the carrier inductor to the total inductors. According to the implemented analysis of transformer based DPA, while using a very output large transformer ratio can improve the back-off efficiency of DPA, and it provides different size of carrier and peaking PAs to optimize the load modulation behavior. In fact, in a symmetrical transformer, the load seen by peaking PA is equal to the carrier PA at a high power level, while for asymmetrical transformer the load seen by peaking PA is smaller than that of carrier one, thus the peaking PA can deliver more power in comparison with carrier PA [132]. Therefore, in the un-even DPA, the peaking PA is optimized for the peak power operation. However, produced distortion by peaking amplifier with high transformer ratio can no longer be compensated, which degrades the linearity of DPA. Moreover, the fully integrated DPAs deliver low back-off efficiency, only because the carrier PA delivers power at back-off.

9. Conclusion

The Doherty PA is conceived as an efficiency enhancement technology for multimode multiband operation due to its low hardware complexity. Although it is ideally linear and highly efficient, the practical DPAs still suffer from nonlinear distortion and low average efficiency. Therefore, a variety of DPA circuit techniques have been adapted to improve the efficiency with extended back-off power, which includes the uneven power drive, gate bias adoption, saturated DPA, modified load modulation network, and digital DPA. Furthermore, in order to avoid the detrimental nonlinear effects of strong saturation, very accurate input splitting design, bias selection, and phase synchronization between two stages are required. Therefore, digital signal processing is often employed to adaptively adjust these parameters. However, every approach that improves the PA efficiency usually has the inherent drawback of complexity. Among the cited advance architectures, DPA employing the envelope tracking technique not only minimizes the power gain reduction and total power dissipation, but also optimizes the efficient operation. Therefore, the ET-DPA is a promising candidate for high-efficiency amplification and an acceptable linearity; howbeit, the penalty of supply modulator complexity should be taken into the account.

On the other hand, DPAs are affected by several bandwidth limiting factors that pose broadband matching problem. This narrowband behavior of the DPA is mostly caused by traditional operating classes for active devices, conventional impedance inverters, offset-lines, and phase compensation network that makes it challenging to apply the DPA for multiband handset applications. A variety of referenced techniques have been carried out with the goal of improving the bandwidth of RF and microwave DPAs in both the analogue and digital domain. In fact, the digital DPAs provide better real-time configurability than the analog DPAs. However, the non-ideal behavior of electrical components and the limitations in fabricated techniques often narrow the number of practical solutions available. Embedding the offset-line within the output matching network, reducing the impedance transformation ratio, and eliminating the influence of output parasites are some introduced practical solutions, which serve to extend the bandwidth and to overcome the manufacturing limitations.

Author Contributions: Conceptualization, M.S. and M.F.B.; methodology, M.S.; formal analysis, M.S.; investigation, M.S.; resources, I.T.E.E. and J.R.; data curation, J.R.; writing—original draft preparation, M.S.; writing—review and editing, J.R. and I.T.E.E.; visualization, R.A.-A.; supervision, J.R.; project administration, J.R.

Funding: This project has received funding from the European Union's Horizon 2020 research and innovation program under grant agreement H2020-MSCA-ITN-2016 SECRET-722424. This work is also funded by the FCT/MEC through national funds and when applicable co-financed by the ERDF, under the PT2020 Partnership Agreement under the UID/EEA/50008/2019 project.

Acknowledgments: This work is supported by the European Union's Horizon 2020 Research and Innovation program under grant agreement H2020-MSCA-ITN-2016-SECRET-722424.

Conflicts of Interest: The authors declare no conflict of interest.

References

1. Kanwal, K.; Safdar, G.A.; Ur-Rehman, M.; Yang, X. Energy Management in LTE Networks. *IEEE Access* **2017**, *5*, 4264–4284. [CrossRef]
2. Qualcomm Technology. Available online: https://www.qualcomm.com/products/rf. (accessed on 1 May 2019).
3. Mohammed, H.A.; Rosdiadee, N.; Mahamod, I. Survey of Green Radio Communications Networks: Techniques and Recent Advances. *J. Comput. Netw. Commun.* **2013**, *2013*, 453893.
4. Theodore, S.; Shu, S.; Rimma, M.; Hang, Z. Millimeter Wave Mobile Communications for 5G Cellular: It Will Work! *IEEE Access* **2013**, *1*, 335–349.
5. Mamta, A.; Abhishek, R. Next Generation 5G Wireless Networks: A Comprehensive Survey. *IEEE Commun. Surv. Tutor.* **2016**, *18*, 1617–1655.
6. Bukhary, A.M.; Abubakar, S.H.; Raed, A.A.; Danjuma, I.M.; Asharaa, A.S.; Issa, T.E.E.; Jonathan, R. Towards a 15.5W Si-LDMOS Energy Efficient Balanced RF Power Amplifier for 5G-LTE Multi-Carrier Applications. *EAI Endorsed Trans. Creat. Technol.* **2018**, *5*, e5.
7. Jayant, B.; Robert, A. Energy consumption in wired and wireless access networks. *IEEE Commun. Mag.* **2011**, *49*, 70–77.
8. Kahn, L.R. Single-Sideband Transmission by Envelope Elimination and Restoration. *Proc. IRE* **1952**, *40*, 803–806. [CrossRef]
9. Komatsuzaki, Y.; Lanfranco, S.; Komlonen, T.; Piirainen, O.; Tanskanen, J.; Sakata, S.; Ma, R.; Shinjo, S.; Yamanaka, K.; Asbeck, P. A High Efficiency 3.6–4.0 GHz Envelope-Tracking Power Amplifier Using GaN Soft-Switching Buck-Converter. In Proceedings of the 2018 IEEE/MTT-S International Microwave Symposium-IMS, Philadelphia, PA, USA, 12–14 June 2018.

10. Tahereh, S.; Raed, A.-A.; Nazar, T.A.; Issa, T.E.E.; Yosif, D.; Anho, N. Linear and Nonlinear Crosstalk in MIMO OFDM Transceivers. In Proceedings of the IEEE International Conference on Electronics, Circuits, and Systems (ICECS), Beirut, Lebanon, 11–14 December 2011; pp. 504–507.

11. Doherty, W.H. A new high-efficiency power amplifier for modulated waves. *Bell Syst. Tech. J.* **1936**, *15*, 469–475. [CrossRef]

12. Wang, Z. Demystifying Envelope Tracking: Use for High-Efficiency Power Amplifiers for 4G and Beyond. *IEEE Microw. Mag.* **2015**, *16*, 106–129. [CrossRef]

13. Chiara, R.; Anna, P.; Roberto, Q.; Vittorio, C.; Marco, P. High Efficiency Power Amplifiers for Modern Mobile Communications: The Load-Modulation Approach. *Electronics* **2017**, *6*, 1–29.

14. Raymond, S.; Simon, M.; James, W. A review of GaN on SiC high electron-mobility power transistors and MMIC. *IEEE Trans. Microw. Theory Tech.* **2012**, *60*, 1764–1783.

15. David, M.P. *Microwave and RF Design of Wireless Systems*; John Wiley & Sons, Inc.: New York, NY, USA, 2000.

16. Nunes, L.C.; Cabral, P.M.; Pedro, J.C. AM/PM distortion in GaN Doherty power amplifiers. In Proceedings of the 2014 IEEE MTT-S International Microwave Symposium (IMS2014), Tampa, FL, USA, 1–6 June 2014; pp. 1–4.

17. Proakis, J.G. *Digital Communications*; McGraw-Hill: New York, NY, USA, 1989.

18. Steve, C. *RF Power Amplifiers for Wireless Communications, Ser*; Artech House Microwave Library: Norwood, MA, USA, 2006.

19. Pengelly, R.; Baker, R. GaN Devices and AMO Technology Enable High Efficiency and Wide Bandwidth. *Microw. J.* **2014**, *57*, 66.

20. Camarchia, V.; Pirola, M.; Quaglia, R.; Jee, S.; Cho, Y.; Kim, B. The Doherty Power Amplifier: Review of Recent Solutions and Trends. *IEEE Trans. Microw. Theory Tech.* **2015**, *63*, 559–571. [CrossRef]

21. Stanley, Ch.; Max, G.; Patrick, B. The next challenge for cellular networks: Backhaul. *IEEE Microw. Mag.* **2009**, *10*, 54–66.

22. Andrei, G. *RF and Microwave Transmitter Design*; John Wiley & Sons, Inc.: New York, NY, USA, 2011.

23. Frederick, H.R.; Peter, A.; Steve, C.; Peter, B.K.; Zoya, B.P.; Nick, P.; John, F.S.; Nathan, O.S. *RF and Microwave Power Amplifier and Transmitter Technologies—Part 2*; High Frequency Electronics LLC.: San Diego, CA, USA, 2003; pp. 22–36.

24. Marian, K.K. *RF Power Amplifiers*; John Wiley and Sons, Ltd.: New York, NY, USA, 2008.

25. Jing, X.; Xiaowei, Z.; Lei, Z.; Yinjin, S. High-efficiency GAN Doherty power amplifier for 100-MHz LTE-advanced application based on modified load modulation network. *IEEE Trans. Microw. Theory Tech.* **2013**, *61*, 2911–2921.

26. Golestaneh, H.; Malekzadeh, F.A.; Boumaiza, S. An Extended-Bandwidth Three-Way Doherty Power Amplifier. *IEEE Trans. Microw. Theory Tech.* **2013**, *61*, 3318–3328. [CrossRef]

27. Maryam, S.; Issa, T.E.E.; Jonathan, R.; Raed, A.A. Modified Symmetric Three-stage Doherty Power Amplifier for 5G. In Proceedings of the EuCAP 13th European Conference on Antennas and Propagation, Krakow, Poland, 31 March–5 April 2019.

28. Amber, S.; Rakesh, K.; Sahab, R. A Doherty Amplifier with Envelope Tracking Technique and DSP for High Efficiency. *Electron. Comput. Sci. Eng.* **2004**, *1*, 147–152.

29. Raymond, O.; Christian, F.; Mustafa, O. Doherty Legacy. *IEEE Microw. Mag.* **2016**, *17*, 41–58.

30. Necip, S.; Simsek, D. Analysis design optimization and performance comparison of bias adapted and asymmetrical Doherty power amplifiers. *Prog. Electromagn. Res.* **2013**, *54*, 337–356.

31. Yunsik, P.; Juyeon, L.; Seunghoon, J.; Seokhyeon, K.; Bumman, K. Gate Bias Adaptation of Doherty Power Amplifier for High Efficiency and High Power. *IEEE Microw. Wirel. Compon. Lett.* **2015**, *25*, 136–138.

32. Sung-Chan, J.; Oualid, H.; Fadhel, M.G. Design Optimization and DPD Linearization of GaN-Based Unsymmetrical Doherty Power Amplifiers for 3G Multicarrier Applications. *IEEE Trans. Microw. Theory Tech.* **2009**, *57*, 2105–2113. [CrossRef]

33. Yong-Sub, L.; Mun-Woo, L.; Yoon-Ha, J. Unequal-Cells-Based GaN HEMT Doherty Amplifier with an Extended Efficiency Range. *IEEE Microw. Wirel. Compon. Lett.* **2008**, *18*, 536–538.

34. Janghoen, K.; Jeonghyeon, C.; Ilda, K.; Bumman, K. Optimum operation of asymmetrical-cells-based linearity Doherty amplifiers—Uneven power drive and power matching. *IEEE Trans. Microw. Theory Tech.* **2005**, *53*, 1802–1809. [CrossRef]

35. Eccleston, K.W.; Smith, K.J.I.; Gough, P.T. Harmonic load modulation in Doherty amplifiers. *Electron. Lett.* **2008**, *44*, 128–129. [CrossRef]

36. Junghwan, M.; Jangheon, K.; Ildu, K.; Jungjoon, K.; Bumman, K. Highly Efficient Three-Way Saturated Doherty Amplifier with Digital Feedback Predistortion. *IEEE Microw. Wirel. Compon. Lett.* **2008**, *18*, 539–541.

37. Ayushi, B.; Karun, R.; Shiban, K.K. A Design Strategy for Bandwidth Enhancement in Three-Stage Doherty Power Amplifier with Extended Dynamic Range. *IEEE Trans. Microw. Theory Tech.* **2018**, *66*, 1024–1033.

38. Yong-Sub, L.; Mun-Woo, L.; Yoon-Ha, J. Linearity improvement of three-way Doherty amplifier using power tracking bias supply method. *Microw. Opt. Technol. Lett.* **2008**, *50*, 728–731.

39. Yong-Sub, L.; Mun-Woo, L.; Yoon-Ha, J. A highly linear and efficient three-way Doherty amplifier using GaN Hemt cells for repeater system. *Microw. Opt. Technol. Lett.* **2009**, *51*, 2895–2898.

40. Choi, J.; Kang, D.; Kim, D.; Kim, B. Optimized Envelope Tracking Operation of Doherty Power Amplifier for High Efficiency over an Extended Dynamic Range. *IEEE Trans. Microw. Tech.* **2009**, *57*, 1508–1515. [CrossRef]

41. Yang, Y.; Cha, J.; Shin, B.; Kim, B. A microwave Doherty amplifier employing envelope tracking technique for high efficiency and linearity. *IEEE Microw. Wirel. Compon. Lett.* **2003**, *13*, 370–372. [CrossRef]

42. Mury, T.; Peter, G. Envelope-tracking-based Doherty power amplifier. *Int. J. Electron.* **2010**, *97*, 525–530.

43. Zhang, Z.; Xin, Z. LTE Doherty Power Amplifier using Envelope Tracking Technique. In Proceedings of the 2014 15th International Conference on Electronic Packaging Technology, Chengdu, China, 12–15 August 2014; pp. 1331–1333.

44. Colantonio, P.; Giannini, F.; Leuzzi, G.; Limiti, E. *On the Class-F Power Amplifier Design*; RF and Microwave CAE 9; John Wiley & Sons: New York, NY, USA, 1999; pp. 129–149.

45. Andrei, G. Highly-efficacy multistage Doherty architecture based on inverse class F power amplifiers for base station applications. In Proceedings of the ARMMS Conference, Oxford, UK, 28 November 2011.

46. Paolo, C.; Franco, G.; Rocco, G.; Luca, P. Theory and Experimental Results of a Class F AB-C Doherty Power Amplifier. *IEEE Trans. Microw. Theory Tech.* **2009**, *57*, 1936–1947.

47. Park, J.C.; Kim, D.; Yoo, C.S.; Lee, W.S.; Yook, J.G.; Hahn, C.K. Efficiency Enhancement of the Doherty Amplifier for 3.5GHz WiMAX Application using Class-F Circuitry. *Microw. Opt. Technol. Lett.* **2010**, *52*, 570–573. [CrossRef]

48. Fan, C.Z.; Zhu, X.W.; Xia, J.; Zhang, L. Efficiency Enhanced Class-F Doherty Power Amplifier at 3.5 GHz for LTE-Advanced Application. In Proceedings of the 2013 Asia-Pacific Microwave Conference Proceedings (APMC), Seoul, Korea, 5–8 November 2013; pp. 707–709.

49. Joonhyang, K. Highly Efficient Asymmetric Class-F−1/F GaN Doherty Amplifier. *IEEE Trans. Microw. Theory Tech.* **2018**, *66*, 4070–4077.

50. Shichang, G.K. High efficiency Class–F RF/Microwave Doherty Amplifiers. *IEEE Microwave Mag.* **2006**, *7*, 40–48.

51. Steven, G.; Alistair, S.; David, S.; Peter, B. A high efficiency Class-F Power Amplifier design technique. *Microwave J.* **2004**. [CrossRef]

52. Barakat, A.; Thian, M.; Fusco, V. A High-Efficiency GaN Doherty Power Amplifier with Blended Class-EF Mode and Load-Pull Technique. *IEEE Trans. Circuits Syst. II Express Briefs* **2019**, *62*, 151–155. [CrossRef]

53. Lee, Y.S.; Lee, M.W.; Jeong, Y.H. High-Efficiency Class-E-Cells-Based GaN HEMT Doherty Amplifier for WCDMA Applications. In Proceedings of the 38th European Microwave Conference, Amsterdam, The Netherlands, 28–30 October 2008; pp. 428–431.

54. Choi, G.W.; Kim, H.J.; Hwang, W.J.; Shin, S.W.; Choi, J.J.; Ha, S.J. High efficiency Class-E tuned Doherty amplifier using GaN HEMT. In Proceedings of the 2009 IEEE MTT-S International Microwave Symposium Digest, Boston, MA, USA, 9–11 June 2009; pp. 925–928.

55. Kim, B.; Kim, I.; Moon, J. Advanced Doherty Architecture. *IEEE Microw. Mag.* **2010**, *11*, 72–86. [CrossRef]

56. Zhou, H.J.; Wu, H.F. Design of an s-band two-way inverted Asymmetrical Doherty power amplifier for long term evaluation application. *Electromagn. Res. Lett.* **2013**, *39*, 73–80. [CrossRef]

57. Ahn, G.; Kim, M.K.; Park, H.-C.; Jung, S.-C.; Van, J.-H.; Cho, H.; Kwon, S.-W.; Jeong, J.-H.; Lim, K.-H.; Kim, J.Y.; et al. Design of a High-Efficiency and High-Power Inverted Doherty Amplifier. *IEEE Trans. Microw. Theory Tech.* **2007**, *55*, 1105–1111. [CrossRef]

58. Kwon, S.; Kim, M.; Jung, S.; Jeong, J.; Lim, K.; Van, J.; Cho, H.; Kim, H.; Nah, W.; Yang, Y. Inverted-load network for High-power Doherty Amplifier. *IEEE Microw. Mag.* **2009**, *10*, 93–98. [CrossRef]

59. Ramzi, D.; Fadhel, M. Digital Doherty Amplifier with Enhanced Efficiency and Extended Range. *IEEE Trans. Microw. Theory Tech.* **2011**, *59*, 2898–2909.

60. Roberto, Q.; Vittorio, C.; Tao, J.; Marco, P.; Simona, D.; Brian, L. K-Band GaAs MMIC Doherty Power Amplifier for Microwave Radio with Optimized Driver. *IEEE Trans.* **2014**, *62*, 2518–2525.

61. Zhao, J.; Wolf, R.; Ellinger, F. Fully Integrated LTE Doherty Power Amplifier. *IEEE Trans.* **2013**, *13*, 1251–1253.

62. Donald, Y.C.; Mayeda, J.C.; Jerry, L. A Review of 5G Power Amplifier Design at Cm-Wave and Mm-Wave Frequencies. *Hindawi Wireless Commun. Mob. Comput.* **2018**. [CrossRef]

63. Zheng, X.; Liu, Y.A.; Yu, C.; Li, S.; Li, J. Design of a Dual-Band Doherty Power Amplifier Utilizing Simplified Phase Offset-Lines. *Prog. Electromagn. Res. C* **2014**, *48*, 21–28. [CrossRef]

64. Wang, F.; Li, T.W.; Wang, H. A Highly Linear Super-Resolution Mixed-Signal Doherty Power Amplifier for High-Efficiency MM-Wave 5G Multi-Gb/s Communications. In Proceedings of the 2019 IEEE International Solid-State Circuits Conference-(ISSCC), San Francisco, CA, USA, 17–21 February 2019; pp. 88–90.

65. Kim, B.; Kim, J.; Kim, I.; Cha, J. The Doherty power amplifier. *IEEE Microw. Mag.* **2006**, *7*, 42–50. [CrossRef]

66. Özen, M.; Andersson, K.; Fager, C. Symmetrical Doherty Power Amplifier with Extended Efficiency Range. *IEEE Trans. Microw. Theory Tech.* **2016**, *64*, 1273–1284. [CrossRef]

67. Colantonio, P.; Giannini, F.; Giofre, R.; Piazzon, L. Doherty Amplifier with Compact Harmonic Traps. In Proceedings of the 2008 IEEE Microwave Integrated Circuit Conference, Amsterdam, Netherlands, 27–28 October 2008; pp. 526–529.

68. Fang, X.H.; Cheng, K.K.M. Extension of High-Efficiency Range of Doherty Amplifier by Using Complex Combining Load. *IEEE Trans. Microw. Theory Tech.* **2014**, *62*, 2038–2047. [CrossRef]

69. Rocco, G.; Luca, P.; Paolo, C.; Franco, G. A Doherty architecture with high feasibility and defined bandwidth behavior. *IEEE Trans. Microw. Theory Tech.* **2013**, *61*, 3308–3317.

70. Kim, S.; Lee, J.; Moon, K.; Park, Y.; Minn, D.; Kim, B. Optimized peaking amplifier of Doherty amplifier using an inductive input second harmonic load. In Proceedings of the 2016 11th European Microwave Integrated Circuits Conference (EuMIC), London, UK, 3–4 October 2016; pp. 129–132.

71. Ma, C.; Pan, W.; Tang, Y.X. Design of Asymmetrical Doherty Power Amplifier with Reduced Memory Effects and Enhanced Back-off Efficiency. *Prog. Electromagn. Res.* **2015**, *56*, 195–203. [CrossRef]

72. Simon, W.; Ray, P.A. High Efficiency Doherty Amplifier with Digital Predistortion for WiMAX. *High Freq. Electron.* **2008**, *7*, 18–28.

73. Fager, C.; Hallberg, W.; Özen, M.; Andersson, K.; Buisman, K.; Gustafsson, D. Design of Linear and Efficient Power Amplifiers by Generalization of the Doherty Theory. *IEEE Trans.* **2017**, *17*, 29–32.

74. Oualid, H.; Sami, B.; Fadhel M., G. A linearized Doherty Amplifier Using Complex Baseband Digital Predistortion Driven by CDMA Signals. In Proceedings of the IEEE Radio and Wireless Conference, Atlanta, GA, USA, 22–22 September 2004; pp. 435–438.

75. Cahuana, J.C.; Landin, P.N.; Gustafsson, D.; Fager, C.; Eriksson, T. Linearization of dual-input Doherty power amplifiers. In Proceedings of the 2014 International Workshop on Integrated Nonlinear Microwave and Millimetre-wave Circuits (INMMiC), Leuven, Belgium, 2–4 April 2014; p. 6815085.

76. Tsai, J.H.; Huang, T.W. A 38–46 GHz MMIC Doherty power amplifier using post-distortion linearization. *IEEE Microw. Wirel. Compon. Lett.* **2007**, *17*, 388–390. [CrossRef]

77. Mike, C.; Parisa, M.; Andrew, Z.; Zoya, P. A 4.2-W 10-GHz GaN MMIC Doherty Power Amplifier. In Proceedings of the 2015 IEEE Compound Semiconductor Integrated Circuit Symposium (CSICS), New Orleans, LA, USA, 11–14 October 2015.

78. Cho, K.J.; Kim, W.J.; Kim, J.H.; Stapleton, S.P. Linearity Optimization of a High Power Doherty Amplifier Based on Post-Distortion Compensation. *IEEE Microw. Wirel. Compon. Lett.* **2005**, *15*, 748–750.

79. Cho, K.J.; Kim, J.H.; Stapleton, S.P. A Highly Efficient Doherty Feedforward Linear Power Amplifier for W-CDMA Base-Station Applications. *IEEE Trans. Microw. Theory Tech.* **2005**, *53*, 292–300.

80. James, W.; Naoki, W.; Andrei, G. Doherty Amplifier Combines High Power and Efficiency. *Microw. RF* **2017**, *63*, 52–56.

81. Sadeghpour, T.; Karkhaneh, H.; Abd-Alhameed, R.A.; Ghorbani, A.; Noras, J.M.; Excell, P.S. Compensation of transmission non-linearity distortion with memory effect for a WLAN802. *11a transmitter. IET Sci. Meas. Technol.* **2012**, *6*, 125–131. [CrossRef]

82. Bathich, K.; Markos, A.Z.; Boeck, G. Wideband GaN Doherty Amplifier with 35 % Fractional Bandwidth. In Proceedings of the 40th European Microwave Conference, Paris, France, 28–30 September 2010; pp. 1007–1009.

83. Khaled, B.; Asdesach, Z.M.; Georg, B. Frequency Response Analysis and Bandwidth Extension of the Doherty Amplifier. *IEEE Trans Microw. Theory Tech.* **2011**, *59*, 934–944.

84. Rubio, J.M.; Fang, J.; Camarchia, V.; Quaglia, R.; Pirola, M.; Ghione, G. 3–3.6-GHz Wideband GaN Doherty Power Amplifier Exploiting Output Compensation Stages. *IEEE Trans.* **2012**, *60*, 2543–2548. [CrossRef]

85. Luca, P.; Paolo, C.; Franco, G.; Rocco, G. 15% Bandwidth 7GHz GaN-MIMIC Doherty Amplifier with Enhanced Axillary chain. *Microw. Opt. Technol. Lett.* **2014**, *56*, 502–504.

86. Watanabe, S.; Takayama, Y.; Ishikawa, R.; Honjo, K. A Broadband Doherty Power Amplifier without a Quarter-Wave Impedance Inverting Network. In Proceedings of the APMC, Kaohsiung, Taiwan, 4–7 December 2012; pp. 361–363.

87. Kurokawa, K. Power waves and the scattering matrix. *IEEE Trans. Microw. Theory Tech.* **1956**, *13*, 194–202. [CrossRef]

88. Gustafsson, D.; Cahuana, D.; Kuylenstierna, I.; Angelov, N.; Fager, C. A wideband and compact GaN MMIC Doherty amplifier for microwave link applications. *IEEE Trans. Microw. Theory Tech.* **2013**, *61*, 922–930. [CrossRef]

89. Bathich, K. Analysis and Design of Efficiency-Enhancement Microwave Power Amplifiers Using the Doherty Technique. Ph.D. Thesis, Berlin University, Berlin, Germany, 2013.

90. Xia, J.; Yang, M.; Guo, Y.; Zhu, A. A Broadband High-Efficiency Doherty Power Amplifier with Integrated Compensating Reactance. *IEEE Trans. Microw. Theory Tech.* **2016**, *64*, 2014–2024. [CrossRef]

91. Fang, X.; Cheng, K.K. Broadband, Wide Efficiency Range, Doherty Amplifier Design Using Frequency -Varying Complex Combining Load. In Proceedings of the 2015 IEEE MTT-S International Microwave Symposium, Phoenix, AZ, USA, 17–22 May 2015.

92. Wang, Z. *High Efficiency Load Modulation Power Amplifiers for Wireless Communications*; Artech House Inc.: Norwood, MA, USA, 2017.

93. Chen, S.; Xue, Q. Optimized Load Modulation Network for Doherty Power Amplifier Performance Enhancement. *IEEE Trans. Microw. Theory Tech.* **2012**, *11*, 3474–3481. [CrossRef]

94. Giofre, R.; Colantonio, P.; Giannini, F.; Piazzon, L. New output combiner for Doherty amplifiers. *IEEE Microw. Wirel. Compon. Lett.* **2013**, *23*, 31–33. [CrossRef]

95. Maryam, S.; Issa, E.; Abubakar, S.H.; Jonathan, R.; Ayman, R.; Raed, A.A. Design of Asymmetrical Doherty GaN HEMT Power Amplifiers for 4G Applications. In Proceedings of the International Conference on Broadband Communications, Networks, and Systems, Faro, Portugal, 19–20 September 2018; pp. 455–465.

96. Piazzon, L.; Giofrè, R.; Quaglia, R.; Camarchia, V.; Pirola, M.; Colantonio, P.; Giannini, F.; Ghione, G. Effect of Load Modulation on Phase Distortion in Doherty Power Amplifiers. *IEEE Microw. Wirel. Compon. Lett.* **2014**, *24*, 505–507. [CrossRef]

97. Abadi, M.N.; Golestaneh, H.; Sarbishaei, H.; Boumaiza, S. An extended bandwidth Doherty power amplifier using a novel output combiner. In Proceedings of the IEEE MTT-S International Microwave Symposium (IMS), Tampa, FL, USA, 1–6 June 2014; pp. 1–4.

98. Meng, F.; Sun, Y.; Tian, L.; Zhu, X.W. A Broadband High-Efficiency Doherty Power Amplifier with Continuous Inverse Class-F Design. In Proceedings of the 32nd URSI GASS, Montreal, QC, Canada, 19–26 August 2017.

99. Sun, G. Broadband Doherty power amplifier via real frequency technique. *IEEE Trans. Microw. Theory Tech.* **2012**, *60*, 99–111. [CrossRef]

100. Akbarpour, M.; Helaoui, M.; Ghannouchi, F.M. A transformer-less load-modulated (TLLM) architecture for efficient wideband power amplifiers. *IEEE Trans. Microw. Theory Tech.* **2012**, *60*, 2863–2874. [CrossRef]

101. Kumar, S.; Tannous, C.; Danshin, T. A multisection broadband impedance transforming branch-line hybrid. *IEEE Trans. Microw. Theory Tech.* **1999**, *43*, 2517–2523. [CrossRef]

102. Neal, T.; Lei, G.; Anding, Z.; Thomas, J.B. A simplified broadband design methodology for linearized high-efficiency continuous Class-F power amplifiers. *IEEE Trans. Microw. Theory Tech.* **2012**, *60*, 1952–1963.

103. Chen, X.; Chen, W.; Ghannouchi, F.M.; Feng, Z.; Liu, Y. A Broadband Doherty Power Amplifier Based on Continuous-Mode Technology. *IEEE Trans. Microw. Theory Tech.* **2016**, *64*, 4505–4517. [CrossRef]

104. Andrei, G.; James, W. A Dual-Band Parallel Doherty Power Amplifier for Wireless Applications. *IEEE Trans. Microw. Theory Tech.* **2012**, *60*, 3214–3222.

105. Quaglia, R.; Camarchia, V.; Rubio, J.J.M.; Pirola, M.; Ghione, G. A 4-W Doherty Power Amplifier in GaN MMIC Technology for 15-GHz Applications. *IEEE Microw. Wirel. Compon. Lett.* **2017**, *27*, 365–367. [CrossRef]

106. Atsush, F.; Hiroshi, O.; Shoichi, N.; Toshio, N. A concurrent Multi-Band Power Amplifier with Compact Matching Network. In Proceedings of the 2011 XXXth URSI General Assembly and Scientific Symposium, Istanbul, Turkey, 13–20 August 2011.

107. Shao, J.; Zhou, R.; Ren, H.; Arigong, B.; Zhou, M.; Kim, H.S.; Zhang, H. Design of GaN Doherty Power Amplifiers for Broadband Applications. *IEEE Microw. Wirel. Compon. Lett.* **2014**, *24*, 248–250. [CrossRef]

108. Luca, P.; Rocco, G.; Paolo, C.; Franco, G. A Method for Designing Broadband Doherty Power Amplifiers. *Prog. Electromagn. Res.* **2014**, *145*, 319–331.

109. Takenaka, K.; Sato, T.; Matsumoto, H.; Kawashima, M.; Nakajima, N. Novel Broadband Doherty Power Amplifier Design for Multiband Handset Applications. In Proceedings of the 2017 IEEE MTT-S International Microwave Symposium (IMS), Honolulu, HI, USA, 4–9 June 2017; pp. 778–781.

110. Shi, W.; He, S.; Zhu, X.; Song, B.; Zhu, Z.; Naah, G.; Zhang, M. Broadband Continuous-Mode Doherty Power Amplifiers with Noninfinity Peaking Impedance. *IEEE Trans. Microw. Theory Tech.* **2018**, *66*, 1034–1046. [CrossRef]

111. Kang, D.; Kim, D.; Cho, Y.; Park, B.; Kim, J.; Kim, B. Design of Bandwidth-Enhanced Doherty Power Amplifiers for Handset Applications. *IEEE Trans. Microw. Theory Tech.* **2011**, *59*, 3474–3483. [CrossRef]

112. Jorge, J.M.R.; Vittorio, C.; Marco, P.; Roberto, Q. Design of an 87% Fractional Bandwidth Doherty Power Amplifier Supported by a Simplified Bandwidth Estimation Method. *IEEE Trans. Microw. Theory Tech.* **2018**, *66*, 1319–1327.

113. Kim, C.H.; Park, B. Fully-Integrated Two-Stage GaN MMIC Doherty Power Amplifier for LTE Small Cells. *IEEE Microw. Wirel. Compon. Lett.* **2016**, *26*, 918–920. [CrossRef]

114. David, Y.T.W.; Slim, B. A Modified Doherty Configuration for Broadband Amplification Using Symmetrical Devices. *IEEE Trans. Microw. Theory Tech.* **2012**, *60*, 3201–3213.

115. Hu, S.; Wang, F.; Wang, H. 2.1 A 28GHz/37GHz/39GHz Multiband linear Doherty Power Amplifier for 5G Massive MIMO Applications. In Proceedings of the 2017 IEEE International Solid-State Circuits Conference (ISSCC), San Francisco, CA, USA, 5–9 February 2017.

116. Zhang, H.; Chen, K.J. A Stub Tapped Branch-Line Coupler for Dual-Band Operations. *IEEE Microw. Wirel. Compon. Lett.* **2007**, *17*, 106–108. [CrossRef]

117. Nghiem, X.A.; Guan, J.; Hone, T.; Negra, R. Design of concurrent multiband Doherty power amplifiers for wireless applications. *IEEE Trans. Microw. Theory Tech.* **2013**, *61*, 4559–4568. [CrossRef]

118. Karun, R.; Fadhel, M.G. Design methodology for dual-band Doherty power amplifier with performance enhancement using dual-band offset lines. *IEEE Trans. Ind. Electron.* **2012**, *59*, 4831–4842.

119. Paul, S.; Paolo, C.; Luca, P.; Franco, G.; Kristoer, A.; Christian, F. Design of a concurrent dual-band 1.8{2.4-GHz GaN-HEMT Doherty power ampler. *IEEE Trans. Microw. Theory Tech.* **2012**, *60*, 1840–1849.

120. Karun, R.; Mohammad, S.H.; Fadhel, M.G. Double the band and optimize. *IEEE Microw. Mag.* **2012**, *13*, 69–82.

121. Colantonio, P.; Feudo, F.; Giannini, F.; Giofrè, R.; Piazzon, L. Design of a dual-band GaN Doherty amplifier. In Proceedings of the 18-th International Conference on Microwaves, Radar and Wireless Communications, Vilnius, Lithuania, 14–16 June 2010; pp. 1–4.

122. Liang, C.; Roblin, P.; Hahn, Y.; Popovic, Z.; Chang, H.C. Novel Outphasing Power Amplifiers Designed with an Analytic Generalized Doherty–Chireix Continuum Theory. *IEEE Trans. Circuit Syst.* **2019**. [CrossRef]

123. Chen, W.; Bassam, S.A.; Li, X.; Liu, Y.; Rawat, K.; Helaoui, M.; Ghannouchi, F.M.; Feng, Z. Design and Linearization of Concurrent Dual Band Doherty PA with Frequency-Dependent Power Ranges. *IEEE Trans. Microw. Theory Tech.* **2011**, *59*, 2537–2546. [CrossRef]

124. Li, X.; Chen, W.; Zhang, Z.; Feng, Z.; Tang, X.; Mouthaan, K. A Concurrent Dual-Band Doherty PA. In Proceedings of the 2010 Asia-Pacific Microwave Conference, Yokohama, Japan, 7–10 December 2010; pp. 654–657.

125. Prathamesh, H.P.; William, H.; Christian, F.; Taylor, W.B. Analysis and Design of a Doherty-Like RF-Input Load Modulated Balanced Amplifier. *IEEE Trans. Microw. Theory Tech.* **2018**, *66*, 5322–5335.

126. Hellberg, R. Composite Power Amplifier. U.S. Patent 7,145,387 B2, 5 December 2017.

127. Christer, M.A.; David, G.; Jessica, C.C.; Richarld, H.; Chiristian, F. A 1–3-GHz digitally controlled Dual-RF input power amplifier design based on a Doherty-Outphasing continuum analysis. *IEEE Trans. Microw. Theory Tech.* **2013**, *61*, 3743–3752.

128. Bumman, K. *Doherty Power Amplifiers, from Fundamentals to Advanced Design Methods*; Elsevier: Amsterdam, The Netherlands, 2018.
129. Haedong, J.; Richard, W.; Tim, C.; David, S.; Christian, S.; Bayaner, A. Self-Outphasing Chireix power amplifier using device input impedance variation. In Proceedings of the 2016 IEEE MTT-S International Microwave Symposium (IMS), San Francisco, CA, USA, 22–27 May 2016; pp. 1–4.
130. Joongjin, N.; Bumman, K. The Doherty Power Amplifier with On-Chip Dynamic Bias Control Circuit for Handset Application. *IEEE Trans. Microw. Theory Tech.* **2007**, *55*, 633–642.
131. Barakat, A.; Thian, M.; Fusco, V. Towards generalized Doherty power amplifier design for multiband operation. Paper presented at International Microwave and Optoelectronics Conference (IMOC 2015), Porto de Galinhas/Pernambuco, Brazil, 3–6 November 2015.
132. Kang, D.; Cho, Y.; Kim, D. Impact of nonlinear CBC on HBT Doherty power amplifiers. *IEEE Trans. Microw. Tech.* **2013**, *61*, 3298–3307. [CrossRef]

Article

Multi-Points Cooperative Relay in NOMA System with *N-1* DF Relaying Nodes in HD/FD Mode for *N* User Equipments with Energy Harvesting

Thanh-Nam Tran [1,2,*] **and Miroslav Voznak** [1]

[1] Faculty of Electrical Engineering and Computer Science, Technical University of Ostrava, 17. listopadu 2172/15, 708 33 Ostrava-Poruba, Czech Republic; miroslav.voznak@vsb.cz

[2] Faculty of Electronics and Telecommunications, Sai Gon University, 220 Tran Binh Trong st., Dict. 5, Ho Chi Minh City, Vietnam

* Correspondence: thanh.nam.tran.st@vsb.cz

Received: 14 December 2018; Accepted: 29 January 2019; Published: 1 February 2019

Abstract: Non-Orthogonal Multiple Access (NOMA) is the key technology promised to be applied in next-generation networks in the near future. In this study, we propose a multi-points cooperative relay (MPCR) NOMA model instead of just using a relay as in previous studies. Based on the channel state information (CSI), the base station (BS) selects a closest user equipment (UE) and sends a superposed signal to this UE as a first relay node. We have assumed that there are N UEs in the network and the N-th UE, which is farthest from BS, has the poorest quality signal transmitted from the BS compared the other UEs. The N-th UE received a forwarded signal from $N-1$ relaying nodes that are the UEs with better signal quality. At the i-th relaying node, it detects its own symbol by using successive interference cancellation (SIC) and will forward the superimposed signal to the next closest user, namely the $(i+1)$-th UE, and include an excess power which will use for energy harvesting (EH) intention at the next UE. By these, the farthest UE in network can be significantly improved. In addition, closed-form expressions of outage probability for users over both the Rayleigh and Nakagami-m fading channels are also presented. Analysis and simulation results performed by Matlab software, which are presented accurately and clearly, show that the effectiveness of our proposed model and this model will be consistent with the multi-access wireless network in the future.

Keywords: cooperative NOMA; multi-points DF relaying nodes; half-duplex; full-duplex; Rayleigh fading channels; Nakagami-m fading channels; energy harvesting

---

1. Introduction

The next-generation network (5G) technology has the advantage of increasing system capacity by superior sharing-spectrum efficiency [1]. Therefore, multiple users in the network can be served in the same frequency band/time slot and various allocation power coefficients by the key technology is called Non-Orthogonal Multiple Access (NOMA). The is fundamentally different from previous orthogonal access methods, e.g., Orthogonal Multiple Access (OMA) [2]. In NOMA system, the users with better channel conditions are allocated less transmitting power coefficients. On the other hand, the users with worse channel conditions are allocated more transmitting power coefficients to guarantee the quality of service (QoS) for all users in the system. After receiving a superposed signal, successive interference cancellation (SIC) is done at the end users [3]. In [4], the authors investigated the impact of imperfect SIC on the analysis performance of NONA system. Their analysis results showed that even though SIC is not perfect, the performance of the NOMA system is still better than the orthogonal system. A down-link NOMA wireless network was studied in [5] by considering using a relay for forwarding signals to combat the fading effect of the transmission channel. Authors applied to dual-hop relaying systems with decode-and-forward (DF) or amplify-and-forward (AF) protocols [6]. Relay full-duplex (FD) model over the Rayleigh fading channels using the DF protocol investigated the performance by optimizing the transmit power factor [7]. The study impacts relay selection of cooperative NOMA on the performance system [8]. The authors in [9] proposed a novel best cooperative mechanism (BCM) for wireless EH and spectrum sharing in the 5G network. The [10–12] include AF and DF relaying. In [12], it showed that a dual-hop power line communication (PLC) system can improve the system capacity compared to direct-link (DL) transmission. And M. Rabie et al. [13] proposed using Multi-hop relay instead of using one hop relay or dual-hop relays. The authors investigated the energy efficiency over PLC channels with assuming log-normal fading. The studies [14,15] analyzed the system performance of multi-hop AF/DF relaying over PLC channels in terms of average bit error and ergodic capacity. These studies showed that the system performance can be improved by increasing the number of relaying. In addition, The authors in [8] studied the impact of relay selection (RS) on system performance. The compared results on two-stage versus max-min RS showed that cooperative NOMA system over Rayleigh fading channels with two-stage RS is better than the max-min one. We hypothesized that there are N users with the N-th user at the far end from BS with the worst channel condition. The QoS of the N-th user can be improved with the cooperation of $N - 1$ users instead of just receiving only a relay cooperation. At each node, one must perform the best neighbor selection to forward the signal next neighbor. The best selection of neighbors is repeated until the signal reaches the destination.

In addition, we also consider EH at UEs. The explosion of the number of wireless devices, radio frequency (RF) EH becomes a potential technology to convert the energy of receiving wireless signal into electricity. Therefore, the MPCR is not only transmitting information but also delivering energy to the users. In Ref. [16–18], only users located close to BS can collect energy. This is because signal reception and energy collection cannot be done simultaneously. Thus, the users need to divide the received signal for EH and information decoding (ID) by using power splitting (PS) or time switching (TS) which was called "received TS" [19,20]. Though the PS approach has been shown to mostly outperform the receive-TS approach, however, the PS is complicated and inefficient for practical implementation. The research results have shown that PS is better than TS, however, PS is more complex and difficult to practical application than TS. In our study, we consider compressing both information and energy in one transmission phase instead of splitting it into two transmission phases as in previous studies. Furthermore, a user faraway from BS can still receive information and collect energy from the nearest relay node. Researchers have made important contributions to the 5G wireless multi-access network. Specially, L. Dai et al. [21] presented the introduction, development process, and recent research trends on NOMA, comprehensively. Because of the potential application of NOMA in the future, there have been many important research contributions [22–29]. These positive research results are motivations for other researchers to continue to study NOMA improvement.

In this study, we focus on MPCR in NOMA network to improve the QoS for the user faraway form BS with poor channel. In terms of contributions in this research, the main contributions include:

- The first, this article proposes a down-link side NOMA network with random N UEs.
- The next, the MPCR model is proposed to improve QoS for the Nth UE with farthest distance from BS among the others users by using $N - 1$ UEs as DF relaying nodes in HD/FD mode. Each $\cdot UE_i$ relaying node receives and forwards a superposed signal to next hop, namely UE_{i+1}, which is nearest from UE_i. This work will loop until the superposed signal is sent to last UE, namely UE_N.
- A algorithm for selecting relay nodes in MPCR is also presented clearly in next section.
- At UE_i with $\forall i > 1$, the received signal has an excess power that is used for EH to charge the battery with assuming unlimited capacity of the battery.
- In additional, this study investigates and finds an outage probability and system throughput for each UE, which are written in closed-form expressions.

- Further, The analysis and simulation results are presented in a clear way by the Monte Carlo simulation (10^6 samples of channels) from the Matlab software to prove our propositions.

This article is presented as following. In the next section, namely Experimental Models, we propose system models and analyse two transmission scenarios which are called $N - 1$ relaying nodes in HD or FD mode. In the third section, we have analyzed the system performance on outage probability and system throughput. In Section 4, we use Matlab software to simulate and results will also be presented in this section. A summary of the results of this study will be presented in Section 5.

Notice: In this study, we use a few notations included as

- $h_{a,b}$ is a channel from source a to destination b.
- α_i is an allocation power coefficient for the i-th UE.
- y_i^Ω is the received signal at the i-th UE with Ω protocol where $\Omega = \{HD, FD\}$.
- $\gamma_{i \to x_j}^\Omega$ is a signal-to-interference-plus-noise-ratios (SINRs) at i-th UE while the i-th UE decodes x_j symbol.
- $\Pr\{.\}$ is a probability.
- $\Re\Theta_i^\Omega$ or $\aleph\Theta_i^\Omega$ is an outage probability of the i-th UE with Ω protocol over Rayleigh or Nakagami-m fading channels, respectively.
- R_i^* is a bit rate threshold of the i-th UE.

2. Experimental Models

In previous studies about NOMA, a direct down-link scenario is considered to serve a number of users in the same time slot. However, in such studies, there are usually a fixed number of users. Therefore, they have not shown the generality of the model. In order to ensure the generality, we have upgraded the model to a random and unpredictable number of users.

2.1. Direct Link Scenario

The authors analyzed different NOMA techniques including power domain and code domain [22]. The role of the power domain is proven to be important in determining the performance of the system through the availability of CSI [23]. The BS send a superposed signal S to all UEs in the same power domain and same time slot as following

$$S = \sqrt{P_0} \sum_{j=1}^{N} \sqrt{\alpha_j} x_j. \tag{1}$$

Thus, the received signal at the i-th UE, $\forall i \in \{1, \ldots, N\}$, would be expressed as following

$$y_i^{Dir} = h_{0,i} \sqrt{P_0} \sum_{j=1}^{N} \sqrt{\alpha_j} x_j + n_i, \tag{2}$$

where $h_{0,i}$ is denoted as the channels from BS to each the i-th UE over Rayleigh or Nakagami-m fading channel. Furthermore, N is a random number of UEs joined to network, α_j in rule with $\sum_{j=1}^{N} \alpha_j = 1$ is an allocation power coefficient for each UE and P_0 is the transmission power of BS. n_i is denoted as the additive white Gaussian noise (AWGN) of the i-th UE, where $n_i \sim CN(0, N_0)$ with zero mean, variance N_0 and $i \in \{1, \ldots, N\}$.

It is important to notice that the channel coefficient from BS to each UE, in paired, is expressed as $h_{0,i}$ in our expressions.

In direct link scenario, the first user in the nearest distance from the BS with the strongest channel conditions was ordered first in the channel gain list. Furthermore, the list is in decreasing order as following

$$h_{0,1} > h_{0,2} > \ldots > h_{0,i} > \ldots > h_{0,N-1} > h_{0,N}. \tag{3}$$

According to the NOMA theory, users with the worst signal quality should be given priority to allocate the highest transmitting power factor. Another assumption in terms of the NOMA characteristics, we have assumed that the BS already owns the CSI of all UEs fully. In a previous study [30], the authors considered that CSI is available to the system and used to determine the decoding order of user's data. The authors in [31] studied how NOMA performance depends on power allocation techniques to ensure fairness for users under instantaneous CSI and average CSI. The superimposed signals are sent to the UEs in the same power domain with different power coefficients, in the hope of ensuring system performance and ensuring service quality fairness for all users. Therefore, the list of allocation power factors is arranged in descending order for each UE in the network as

$$\alpha_1 < \alpha_2 < \ldots < \alpha_i < \ldots < \alpha_{N-1} < \alpha_N. \tag{4}$$

In Figure 1, the UE_N is farthest from the BS. Thus, the x_N symbol is allocated the strongest power factor. Therefore, x_N symbol will be first decoded at all UEs in the network by applying SIC [3]. Furthermore, the order of decoding is done sequentially according to the reversed list of power factor allocations presented in (4) expression. The Signal-to-interference-plus-noise ratios (SINRs) of all UEs have been expressed as

$$\gamma_{i \to x_j}^{Dir} = \frac{\left| h_{0,i} \right|^2 \rho_0 \alpha_j}{\left| h_{0,i} \right|^2 \rho_0 \sum\limits_{k=1}^{j-1} \alpha_k + 1}, \tag{5}$$

where $i \in \{1, \ldots, N\}$ and $j \in \{N, \ldots, i\}$.

In a special case at the UE_1, after it decoded x_j symbols with $j \in \{N, \ldots, 2\}$ by using (5), UE_1 decodes its own symbol x_1 with only AWGN n_1 as

$$\gamma_{1 \to x_1}^{Dir} = \left| h_{0,1} \right|^2 \rho_0 \alpha_1. \tag{6}$$

Furthermore, ρ_0 in (5) or (6) is signal-to-noise ratio (SNR) which can be calculated by

$$\rho_i = \frac{P_i}{N_0}, \tag{7}$$

where $i \in \{0, \ldots, N-1\}$, e.g., $\rho_0 = P_0 / N_0$ with P_0 is the transmitting power of the BS.

The achievable instantaneous bit rate of the i-th UE when it decodes x_j symbol with $x_j \in \{x_N, \ldots, x_i\}$ is shown by

$$R_{i \to x_j}^{Dir} = \frac{1}{2} \log_2 \left(1 + \gamma_{i \to x_j}^{Dir} \right), \tag{8}$$

where $i \in \{1, \ldots, N\}$ and $j \in \{N, \ldots, i\}$. If $i \neq j \neq 1$, and $\gamma_{i \to x_j}^{Dir}$ is given by (5) then. Else if $i = j = 1$, and $\gamma_{i \to x_j}^{Dir}$ is given by (6) then.

2.2. $N - 1$ DF Relaying Nodes Scenario

On the other hand, the system model in [13] has only one relaying to improve the QoS of UEs which are faraway from the BS. We propose a improved model with using a MPCR model instead of using only one user as a relay device. See in Figure 1, there are N users in the network with descending order channel conditions with the N-th UE has the poorest signal compared to the other UEs. The Figure 1a,b are $N - 1$ HD relaying nodes model and $N - 1$ FD relaying nodes models, respectively. In FD mode, the relays are impacted by the loop interference channels, which themselves affected the system's performance. This study investigates the system performance on MPCR in HD or FD mode for N users over Rayleigh or Nakagami-m fading channels. Previous studies on the NOMA system used a cooperative relay to improve system performance compared to a direct transmission system. The contributions of previous studies [30–32] are the motivation for this research to continue to improve system performance.

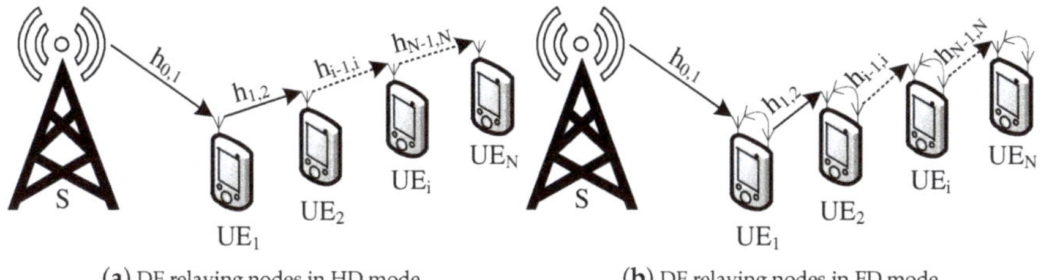

(**a**) DF relaying nodes in HD mode.　　　　(**b**) DF relaying nodes in FD mode.

Figure 1. The NOMA system with $N-1$ relaying nodes in HD/FD mode.

Z. Ding et al. [8] proposed the relay selection method to choose the best relay with the best channel condition by using two-stage relay selection protocol which outperforms versus max-min relay selection protocol. There is a difference compared model in [8] versus our model. The authors consider selection a best relay in N relays to serve for two other users [8]. In our proposed model, Figure 1, all of the $N-1$ UEs can be selected for relaying node. A selected relay node set is initialized empty $\omega = \varnothing$, and a first relaying node can be selected by

$$\omega_1 = \max\left\{ R^{\Omega}_{i \to x_1} > R^*_1 \right\}, \tag{9}$$

where $R^{\Omega}_{i \to x_1}$ is given by (22), and ω_1 has been added into $\omega = \omega \cup \omega_1$ then.

BS sends a superposed signal to the closest distance user with strongest channel condition, namely UE_1 in the Figure 1a,b, after BS selected UE_1 as a relay successfully. It is important to point out the difference. In this study, each relay node has a single or a twin antenna and works in HD or FD mode.

The received signals at the UE_1 in HD or FD mode are respectively the same like (2) or (10) as

$$y_1^{FD} = h_{0,1}\sqrt{P_0} \sum_{j=1}^{N} \sqrt{\alpha_j} x_j + h_{LI,1}\sqrt{P_0}\tilde{x}_1 + n_1, \tag{10}$$

where $h_{LI,1}$ is the loop interference channel generated by the itself transmitter antenna, and n_1 is the AWGN noise of the device UE_1.

In case the UE_1 is working in HD relaying mode, UE_1 decodes its own symbol by applying (5) and (6), respectively. On the other hand, the UE_1 is working in FD relaying mode, UE_1 decodes x_j symbol with $j \in \{N, \ldots, 2\}$ or $j = 1$ by applying SINRs in (11a) or (11b), respectively,

$$\gamma_{1 \to x_j}^{FD} \triangleq \frac{|h_{0,1}|^2 \rho_0 \alpha_j}{|h_{0,1}|^2 \rho_0 \sum\limits_{k=1}^{j-1} \alpha_k + |h_{LI,1}|^2 \rho_1 + 1} \tag{11a}$$

$$\triangleq \frac{|h_{0,1}|^2 \rho_0 \alpha_1}{|h_{LI,1}|^2 \rho_1 + 1}. \tag{11b}$$

Then, the UE_1 sends a mixed signal, namely S_1 in (13), to the next UE which is next nearest relay node, namely UE_2. The second relay node can be selected by applying (9) as

$$\omega_2 = \max\left\{R^{\Omega}_{i\to x_2} > R^*_2, i = \{1,\ldots,N\}, i \notin \omega\right\},\tag{12}$$

where R^{Ω}_i is also given by (22) and not being contained in ω set which is a selected relay nodes set. We removed UE_i with $i \in \omega$ from the relays selection because the signal could be sent back to the previous relay node and the superposed signal is unable to send to the UE_N. Furthermore, the ω_2 is also added into ω then. Note that the nearest neighbor represented in [33,34] are neighbors closest to the BS. However, the authors in [35] have extended the definition of nearest neighbor as the device can set up the transmission channel in the best condition compared to the other devices.

A mixed signal is sent to the next relay node as expressed

$$S_1 = \sqrt{P_1}\left(\sqrt{\alpha_1}x_\varnothing + \sum_{j=2}^{N}\sqrt{\alpha_j}x_j\right),\tag{13}$$

where $x_\varnothing$ is an empty information symbol which was also namely x_1 decoded at the UE_1.

The received signals at the UE_2 in both HD and FD relaying modes are expressed as, respectively,

$$y_2^{HD} = h_{1,2}\sqrt{P_1}\left(\sqrt{\alpha_1}x_\varnothing + \sum_{j=2}^{N}\sqrt{\alpha_j}x_j\right) + n_2,\tag{14}$$

and

$$y_2^{FD} = h_{1,2}\sqrt{P_1}\left(\sqrt{\alpha_1}x_\varnothing + \sum_{j=2}^{N}\sqrt{\alpha_j}x_j\right)$$
$$+ h_{LI,2}\sqrt{P_2}\tilde{x}_2 + n_2,\tag{15}$$

where $h_{1,2}$ is the channel from UE_1 to UE_2, P_1 is denoted as transmitting power at UE_1, and $h_{LI,2}$ is loop interference channel from transmitting antenna to receiving one at UE_2. Specially, the x_1 symbol existed in (2) and (10) but it was replaced by $x_\varnothing$ symbol in (14) and (15). Because x_1 was previously decoded and removed from the mixed signal by UE_1. Therefore, the $x_\varnothing$ symbol does not contain information and becomes a redundancy in the mixed signal. This paper will use excess power of $x_\varnothing$ symbol for EH purposes as is described in the next section.

The SINRs for decoding x_j symbol and its own x_2 symbol at UE_2 in both HD and FD relaying modes can be expressed, respectively, as following

$$\gamma_{2\to x_j}^{HD} \triangleq \frac{|h_{1,2}|^2\rho_1\alpha_j}{|h_{1,2}|^2\rho_1\sum_{k=2}^{j-1}\alpha_k + 1}\tag{16a}$$

$$\triangleq |h_{1,2}|^2\rho_1\alpha_2,\tag{16b}$$

and

$$\gamma_{2\to x_j}^{FD} \triangleq \frac{|h_{1,2}|^2\rho_1\alpha_j}{|h_{1,2}|^2\rho_1\sum_{k=2}^{j-1}\alpha_k + |h_{LI,2}|^2\rho_2 + 1}\tag{17a}$$

$$\triangleq \frac{|h_{1,2}|^2\rho_1\alpha_2}{|h_{LI,2}|^2\rho_2 + 1},\tag{17b}$$

where (16a) and (17a) with $j \in \{N,\ldots,3\}$, or (16b) and (17b) with $j = 2$.

After UE_2 decoded its own symbol, it selects a next relay node and sends a new superposed signal to next nearest UE, namely UE_3. This work will loop until a superposed signal is sent to the farthest UE, namely UE_N in Figure 1.

Proposition 1. *In this study, we propose a EH model to use excess power in the mixed signals for purposing EH as Figure 2. As expressing in (18) and (19), the received signals at the i-th UE, where $i \in \{2, \ldots, N\}$, have an empty $x_\oslash$ symbol with no information. Thus, the transmit power coefficients of each empty symbol can be harvested. In previous studies, the power for EH was transmitted to users on different time slots or on different antennas on the receivers. However, in this study, we use only one antenna for receiving both signals and energy from the transmitter.*

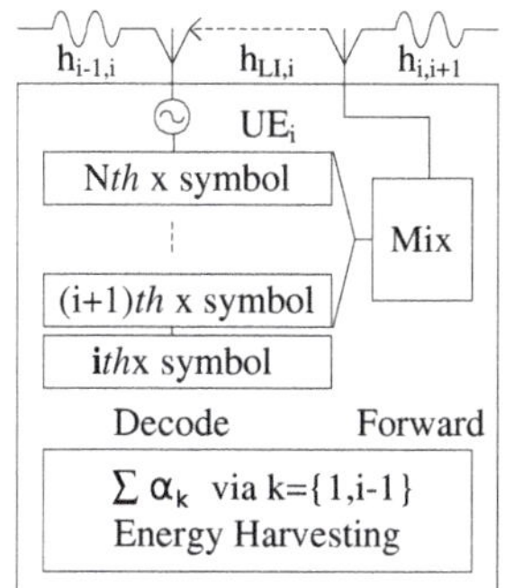

Figure 2. DF protocol and EH protocol at the *i*-th UE node.

In general, the received signals at the UE_i in both HD and FD relaying nodes can be rewritten by, respectively

$$y_i^{HD} = h_{i-1,i}\sqrt{P_{i-1}}\left(\sum_{l=1}^{i-1}\sqrt{\alpha_l}x_\oslash + \sum_{k=i}^{N}\sqrt{\alpha_k}x_k\right) + n_i, \tag{18}$$

and

$$y_i^{FD} = h_{i-1,i}\sqrt{P_{i-1}}\left(\sum_{l=1}^{i-1}\sqrt{\alpha_l}x_\oslash + \sum_{k=i}^{N}\sqrt{\alpha_k}x_k\right)$$
$$+ h_{LI,i}\sqrt{P_i}\tilde{x}_i + n_i, \tag{19}$$

where y_i^{HD} and y_i^{FD} are denoted as receiving signals at the UE_i node, $h_{i-1,i}$ is the channel from previous node to current node, P_{i-1} and P_i are transmitting power of previous UE and current UE, respectively. It is important to notice that $\sum_{l=1}^{i-1}\alpha_l + \sum_{k=i}^{N}\alpha_k = 1$.

The SINRs of each the *i*-th UE relaying node for detecting x_j symbol in HD and FD modes are expressed as, respectively

$$\gamma_{i\to x_j}^{HD} \triangleq \frac{|h_{i-1,i}|^2\rho_{i-1}\alpha_j}{|h_{i-1,i}|^2\rho_{i-1}\sum_{k=i}^{j-1}\alpha_k + 1}, \tag{20a}$$

$$\triangleq |h_{i-1,i}|^2\rho_{i-1}\alpha_i, \tag{20b}$$

and

$$\gamma^{FD}_{i \to x_j} \triangleq \frac{|h_{i-1,i}|^2 \rho_{i-1} \alpha_j}{|h_{i-1,i}|^2 \rho_{i-1} \sum\limits_{k=i}^{j-1} \alpha_k + |h_{LI,i}|^2 \rho_i + 1}, \tag{21a}$$

$$\triangleq \frac{|h_{i-1,i}|^2 \rho_{i-1} \alpha_i}{|h_{LI,i}|^2 \rho_i + 1}, \tag{21b}$$

where both (20a) and (21a) are with $i \in \{1, \ldots, N\}$ and $j \in \{N, \ldots, i+1\}$. Furthermore, both (20b) and (21b) are with $i = j$.

In NOMA theory, reachable instantaneous bit rate can be calculated by

$$R^{\Omega}_{i \to x_j} = \frac{1}{2} \log_2 \left(1 + \gamma^{\Omega}_{i \to x_j}\right), \tag{22}$$

where $\Omega = \{HD, FD\}$, $i \in \{1, \ldots, N\}$, and $j \in \{N, \ldots, i\}$. If $i \neq j$, and $\gamma^{\Omega}_{i \to x_j}$ is given by (20a) or (21a) then. Else if $i = j$, and $\gamma^{\Omega}_{i \to x_j}$ is given by (20b) or (21b) then.

A selected relay node can be performed by

$$\omega_i = \max \left\{ R^{\Omega}_{i \to x_j} > R^*_j, i \in \{1, \ldots, N\}, i \notin \omega \right\}. \tag{23}$$

Furthermore, a selected relay nodes set ω after the signal has been sent to the UE_N included

$$\omega = \omega_1 \cup \omega_2 \cup \ldots \cup \omega_{N-1}. \tag{24}$$

3. The System Performance Analysis

In this section, we evaluate the performance of the system that we have proposed based on outage probability and system throughput, in order.

3.1. Outage Probability

In terms of investigating outage probability, the outage probability is defined as the occurrence of the stop transmitting event if any instantaneous bit rate in (8) or (22) cannot reach minimum bit rate thresholds.

The probability density function (PDF) and cumulative distribution function (CDF) of Rayleigh distribution are shown by, respectively,

$$f_{|h_{a,b}|^2}(x) = \frac{1}{\sigma^2_{a,b}} e^{-\frac{x}{\sigma^2_{a,b}}} dx, \tag{25}$$

and

$$F_{|h_{a,b}|^2}(x) = 1 - e^{-\frac{x}{\sigma^2_{a,b}}}, \tag{26}$$

where $|h_{a,b}|^2$ are random independent variables namely x in PDF and CDF, respectively, with a and b are source and destination of channels, and $\sigma^2_{a,b}$ is mean of channel with $\sigma^2_{a,b} = E\left[|h_{a,b}|^2\right]$.

In general, the PDF and CDF over nakagami-m fading channels can be expressed, respectively,

$$f_{|h_{a,b}|^2}(x) = \left(\frac{m}{\sigma^2_{a,b}}\right)^m \frac{x^{m-1}}{\Gamma(m)} e^{-\frac{mx}{\sigma^2_{a,b}}}, \tag{27}$$

and

$$F_{|h_{a,b}|^2}(x) = \frac{\gamma\left(m, \frac{mx}{\sigma_{a,b}^2}\right)}{\Gamma(m)}$$

$$= 1 - e^{-\frac{mx}{\sigma_{a,b}^2}} \sum_{j=0}^{m-1} \left(\frac{mx}{\sigma_{a,b}^2}\right)^j \frac{1}{j!}. \tag{28}$$

In direct link scenario, outage event occurs if UE_i, where $i \in \{1, \ldots, N\}$, cannot decode x_j symbol, where $j \in \{N, \ldots, i\}$. The outage probability for each of the joining UE in NOMA system is expressed as

$$\Theta_i^{Dir} = 1 - \prod_{j=N}^{i} \Pr\left(R_{i \to x_j}^{Dir} > R_j^*\right). \tag{29}$$

where $R_{i \to x_j}^{Dir}$ is given by (8) and R_j^* is bit rate threshold of UE_j.

By applying the CDF in (25) and (27), the (29) is solved and it can be rewritten in closed-form as

$$\Re\Theta_i^{Dir} = 1 - \prod_{j=N}^{i} e^{-\frac{R_j^{**}}{x_j \rho_0 \sigma_{0,i}^2}}, \tag{30}$$

and

$$\aleph\Theta_i^{Dir} = 1 - \prod_{j=N}^{i} \left[\frac{\left(\frac{m}{\sigma_{0,i}^2}\right)^m \left(\left(\frac{m}{\sigma_{0,i}^2}\right)^{-m}\Gamma(m) + \left(\frac{R_j^{**}}{x_j \rho_0}\right)^m \left(\frac{mR_j^{**}}{x_j \rho_0 \sigma_{0,i}^2}\right)^{-m}\left(\Gamma\left(m, \frac{mR_j^{**}}{x_j \rho_0 \sigma_{0,i}^2}\right) - \Gamma(m)\right)\right)}{\Gamma(m)}\right], \tag{31}$$

where $\Gamma(.)$ and $\Gamma(.,.)$ are gamma function and gamma incomplete function, respectively. Furthermore, $R_j^{**} = 2^{2R_j^*} - 1$. It is important to notice that (30) and (31) are with the users over Rayleigh and Nakagami-m fading channels, respectively. In addition, χ_j in both (30) and (31) is given by

$$\chi_j \overset{\Delta}{=} \alpha_j - R_j^{**} \sum_{k=1}^{j-1} \alpha_k \tag{32a}$$

$$\chi_j \overset{\wedge}{=} \alpha_1, \tag{32b}$$

where (32a) is with $\forall i$, and $j \in \{N, \ldots, 2\}$ then. Furthermore, Equation (32b) is with $i = j = 1$ then.

Remark 1. *Base on the proposed model with $N - 1$ relaying nodes as in Figure 1, this study investigates the outage probabilities of N UE nodes in both HD and FD modes as*

$$\Theta_i^{\Omega} = \left(1 - \underbrace{\prod_{l=1}^{i-1} \Pr\left(R_{l \to x_i}^{\Omega} > R_i^*\right)}_{\eta}\right) and \left(1 - \underbrace{\prod_{j=N}^{i} \Pr\left(R_{i \to x_j}^{\Omega} > R_j^*\right)}_{\mu}\right), \tag{33}$$

where η is the successful probability to detect x_i symbol at previous UEs and μ is the successful probability to detect x_j symbol at the i-th UE. In a special case of the i-th UE with $i = 1$, It is important to notice that η in (33) is equal with zero and the (33) becomes the same with (29). In (33), η and μ are also solved by applying the CDF and gotten closed-form outage probability of each UE node over Rayleigh fading channel on both HD and FD modes as, respectively,

$$\Re\Theta_i^{HD} = \underbrace{\left(1 - \underbrace{\prod_{l=1}^{i-1} e^{-\frac{R_i^{**}}{\psi_i \rho_{l-1} \sigma_{l-1,l}^2}}}_{\eta}\right)}_{A_1} \underbrace{\left(1 - \underbrace{\prod_{j=N}^{i} e^{-\frac{R_j^{**}}{\chi_j \rho_i \sigma_{i-1,i}^2}}}_{\mu}\right)}_{A_2}, \tag{34}$$

and

$$\Re\Theta_i^{FD} = \underbrace{\left(1 - \underbrace{\prod_{l=1}^{i-1} \left(e^{-\frac{R_i^{**}}{\psi_i \rho_{l-1} \sigma_{l-1,l}^2}} \frac{\psi_i \rho_{l-1} \sigma_{l-1,l}^2}{\psi_i \rho_{l-1} \sigma_{l-1,l}^2 + R_i^{**} \rho_l \sigma_{h_{LI,l}}^2}\right)}_{\eta}\right)}_{B_1}$$

$$\underbrace{\left(1 - \underbrace{\prod_{j=N}^{i} \left(e^{-\frac{R_j^{**}}{\chi_j \rho_{i-1} \sigma_{i-1,i}^2}} \frac{\chi_j \rho_{i-1} \sigma_{i-1,i}^2}{\chi_j \rho_{i-1} \sigma_{i-1,i}^2 + R_j^{**} \rho_i \sigma_{LI,i}^2}\right)}_{\mu}\right)}_{B_2}. \tag{35}$$

To be clearer, here is some information that should be clearly explained. We denoted $\Re\Theta_i^{\Omega}$, where $i \in \{1,\ldots,N\}$ and $\Omega = \{HD, FD\}$, is the outage probability of UE_i over Rayleigh fading channels. The η symbol in both (34) and (35) is the successful detected x_i symbol at UE_l probability with $l \in \{1,\ldots,i-1\}$. Similarly, the μ symbol in both (34) and (35) is the successful detected x_j symbol with $j \in \{N,\ldots,i\}$ at the UE_i. Here are two cases such as:

- First case with $i = 1$, $\eta = 0$ in both (34) and (35) then. Furthermore, the outage probability of the UE_1 in HD/FD mode is $\Re\Theta_i^{\Omega} = \{A_2, B_2\}$.
- In addition, second case with $\forall i > 1$, the (34) and (35) are with $\Re\Theta_i^{\Omega} = \{A_1.B_1, A_2.B_2\}$.

In only the second case: ψ_i in both (34) and (35) is given by

$$\Psi_i = \left(\alpha_i - R_i^{**} \sum_{k=l}^{i-1} \alpha_k\right). \tag{36}$$

In both cases: χ_j is given by (32a) or (32b) after it has been rewritten as following, respectively,

$$\left[\begin{array}{l} \chi_j \overset{\Delta}{=} \alpha_j - R_j^{**} \sum_{k=i}^{j-1} \alpha_k \\[6pt] \chi_j \overset{\wedge}{=} \alpha_i \end{array}\right. \tag{37}$$

Remark 2. *The presented results of the studies [8,36] have firmly contributed to the role of NOMA system over the Rayleigh fading channels. However, studies on the NOMAn system over the Nakagami-m fading channels have received little attention because of its complexity. Therefore, we investigate the outage probability of each UE over Nakagami-m fading channels with $m = 2$ on both $N - 1$ HD/FD relaying nodes. Furthermore, the (33) can be solved by applying the PDF in (27) which is expressed in closed-form, respectively, as this research contributes.*

$$\aleph\Theta_i^{HD,m=2} = \underbrace{\left(1 - \underbrace{\prod_{l=1}^{i-1}\left[e^{-\frac{2R_i^{**}}{\psi_i\rho_{l-1}\sigma_{l-1,l}^2}}\frac{2R_i^{**}+\psi_i\rho_{l-1}\sigma_{l-1,l}^2}{\psi_i\rho_{l-1}\sigma_{l-1,l}^2}\right]}_{\eta} \right)}_{C_1}$$

$$\underbrace{\left(1 - \underbrace{\prod_{j=N}^{i}\left[e^{-\frac{2R_j^{**}}{\chi_j\rho_{i-1}\sigma_{i-1,i}^2}}\frac{2R_j^{**}+\chi_j\rho_{i-1}\sigma_{i-1,i}^2}{\chi_j\rho_{i-1}\sigma_{i-1,i}^2}\right]}_{\mu} \right)}_{C_2}, \tag{38}$$

and

$$\aleph\Theta_i^{FD,m=2} = \underbrace{\left(1 - \underbrace{\prod_{l=1}^{i-1}\left[\frac{e^{-\frac{2R_i^{**}}{\psi_i\rho_{l-1}\sigma_{l-1,l}^2}}}{\psi_i\rho_{l-1}\sigma_{l-1,l}^2\left(\psi_i\rho_{l-1}\sigma_{l-1,l}^2\left(\psi_i\rho_{l-1}\sigma_{l-1,l}^2+2R_i^{**}\right)+\rho_l\sigma_{LI,l}^2 R_i^{**}\left(3\psi_i\rho_{l-1}\sigma_{l-1,l}^2+2R_i^{**}\right)\right)}{\left(\psi_i\rho_{l-1}\sigma_{l-1,l}^2+\rho_l\sigma_{LI,l}^2 R_i^{**}\right)^3}}\right]}_{\eta} \right)}_{D_1}$$

$$\underbrace{\left(1 - \underbrace{\prod_{j=N}^{i}\left[\frac{e^{-\frac{2R_i^{**}}{\chi_j\rho_{i-1}\sigma_{i-1,i}^2}}}{\chi_j\rho_{i-1}\sigma_{i-1,i}^2\left(\chi_j\rho_{i-1}\sigma_{i-1,i}^2\left(\chi_j\rho_{i-1}\sigma_{i-1,i}^2+2R_j^{**}\right)+\rho_i\sigma_{LI,i}^2 R_j^{**}\left(3\chi_j\rho_{i-1}\sigma_{i-1,i}^2+2R_j^{**}\right)\right)}{\left(\chi_j\rho_{i-1}\sigma_{i-1,i}^2+\rho_i\sigma_{LI,i}^2 R_j^{**}\right)^3}}\right]}_{\mu} \right)}_{D_2}. \tag{39}$$

There have been two cases described above. It is not necessary to represent these cases. The analysis results will be presented in the next section. See Appendix A for proofing of remarks.

3.2. System Throughput

The total achievable received data rate at UE_i, which is denoted as system throughput P_{sum}^Ω, is the sum of throughput results of all UEs in system shown by

$$P_{sum}^\Omega = \sum_{i=1}^{N} P_i^\Omega = \sum_{i=1}^{N}\left(1-\Theta_i^\Omega\right) R_i^*. \tag{40}$$

3.3. A Proposal for Energy Harvesting

Proposition 2. *In (18) and (19), the received signals at UE_i, with $\forall i > 1$, include two parts which are x_k data symbol and $x_\varnothing$ empty symbol where $k \in \{i,\ldots,N\}$ and $l \in \{1,\ldots,i-1\}$. The $x_\varnothing$ does not contain information. Therefore, we proposed collecting the energy of allocating power coefficient of the $x_\varnothing$ symbol for charging the battery. Another assumption is that the battery is not limited by capacity. Thus, the EH for each UE in both HD and FD scenarios are expressed by, respectively*

$$EH_i = \xi \sqrt{\sum_{l=1}^{i-1} \alpha_l \rho_{i-1} |h_{i-1,i}|^2}, \tag{41}$$

where $i \in \{2, \ldots, N\}$ and ξ is collection coefficient.

3.4. A Proposed Algorithm for $N-1$ Relaying Nodes

Proposition 3. *In this section, an algorithm for processing with $N-1$ relaying nodes as shown in Figure 1 is proposed. The treatment flow is done in the waterfall pattern in the order shown in Figure 2.*

1. *Generate a random N UEs in the network with N channels from BS to UEs.*
2. *Creating a list of channels in descending order with the element at the top of the list is the best channel. Upon completion of the arrangement, BS will know which user is best chosen to use for first hop relaying node.*
3. *Through the results of the analysis [30], the authors have found that the performance of the NOMA system depends on the efficiency of the power allocation and the selection of the bit rate threshold, accordingly. Lack of CSI may affect the performance of the NOMA system. We have assumed that at BS and at each UE, there is full CSI of the UEs. Based on ordering of SCI as shown in (3), allocate the power coefficients and select the bit rate threshold for the UEs as, respectively*

$$\alpha_i = \frac{\min\left(\sigma_{0,j}^2\right)}{\sum\limits_{k=i}^{N} \sigma_{0,k}^2}, \tag{42}$$

and

$$R_i^* = \frac{\max\left(\sigma_{0,i}^2\right)}{\sum\limits_{k=i}^{N} \sigma_{0,k}^2}, \tag{43}$$

where in ordering and paring $i \in \{1, \ldots, N\}$ and $j \in \{N, \ldots, 1\}$. After the BS allocates the transmit power factor to the UEs, logically, a superposed signal will be sent to the nearest UE which is selected as the first hop relaying node, namely UE_1.

4. *The UE_1 receives and decodes x_j symbol with $j \in \{N, \ldots, i\}$ by (20a)–(21b), and excess power is collected by the UE for recharging. The UE_1 will select a next relay node by (23) and send a superposed signal as (18) or (19) to next hop relaying node after UE_1 detects its own symbol, namely x_1, successfully. This work (step 4) will be repeated until the superposed signal will be transmitted to the last UE, namely UE_N in model. The outage probability will occur when x_j, where $j \in \{N, \ldots, i\}$, cannot be detected successfully at UE_i with $i \in \{1, \ldots, N\}$.*

4. Numerical Results and Discussion

It is important to announce that all of our analysis results are simulated by the Matlab software and are presented accurately and clearly. We undertake no reproduction of any prior research results. Furthermore, this study does not use any given data set, channels were generated randomly during the simulation of a rule. e.g., if there are random N users, the random channels are arranged according to the rule $h_{0,1} > h_{0,2} > \ldots > h_{0,i} > \ldots > h_{0,N-1} > h_{0,N}$ and the corresponding channel coefficients $1/1 > 1/2 > \ldots > 1/i > \ldots > 1/(N-1) > 1/N$.

For the results to be clear and accurate, we have performed the Monte Carlo simulation with 10^6 random samples of each $h_{a,b}$ channel.

4.1. Numerical Results and Discussion for Outage Probability

It is important to note that the outage probability results of Dir, HD and FD scenarios are presented by black dashed lines, red dash-dot lines, and blue solid lines, respectively, as shown in Figure 3a,b.

In the first case, we assume that there are only three users connected in the network at t-th time slot. We analyzed the performance of the system based on the outage probability of each user in three different scenarios such as Dir, HD and FD schemes. There are some simulation parameters, e.g., the channel coefficients $h_{0,1} = 1$, $h_{0,2} = 1/2$, and $h_{0,3} = 1/3$ are in accordance with the earlier presented assumptions. Based on the transmission channel coefficients of the users, we can allocate power factors for the users UE_1, UE_2, and UE_3 with $\alpha_1 = 0.1818$, $\alpha_2 = 0.2727$, $\alpha_3 = 0.5455$, respectively, with $\sum_{i=1}^{3} \alpha_i = 1$ by applying (42). Because the third user, namely UE_3, has the poorest signal quality, it is prioritized to allocate the biggest power factor among the users. Our analysis results showed that users who are far from BS with poor signal quality have better results, e.g., the outage probability results of the UE_2 and the UE_3 are better than the UE_1, although their signal qualities are weaker than the first one. In addition, Figure 3a showed that UE_3 has the outage probability results which were marked with diamond marker, which are the best results compared to the other ones, although UE_3 has the weakest signal quality $h_{0,3} = 1/3$. Because UE_3 receives cooperation from the other UEs, the UE_3's QoS has been improved and is better than the other ones. These results demonstrate the effectiveness of the proposed MPCR model. In addition, the outage probability results of the first user, namely UE_1, has worse results than the other UEs, and U_1's outage probability results approximate to each other in all three scenarios, namely Dir, HD and FD relaying scenarios. The UE_1 with the strongest channel coefficient $h_{0,1} = 1$ has been allocated the worst power coefficient $\alpha_1 = 0.1818$ compared to the others. A previous study of FD relay [37,38] and the results of comparison between FD and HD [27] showed that the outage probability results of the relaying in FD mode was worse than the HD one. There is a similarity in these research results. The system performance efficiency of the MPCR model with $N - 1$ FD relaying nodes has resulted in approximation with $N - 1$ HD relaying nodes in the low dB SNRs. However, as ascending the SNRs, the performance of the MPCR system with $N - 1$ HD relaying nodes becomes better demonstrated by the red dash-dot lines in Figure 3a. Specifically, the first user's outage probability results in the FD scenario are the worst. However, there is not much difference compared to the other scenarios, such as Dir and HD scenarios. The reason is that the first relaying node in FD mode is affected by its own antenna channel noise, whereas in the direct and HD transmission scenarios with one antenna there are no loop interference channels.

To be more clear, we increased the number of users in the network to $N = 4$ users with the channel coefficient of UE_4 was $h_{0,4} = 1/4$ at $(t+1)$-th time slot. In addition, the outage probability of the users are presented in Figure 3b. This is because the system has a new joined user, namely UE_4, involved in the network with very weak signal quality. Therefore, we reused (42) to reallocate the transmit power factors to the users with $\alpha_1 = 0.12$, $\alpha_2 = 0.16$, $\alpha_3 = 0.24$, $\alpha_4 = 0.48$ as showing in Table 2. This is also because the power distribution coefficients have been changed. As a result, the instantaneous bit rate thresholds of users have been changed accordingly. The instantaneous bit rate thresholds of the user are $R_i^* = \{0.48, 0.24, 0.16, 0.12\}$ bps/Hz with $i \in \{1, \ldots, 4\}$. In this case, to ensure the QoS to the fourth user with the poorest signal quality, we have allocated to this user the biggest power factor, namely $\alpha_4 = 0.48$, and the lowest threshold, namely $R_4^* = 0.12$ bps/Hz, compared with the other users in the network. In addition, the other users must share power coefficient to UE_4 in the same power domain. The compared row contents in Tables 1 and 2 correspondingly, both α_i and R_i^* with $i = \{1, 2, 3\}$ are reduced for sharing power and bit rate to UE_4. As showing in Figure 3b, although the UE_4 has the poorest signal quality, it has the best outage probability results. This demonstrates that the MPCR combines with allocating power factor and instantaneous bit rate threshold selection are effective. In particular, the outage probability results in both HD and FD scenarios using $N - 1$ relaying nodes always outperform the scheme with no relaying.

Table 1. 3 UEs in NOMA system at t-th time slot.

UEs	Channels	Allocation Power Coefficients	Bit Rate Thresholds
UE_1	$h_{0,1} = 1$	$\alpha_1 = 0.1818$	$R_1^* = 0.5455$
UE_2	$h_{0,2} = 0.5$	$\alpha_2 = 0.2727$	$R_2^* = 0.2727$
UE_3	$h_{0,3} = 0.3333$	$\alpha_3 = 0.5455$	$R_3^* = 0.1818$

Table 2. 4 UEs in NOMA system at $(t+1)$th time slot.

UEs	Channels	Allocation Power Coeffecients	Bit Rate Thresholds
UE_1	$h_{0,1} = 1$	$\alpha_1 = 0.1200$	$R_1^* = 0.4800$
UE_2	$h_{0,2} = 0.5$	$\alpha_2 = 01600$	$R_2^* = 0.2400$
UE_3	$h_{0,3} = 0.3333$	$\alpha_3 = 0.2400$	$R_3^* = 0.1600$
UE_4	$h_{0,4} = 0.2500$	$\alpha_4 = 0.4800$	$R_4^* = 0.1200$

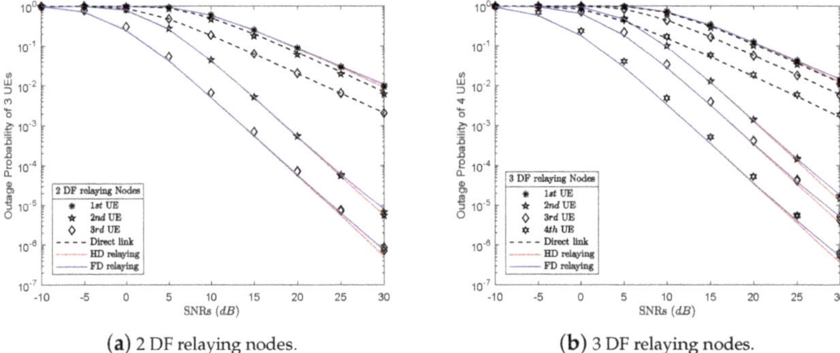

(**a**) 2 DF relaying nodes.   (**b**) 3 DF relaying nodes.

Figure 3. The outage probability results of $N = \{3, 4\}$ UEs over Rayleigh fading channels.

Furthermore, this study investigates the impact of both allocation power coefficient and SNRs affecting user's service quality, especially weak users. In Figure 3b, the weakest user, namely UE_4, has been assigned a fixed power factor $\alpha_4 = 0.48$. This study considers if the power allocation coefficient for UE_4 increases or decreases, the quality of service of UE_4 is varied over the corresponding SNRs. For simplicity, we assume that user UE_4 and the other users are over the Rayleigh fading channel. On the other hand, the UE_4 and the other users that are over Nakagami-m fading channels will be analyzed later. This study has assumed that the fourth user can be allocated a variable power factor $\alpha_4 \in \{0.1, \ldots, 0.9\}$. The Figure 4 shows the outage probability of the UE_4 with the allocation power factor which can be variable by one-by-one submitting each variable value α_4 into (34), (35), (38), and (39). It is important to notice that the outage probability results of UE_4 in direct, HD relaying, FD relaying scenario are presented by solid grid, dashed grid, and dash-dot grid, respectively. The Figure 4 showed that the outage probability results of UE_4 with the cooperation of 3 HD relaying nodes and 3 FD relaying nodes in MPCR scenarios are better than the UE_4's results in direct scenario. Specially, the outage probability results of UE_4 in MPCR system with $N - 1$ HD/FD relaying nodes are also approximations in all SNRs. These results are consistent with the UE_4's results presented earlier in Figure 3b.

In addition, this study investigates the outage probability of the users over Nakagami-m fading channels scenario versus the ones over Rayleigh fading channels scenario as shown in Figure 5. To ensure that this comparison is fair, the simulation parameters in the Nakagami-m fading channels scenario are the same as the simulation parameters shown in Table 1. Therefore, it is not necessary to represent these simulation parameters. In low SNRs, the outage probability results of the users over Rayleigh fading channels and Nakagami-m fading channels are approximated. However, when the

SNRs are increased, the outage probability results of the users over the Nakagami-m scenario are greatly improved.

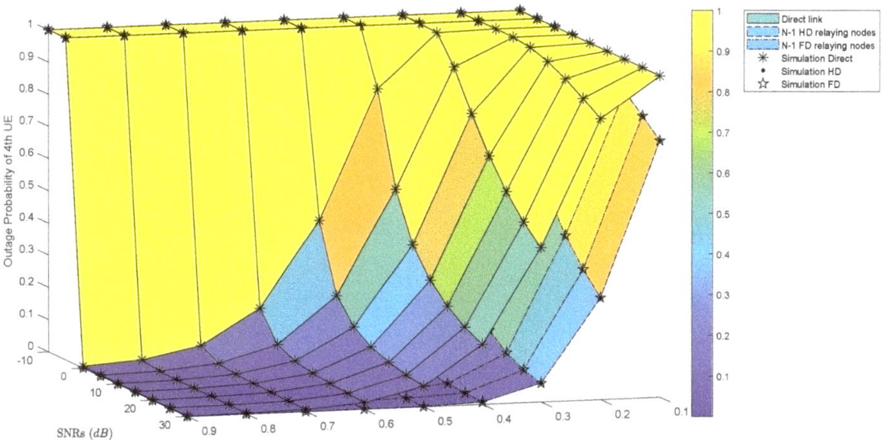

Figure 4. The outage probability results of $4th$ UE with $\alpha_4 = \{0.1, \ldots, 0.9\}$ and $SNRs = \{-10, \ldots, 30\}$.

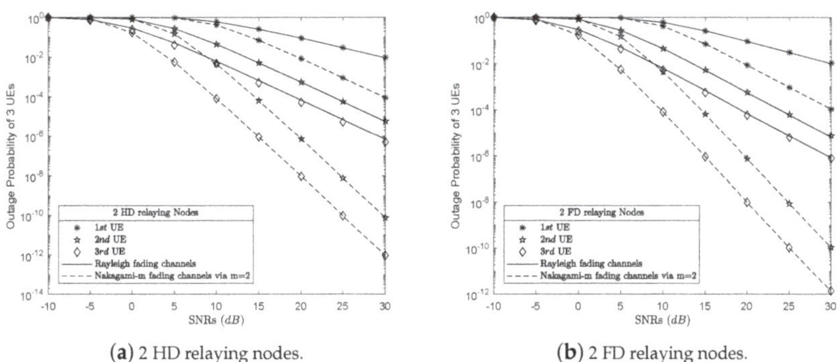

(**a**) 2 HD relaying nodes.

(**b**) 2 FD relaying nodes.

Figure 5. The outage probability results of three UEs over Rayleigh fading channels versus Nakagami-m fading channels via $m = 2$.

4.2. Numerical Results and Discussion for System Throughput

In system performance evaluation, system throughput is an important criterion that is known as the sum of instantaneous achievable bit rate of each user in the system. We reuse the simulation parameters as described in the evaluation of the outage probability shown in Tables 1 and 2. Therefore, we do not restate these parameters. The system throughput of each user with $N = 3$ UEs and $N = 4$ ones are presented in Figure 6a,b, respectively. It is important to notice that the solid lines, dash-dot lines and dashed lines are the system throughput of the users in Dir, HD and FD scenarios, respectively. This is because the outage probability results of the users in HD and FD scenarios are approximately equal. As a result, the throughput results of these users are also approximately equal. Thus, the dash-dot lines and dashed ones are overlapped in both Figure 6a,b. The analysis results showed that the system throughput of users in the $N - 1$ HD/FD relaying nodes scenarios are always better than the system throughput of the ones in the non-relay scenario. Specifically, the first UE's system throughput is approximate in all three scenarios. At SNR in 30 dB, all users in three scenarios reach their bit rate thresholds R_i^*.

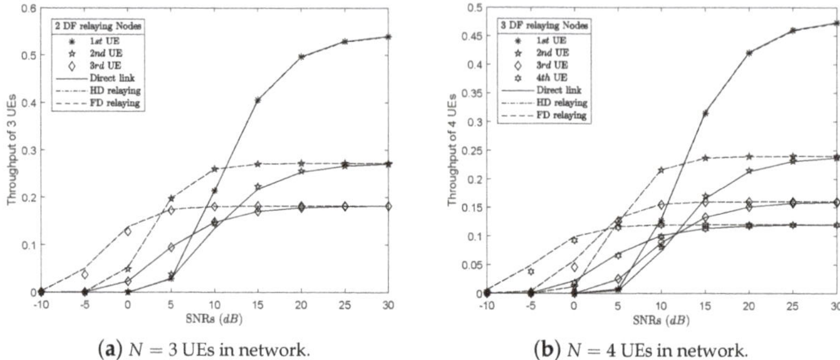

(**a**) $N = 3$ UEs in network. (**b**) $N = 4$ UEs in network.

Figure 6. The system throughput results of the users over Rayleigh fading channels.

On the other hand, this study analyzes the impact of the allocation power factor α_4 on the fourth user's throughput with variable $\alpha_4 \in \{0.1, ..., 0.9\}$ values instead fixing $\alpha_4 = 0.48$. As shown in Figure 7, higher grid lines are better results than the other ones. In this case, the instantaneous bit rate threshold of UE_4 is $R_4^* = 0.12$ bps/Hz. In low SNRs, e.g., $SNR = 0$ db, the system throughput results in all scenarios being approximately zero. On the other hand, although the SNRs have been increased, e.g., $SNR = 10$ dB, the system throughput results are still approximately zero if the power factor, namely α_4, is still in low, e.g., $\alpha_4 = 0.1$. However, with $\alpha_4 = 0.4$ and SNR is still held in 10 dB, the system throughput results of UE_4 in both three HD relaying nodes and three FD relaying nodes in MPCR scenarios are improved and reach their bit rate threshold. The Figure 6b showed that at SNR in 10 dB and $\alpha_4 = 0.48$, the UE_4 reach its bit rate threshold, approximately. Another e.g., in paired $\alpha_4 = 0.5$ and $SNR = 0$ dB, UE_4 also reach its bit rate threshold in Figure 7. By this analysis, we can find pairs of values α_4 and SNR where UE_4 can reach the threshold $R_4^* = 0.12$ bps/Hz.

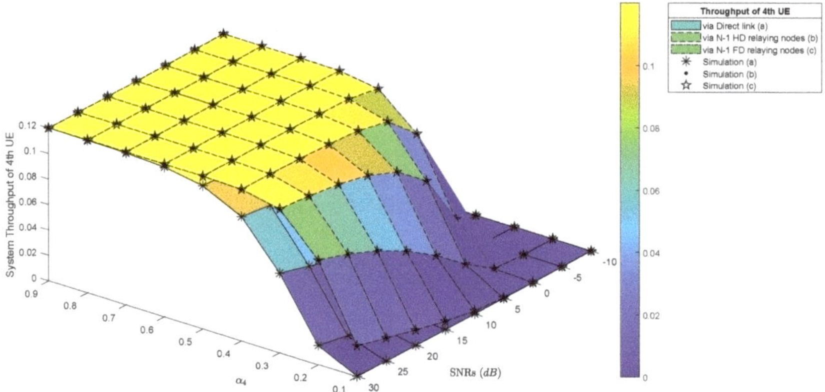

Figure 7. The throughput of the 4th UE over Rayleigh fading channels with $\alpha_4 = \{0.1, \ldots, 0.9\}$ and $SNRs = \{-10, \ldots, 30\}$ dB.

The system throughput of the users in $N - 1$ HD relaying nodes over both Rayleigh and Nakagami-m scenarios were analyzed, compared and presented in Figure 8a. In Figure 8a, there are $N = 3$ UEs over Rayleigh fading channels and Nakagami-m fading channels with solid lines and dashed ones, respectively. This is because of the results of $\Theta_1^{HD} > \Theta_2^{HD} > \Theta_3^{HD}$ as shown in Figure 5a. By applying (40), we get $P_1^{HD} < P_2^{HD} < P_3^{HD}$ with low SNRs. With increasing SNRs, the system throughput of each UE changes, e.g., $SNR = 30$ dB, $P_1^{HD} > P_2^{HD} > P_3^{HD}$ and reach their bit rate thresholds R_i^*.

The similarly results also happen in $N - 1$ FD relaying nodes scheme as shown in Figure 8b. Specifically, because the users over Nagami-m fading channels have better outage probability results than the ones over the Rayleigh fading channels as shown in Figure 5b, in some SNRs, e.g., $SNR = 10\,\text{dB}$ then $\aleph\Theta_i^{FD} < \Re\Theta_i^{FD}$. Therefore, $\aleph P_i^{FD} > \Re P_i^{FD}$ where $\aleph$ and $\Re$ were denoted as Nakagami-m and Rayleigh fading channels, respectively, after applying (40). These results proved that the Nakagami-m channel is better than the Rayleigh channel. However, when SNRs are increasing, the users have the throughput results approximately and close to the thresholds $\aleph P_i^{HD} \approx \Re P_i^{HD} \approx R_i^*$.

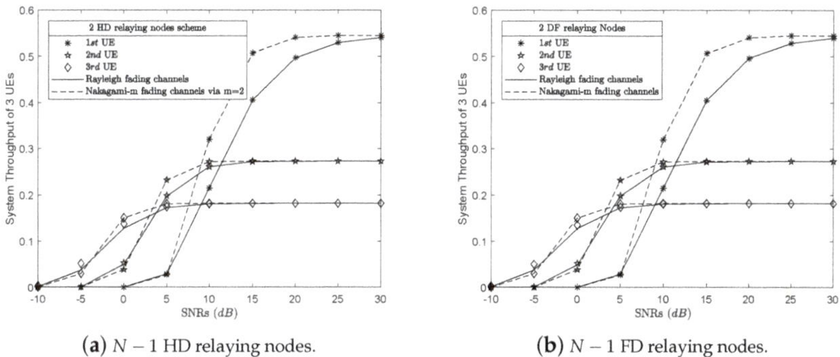

(**a**) $N - 1$ HD relaying nodes.　　(**b**) $N - 1$ FD relaying nodes.

Figure 8. Comparison of the system throughput results of Rayleigh versus Nakagami-m via $m = 2$.

4.3. N UEs with N − 1 HD/FD Relaying Nodes

As modeling Figure 1a,b, the proposed Proposition 3 can investigate the system performance with N UEs where N is a random and big number. Because of the limited power of our personal computers, this study only investigates and presents cases where there are only three or four users, $N = \{3,4\}$, in the system. However, the results presented do not show all the advantages of proposing algorithm. Thus, we are increasing the limit the number user with bigger number N. As shown in Figure 9a,b, there are 9 UEs in the network. By applying Proposition 3, we investigated the outage probability of the UEs in the network over both Rayleigh and Nakagami-m fading channels. For e.g., in $N - 1$ HD relaying nodes scenario, the outage probability of the first UE, namely UE_1, can be calculated by (34) or (28) over Rayleigh or Nakagami-m fading channels with $m = 2$, respectively, where $\eta = 0$. Another e.g., in FD scenario, the outage probability of last UEs, namely UE_9, over Rayleigh or Nakamagmi-m fading channels can be computed by (35) or (39), respectively. With the number of users is greater than nine UEs, $N > 9$, the results of the analysis are difficult to observe in the figure and it needs more time for the simulation. Therefore, we end the investigation with up to nine users in network.

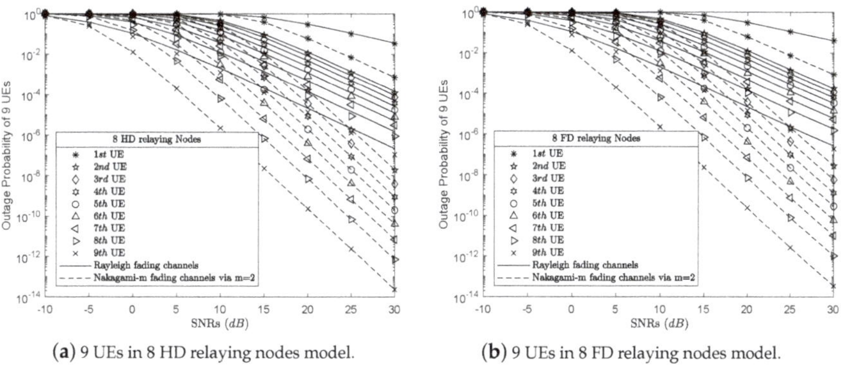

(**a**) 9 UEs in 8 HD relaying nodes model.　　(**b**) 9 UEs in 8 FD relaying nodes model.

Figure 9. Comparison of the outage probability results of Rayleigh versus Nakagami-m fading channels.

5. Conclusions

In this study, we proposed a novel NOMA network model with $N - 1$ relaying nodes instead of using only one relay as in previous studies. A superposed signal would be sent through $N - 1$ relaying nodes before it reaches the farthest UE which is denoted by UE_N. The closed-form expressions of $N - 1$ HD/FD relaying nodes scenarios over Rayleigh/Nakagami-m fading channels are also presented along with an explanation for the corresponding processing. By presenting results in the figures, our proposed models with $N - 1$ HD/FD relaying nodes are effective for applying to the cooperator NOMA network in the next generation of wireless telecommunications.

Author Contributions: T.-N.T. is the first author who proposed the main idea, analyzed and simulated the system, and presented the writing—original draft preparation, writing—review and editing, visualization. M.V. is the second who has experience in wireless communication research. He has made a supervision, review, and given the first author some useful comments and funding acquisition for this research. All authors read and approved the final manuscript.

Funding: This research received no external funding.

Acknowledgments: We would like to extend special thanks to the Reviewers for their comments and suggestions to improve this article.

Conflicts of Interest: We declare no conflict of interest. The funders had no role in the design of the study; in the collection, analyses, or interpretation of data; in the writing of the manuscript, or in the decision to publish the results.

Abbreviations

The following abbreviations are used in this manuscript:

No.	Abbreviations	Full description
1	AWGNs	Additive white Gaussian noises
2	BS	Base station
3	CDF	Cummuative distribution function
4	CSI	Channel state information
5	FD	Full-duplex
6	Fig.	Figure
7	HD	Half-duplex
8	MPCR	Multi-Point Cooperative Relay
9	NOMA	non-orthogonal multiple access
10	PDF	Probability density function
11	QoS	Quality of service
12	S	Source
13	SIC	Successive interference cancellation
14	SINR	Signal-to-interference-plus-noise ratio
15	SNR	Signal-to-noise ratio
16	UEs	User Equipments

Appendix A

Proof of $N - 1$ HD relaying nodes scenario: The condition for occurrence of the outage events has been presented in (33). By submitting (22), where $\Omega = HD$, into (33), we can get a expression for computing the outage probability of each UE in $N - 1$ HD relaying nodes scenario as following

$$\Theta_i^{HD} = \left(1 - \prod_{l=1}^{i-1} \Pr\left(|h_{l-1,l}|^2 > \frac{R_i^{**}}{\chi_i \rho_{l-1}} \right) \right) and \left(1 - \prod_{j=N}^{i} \Pr\left(|h_{i-1,i}|^2 > \frac{R_j^{**}}{\chi_j \rho_{i-1}} \right) \right). \tag{A1}$$

The (A1) can be rewritten in experimental integral by applying the PDF (25) of Rayleigh distributions as

$$\Re\Theta_i^{HD} = \left(1 - \prod_{l=1}^{i-1} \int_{\frac{R_i^{**}}{\chi_i\rho_{l-1}}}^{\infty} \frac{1}{\sigma_{l-1,l}^2} e^{-\frac{x}{\sigma_{l-1,l}^2}}\, dx\right) and \left(1 - \prod_{j=N}^{i} \int_{\frac{R_j^{**}}{\chi_j\rho_{i-1}}}^{\infty} \frac{1}{\sigma_{i-1,i}^2} e^{-\frac{x}{\sigma_{i-1,i}^2}}\, dx\right). \tag{A2}$$

The (A2) can be solved and expressed as (34).

On the other hand, the (A2) can be written with the PDF (27) of Nakagami-*m* fading channels as following

$$\aleph\Theta_i^{HD} = \left(1 - \prod_{l=1}^{i-1} \int_{\frac{R_i^{**}}{\chi_i\rho_{l-1}}}^{\infty} \left(\frac{m}{\sigma_{l-1,l}^2}\right)^m \frac{x^{m-1}}{\Gamma(m)} e^{-\frac{mx}{\sigma_{l-1,l}^2}}\, dx\right) and \left(1 - \prod_{j=N}^{i} \int_{\frac{R_j^{**}}{\chi_j\rho_{i-1}}}^{\infty} \left(\frac{1}{\sigma_{i-1,i}^2}\right)^m \frac{x^{m-1}}{\Gamma(m)} e^{-\frac{mx}{\sigma_{i-1,i}^2}}\, dx\right). \tag{A3}$$

and after the (A3) was solved, it can be expressed as (38). □

Proof of $N-1$ FD relaying nodes scenario: Similarly, by submitting (22) with $\Omega = FD$ into (33), we can get an expression for computing the outage probability of each UE in $N-1$ FD relaying nodes scenario

$$\Theta_i^{FD} = \left(1 - \prod_{l=1}^{i-1} \Pr\left(|h_{l-1,l}|^2 > \frac{R_i^{**}\left(|h_{Li,l}|^2\rho_l+1\right)}{\chi_i\rho_{l-1}}, |h_{Li,l}|^2 > 0\right)\right) \left(1 - \prod_{j=N}^{i} \Pr\left(|h_{i-1,i}|^2 > \frac{R_j^{**}\left(|h_{Li,i}|^2\rho_i+1\right)}{\chi_j\rho_{i-1}}, |h_{Li,i}|^2 > 0\right)\right). \tag{A4}$$

The (A4) is also rewritten in experimental integral by applying the PDF of Rayleigh or Nakagami-*m* fading which are respectively (25) or (27), respectively, as

$$\Re\Theta_i^{HD} = \left(1 - \prod_{l=1}^{i-1} \int_0^{\infty} \int_{\frac{R_i^{**}(y\rho_l+1)}{\chi_i\rho_{l-1}}}^{\infty} \frac{1}{\sigma_{l-1,l}^2\sigma_{Ll,l}^2} e^{-\left(\frac{x}{\sigma_{l-1,l}^2}+\frac{y}{\sigma_{Ll,l}^2}\right)}\, dxdy\right) \left(1 - \prod_{j=N}^{i} \int_0^{\infty} \int_{\frac{R_j^{**}(y\rho_i+1)}{\chi_j\rho_{i-1}}}^{\infty} \frac{1}{\sigma_{i-1,i}^2\sigma_{Ll,i}^2} e^{-\left(\frac{x}{\sigma_{i-1,i}^2}+\frac{y}{\sigma_{Ll,i}^2}\right)}\, dxdy\right), \tag{A5}$$

and

$$\aleph\Theta_i^{FD} = \left(1 - \prod_{l=1}^{i-1} \int_0^{\infty} \int_{\frac{R_i^{**}(y\rho_l+1)}{\chi_i\rho_{l-1}}}^{\infty} \left(\frac{m^2}{\sigma_{l-1,l}^2\sigma_{Ll,l}^2}\right)^m \frac{(xy)^{m-1}}{(\Gamma(m))^2} e^{-m\left(\frac{x}{\sigma_{l-1,l}^2}+\frac{y}{\sigma_{Ll,l}^2}\right)}\, dxdy\right) \left(1 - \prod_{j=N}^{i} \int_0^{\infty} \int_{\frac{R_j^{**}(y\rho_i+1)}{\chi_j\rho_{i-1}}}^{\infty} \left(\frac{m^2}{\sigma_{i-1,i}^2\sigma_{Ll,i}^2}\right)^m \frac{(xy)^{m-1}}{(\Gamma(m))^2} e^{-m\left(\frac{x}{\sigma_{i-1,i}^2}+\frac{y}{\sigma_{Ll,i}^2}\right)}\, dxdy\right). \tag{A6}$$

For e.g., $m = 2$, the (A5) and (A6) are solved and expressed as (38) and (39), respectively. End of proof. □

References

1. Saito, Y.; Benjebbour, A.; Kishiyama, Y.; Nakamura, T. System-level performance evaluation of downlink non-orthogonal multiple access (NOMA). In Proceedings of the 2013 IEEE 24th Annual International Symposium on Personal, Indoor, and Mobile Radio Communications (PIMRC), London, UK, 8–11 September 2013.
2. Ding, Z.; Yang, Z.; Fan, P.; Poor, H. On the Performance of Non-Orthogonal Multiple Access in 5G Systems with Randomly Deployed Users. *IEEE Signal Process. Lett.* **2014**, *21*, 1501–1505.
3. Higuchi, K.; Benjebbour, A. Non-orthogonal Multiple Access (NOMA) with Successive Interference Cancellation for Future Radio Access. *IEICE Trans. Commun.* **2015**, *98*, 403–414.
4. Men, J.; Ge, J.; Zhang, C. Performance Analysis for Downlink Relaying Aided Non-Orthogonal Multiple Access Networks with Imperfect CSI Over Nakagami-*m* Fading. *IEEE Access* **2017**, *5*, 998–1004. [CrossRef]
5. Ding, Z.; Peng, M.; Poor, H. Cooperative Non-Orthogonal Multiple Access in 5G Systems. *IEEE Commun. Lett.* **2015**, *19*, 1462–1465.
6. Xiao, Y.; Hao, L.; Ma, Z.; Ding, Z.; Zhang, Z.; Fan, P. Forwarding Strategy Selection in Dual-Hop NOMA Relaying Systems. *IEEE Commun. Lett.* **2018**, *22*, 1644–1647.
7. Davoodi, A.; Emadi, M.; Aref, M. Analytical power allocation for a full duplex decode-and-forward relay channel. In Proceedings of the 2013 Iran Workshop on Communication and Information Theory, Tehran, Iran, 8–9 May 2013.
8. Ding, Z.; Dai, H.; Poor, H. Relay Selection for Cooperative NOMA. *IEEE Wirel. Commun. Lett.* **2016**, *5*, 416–419.
9. Gao, H.; Ejaz, W.; Jo, M. Cooperative Wireless Energy Harvesting and Spectrum Sharing in 5G Networks. *IEEE Access* **2016**, *4*, 3647–3658.
10. Tonello, A.; Versolatto, F.; D'Alessandro, S. Opportunistic Relaying in In-Home PLC Networks. In Proceedings of the 2010 IEEE Global Telecommunications Conference GLOBECOM 2010, Miami, FL, USA, 6–10 December 2010.
11. Lampe, L.; Vinck, A. Cooperative multihop power line communications. In Proceedings of the 2012 IEEE International Symposium on Power Line Communications and Its Applications, Beijing, China, 27–30 March 2012; pp. 1–6.
12. Cheng, X.; Cao, R.; Yang, L. Relay-aided amplify-and-forward powerline communications. *IEEE Trans. Smart Grid* **2013**, *4*, 265–272.
13. Rabie, K.; Adebisi, B.; Gacanin, H.; Nauryzbayev, G.; Ikpehai, A. Performance evaluation of multi-hop relaying over non-gaussian PLC channels. *J. Commun. Netw.* **2017**, *19*, 531–538.
14. Dubey, A.; Mallik, R.K.; Schober, R. Performance analysis of a multi-hop power line communication system over log-normal fading in presence of impulsive noise. *IET Commun.* **2015**, *9*, 1–9.
15. Dubey, A.; Mallik, R.K. PLC system performance with AF relaying. *IEEE Trans. Commun.* **2015**, *63*, 2337–2345.
16. Lu, X.; Wang, P.; Niyato, D.; Kim, D.I.; Han, Z. Wireless networks with RF energy harvesting: A contemporary survey. *IEEE Commun. Surv. Tutor.* **2015**, *17*, 757–789.
17. Nasir, A.A.; Tuan, H.D.; Ngo, D.T.; Durrani, S.; Kim, D.I. Path-following algorithms for beamforming and signal splitting in RF energy harvesting networks. *IEEE Commun. Lett.* **2016**, *20*, 1687–1690.
18. Nguyen, V.D.; Duong, T.Q.; Tuan, H.D.; Shin, O.S.; Poor, H.V. Spectral and energy efficiencies in full-duplex wireless information and power transfer. *IEEE Trans. Commun.* **2017**, *65*, 2220–2233.
19. Tam, H.H.M.; Tuan, H.D.; Nasir, A.A.; Duong, T.Q.; Poor, H.V. MIMO energy harvesting in full-duplex multi-user networks. *IEEE Trans. Wirel. Commun.* **2017**, *16*, 3282–3297.
20. Nasir, A.A.; Tuan, H.D.; Duong, T.Q.; Poor, H.V. Secrecy rate beamforming for multicell networks with information and energy harvesting. *IEEE Trans. Signal Process.* **2017**, *65*, 677–689.
21. Dai, L.; Wang, B.; Ding, Z.; Wang, Z.; Chen, S.; Hanzo, L. A Survey of Non-Orthogonal Multiple Access for 5G. *IEEE Commun. Surv. Tutor.* **2018**, *20*, 2294–2323.
22. Islam, S.; Avazov, N.; Dobre, O.; Kwak, K. Power-Domain Non-Orthogonal Multiple Access (NOMA) in 5G Systems: Potentials and Challenges. *IEEE Commun. Surv. Tutor.* **2017**, *19*, 721–742.
23. Islam, S.; Zeng, M.; Dobre, O.; Kwak, K. Resource Allocation for Downlink NOMA Systems: Key Techniques and Open Issues. *IEEE Wirel. Commun.* **2018**, *25*, 40–47.
24. Shi, S.; Yang, L.; Zhu, H. Outage Balancing in Downlink Non-Orthogonal Multiple Access with Statistical Channel State Information. *IEEE Trans. Wirel. Commun.* **2016**, *15*, 4718–4731.

25. Di, B.; Song, L.; Li, Y. Sub-Channel Assignment, Power Allocation, and User Scheduling for Non-Orthogonal Multiple Access Networks. *IEEE Trans. Wirel. Commun.* **2016**, *15*, 7686–7698.

26. Lei, L.; Yuan, D.; Ho, C.; Sun, S. Power and Channel Allocation for Non-Orthogonal Multiple Access in 5G Systems: Tractability and Computation. *IEEE Trans. Wirel. Commun.* **2016**, *15*, 8580–8594.

27. Fang, F.; Zhang, H.; Cheng, J.; Leung, V. Energy-Efficient Resource Allocation for Downlink Non-Orthogonal Multiple Access Network. *IEEE Trans. Commun.* **2016**, *64*, 3722–3732.

28. Hanif, M.; Ding, Z.; Ratnarajah, T.; Karagiannidis, G. A Minorization-Maximization Method for Optimizing Sum Rate in the Downlink of Non-Orthogonal Multiple Access Systems. *IEEE Trans. Signal Process.* **2016**, *64*, 76–88.

29. Ding, Z.; Adachi, F.; Poor, H. The Application of MIMO to Non-Orthogonal Multiple Access. *IEEE Trans. Wirel. Commun.* **2016**, *15*, 537–552.

30. Wan, D.; Wen, M.; Ji, F.; Liu, Y.; Huang, Y. Cooperative NOMA Systems with Partial Channel State Information Over Nakagami-*m* Fading Channels. *IEEE Trans. Commun.* **2018**, *66*, 947–958.

31. Timotheou, S.; Krikidis, I. Fairness for Non-Orthogonal Multiple Access in 5G Systems. *IEEE Signal Process. Lett.* **2015**, *22*, 1647–1651.

32. Liu, H.; Ding, Z.; Kim, K.; Kwak, K.; Poor, H. Decode-and-Forward Relaying for Cooperative NOMA Systems with Direct Links. *IEEE Trans. Wirel. Commun.* **2018**, *17*, 8077–8093.

33. Sadek, A.K.; Han, Z.; Liu, K.J.R. A distributed relay-assignment algorithm for cooperative communications in wireless networks. In Proceedings of the 2006 IEEE International Conference on Communications, Istanbul, Turkey, 11–15 June 2006.

34. Sreng, V.; Yanikomeroglu, H.; Falconer, D.D. Relay selection strategies in cellular networks with peer-to-peer relaying. In Proceedings of the 2003 IEEE 58th Vehicular Technology Conference, VTC 2003-Fall (IEEE Cat. No.03CH37484), Orlando, FL, USA, 6–9 October 2003.

35. Jing, Y.; Jafarkhani, H. Single and multiple relay selection schemes and their achievable diversity orders. *IEEE Trans. Wirel. Commun.* **2009**, *8*, 1414–1423.

36. Kim, J.B.; Lee, I.H. Capacity Analysis of Cooperative Relaying Systems Using Non-Orthogonal Multiple Access. *IEEE Commun. Lett.* **2015**, *19*, 1949–1952.

37. Thanh-Nam, T.; Dinh-Thuan, D.; Voznak, M. Full-duplex Cognitive Radio NOMA Networks: Outage and Throughput Performance Analysis. *Int. J. Electron. Telecommun.* **2019**, in processing.

38. Thanh-Nam, T.; Dinh-Thuan, D.; Voznak, M. On Outage Probability and Throughput Performance of Cognitive Radio Inspired NOMA Relay System. *Adv. Electr. Electron. Eng.* **2018**, *16*, 501–512.

 *electronics*

Article

Wireless-Powered Cooperative MIMO NOMA Networks: Design and Performance Improvement for Cell-Edge Users

Chi-Bao Le [1], Dinh-Thuan Do [2,*] and Miroslav Voznak [3]

[1] Faculty of Electronics Technology, Industrial University of Ho Chi Minh City (IUH),
   Ho Chi Minh City 700000, Vietnam; lechibao@iuh.edu.vn
[2] Wireless Communications Research Group, Faculty of Electrical and Electronics Engineering,
   Ton Duc Thang University, Ho Chi Minh City 700000, Vietnam
[3] Department of Telecommunications, Faculty of Electrical Engineering and Computer Science,
   VSB—Technical University of Ostrava, 17. listopadu 15/2172, 708 33 Ostrava, Czech Republic;
   miroslav.voznak@vsb.cz
* Correspondence: dodinhthuan@tdtu.edu.vn

Received: 18 February 2019; Accepted: 12 March 2019; Published: 16 March 2019

Abstract: In this paper, we study two transmission scenarios for the base station (BS) in cellular networks to serve the far user, who is located at the cell-edge area in such a network. In particular, we show that wireless-powered non-orthogonal multiple access (NOMA) and the cell-center user in such a model can harvest energy from the BS. To overcome disadvantages of the cell-edge user due to its weak received signal, we fabricate a far NOMA user with multiple antennas to achieve performance improvement. In addition, the first scenario only considers a relay link deployed to forward signals to a far NOMA user, while both direct links and relay links are generally enabled to serve a far user in the second scenario. These situations, together with their outage performance, are analyzed and compared to provide insights in the design of a real-multiple-antenna NOMA network, in which the BS is also required to equip multiple antennas for robust quality of transmission. Higher complexity in computations is already known in consideration of outage metrics with respect to performance analysis, since the system model employs multiple antennas. To this end, we employ a transmit antenna selection (TAS) policy to formulate closed-form expressions of outage probability that satisfies the quality-of-service (QoS) requirements in the NOMA network. Our simulation results reveal that the performance of the considered system will be improved in cases of higher quantity of transmit antennas in dedicated devices. Finally, the proposed design in such a NOMA system cannot only ensure a downlink with higher quality to serve a far NOMA user, but also provide significant system performance improvement compared to a traditional NOMA networks using a single antenna.

Keywords: non-orthogonal multiple access; multiple antenna; transmit antenna selection; outage probability

---

1. Introduction

Recently, to deploy the next generation of wireless networks, one of the potential technologies referred to as a promising application in 5G is Non-Orthogonal Multiple Access (NOMA) [? ? ?]. With more advantages compared to Orthogonal Multiple-Access (OMA) schemes, NOMA allows a superior number of users to be connected to a wireless network concurrently [?]. Furthermore, cooperative NOMA is considered to be extended work, and it can be developed from traditional relaying networks [? ? ? ?] with a NOMA scheme. More specifically, outage performance is examined since the conditions of extended coverage and energy harvesting are guaranteed [? ?]. To investigate the further challenges in cooperative NOMA, a full-duplex (FD), together with device-to-device, is proposed for a cooperative NOMA scheme, as studied in [?], in which the outage performance can be enhanced with respect to satisfying the quality requirements of NOMA users. A decoding and forward relaying scheme was proposed, and two metrics, including the outage probabilities and average throughput of the paired users, are derived in closed-form [?]. To achieve wireless-powered NOMA, the degraded performance of NOMA as a result of the of inter-cell interference can be examined [?]. In particular, the authors in [?] showed the impacts of relay selection schemes in the analysis of physical-layer security for such a NOMA. In other system models, the concept of both downlink and uplink NOMA (termed as DU-CNOMA) is proposed, which considered system performance [?]. In [?], the authors investigated ergodic sum capacity, outage probability, and outage sum capacity. In [?], an optimum joint user and relay selection procedure was suggested to employ dual-hop transmission in cooperative NOMA networks, in which multiple Amplify-and-Forward (AF) relays forward signals from multiple users to two terminals. Other emerging trends need to be considered, i.e., the authors in [?] considered a two-tier heterogeneous network (HetNet) with non-uniform small cell deployment for cooperative NOMA to establish a NOMA-based HetNet model. In particular, critical performance metrics are analyzed, such as coverage probability and achievable rate.

More recently, when considering the random location of relay in NOMA systems, some stochastic geometry models can be applied for performance evaluation. For instance, in [?], to improve the security of a random network, large-scale NOMA systems were examined in terms of physical-layer security. In addition, the authors in [?] proposed a secured zone containing the source node. More specifically, the emerging techniques, including NOMA and energy harvesting, have been implemented for next generation wireless networks [?]. By allowing energy harvesting, the system maintains distinct power levels when the users commence NOMA transmissions [?]. It is worth noting that harvested energy is re-used for signal processing in a 3-phase scheme of energy-harvesting NOMA [?]. In particular, the authors in [?] presented the case in which the NOMA user harvests energy from the received downlink signals to further the process on the uplink, which leads to improving the average ergodic rate of the system. In recent work, a low-complexity iterative algorithm has been proposed to maximize the energy efficiency of the D2D pair in an energy harvesting-enabled device-to-device (D2D) NOMA [?]. The main result is that such a system is discovered in the case of the existence of D2D communications underlaying a NOMA-based cellular network while guaranteeing the quality of service of cellular users [?]. In other metrics of multi-objective resource optimization problems, simultaneous wireless information and power transfer, together with a NOMA cognitive radio (CR-NOMA) network, is investigated under a practical non-linear energy-harvesting model [?]. Furthermore, there is an information–energy trade-off in CR-NOMA, and the analytical results confirmed that CR-NOMA can outperform OMA if the channel power gains of users are sufficiently different [?].

In other trends of research on NOMA, multiple-input multiple-output (MIMO) is incorporated with NOMA to introduce a new scheme with favorable results [? ? ? ?]. In particular, a user-pairing algorithm classifying users into clusters is proposed. For instance, Ding et al. [?] and Al-Abbasi et al. [?] considered a multi-user MIMO downlink channel. The authors used a beamforming technique to avoid inter-cluster interference, whereas NOMA is used to manage intra-cluster interference. In [?], the authors developed a novel transceiver for the MIMO NOMA system and showed its efficiency in terms of power consumption. More general frameworks in NOMA technique using Successive Interference Cancellation (SIC) for MIMO systems are provided in [?]. It was assumed that the instantaneous channel state information (CSI) is available at the base station (*BS*). They also assumed that each user is equipped with several antennas equal to or higher than the number of antennas at the base station (BS) so that users can eliminate inter-cluster interference and manage intra-cluster interference using SIC. It was shown that NOMA enhances the performance of MIMO systems in terms of the spectrum efficiency and user fairness.

Regarding multiple-input single-output (MISO) NOMA systems, the authors in [?] recommended a sub-optimal precoding design for minimization of the transmit power. Regarding transmit antenna selection, the performance of MISO NOMA systems was presented in [?]. In [?], MIMO using a single carrier (SC) applied in NOMA systems is proposed to achieve a substantially higher spectral efficiency compared to the traditional MIMO SC-OMA systems, by exploiting the degrees of freedom (DoF) offered in both the spatial domain and the power domain. On the other hand, the application of NOMA to improve the fairness and spectrum use in multi-carrier (MC) systems was studied in [? ? ?]. In [?], to consider the weighted system throughput in single-antenna MC-NOMA systems, the authors developed a sub-optimal sub-carrier and power allocation algorithm. Optimal sub-carrier and power allocation algorithms for minimization of the total transmitted power and maximization of the weighted system throughput in MC-NOMA systems were proposed in [? ?], respectively. Most of the above works on NOMA systems have considered a single-antenna NOMA user. However, recent studies have shown that outage performance can be improved effectively in the spatial domain, if multiple antennas are used [?]. These recent analyses motivate us to consider situations where multiple antennas are deployed at the BS and destination. In particular, it is important to study outage performance regarding the multi-antenna NOMA user in cooperative NOMA networks.

Our main contributions are summarized as follows:

- Extending our previous work in terms of single-input single-output (SISO) NOMA strategy [?], we introduce a realistic scenario with multiple antennas which is equipped at the far NOMA user in the considered NOMA. This model also employs multiple-antenna BS. To provide the capability for energy harvesting, the near NOMA user can re-use the harvested power to serve the far NOMA user, who has a weaker channel condition. Two schemes are investigated with or without the existence of a direct link between the BS, and hence performance of far NOMA user is determined.

- We first examine outage performance at the near user, who has a single antenna. Then, we derive outage probability expressions for the near NOMA user and the outage comparison is exhibited with the far NOMA user. The number of deployed antennas or location arrangement of the BS, relay, and destination node are examined as crucial impacts on the considered outage performance.

- In addition, to extract further metrics and highlight the system behavior, throughput performance of these users is presented. Targeting the threshold signal-to-noise ratio (SNR), optimal throughput can be achieved via a numerical method. Such an evaluation is presented in the numerical results section.

- Our findings reveal that a higher number of transmit antennas at the BS provides a superior outage probability for both the near and far users compared to the traditional model. In addition, outage performance of the far NOMA user will be improved when increasing the number of its received antennas. Moreover, comparing the proposed multiple-antenna NOMA system with different locations of the user and energy-harvesting time, we provide detailed guidelines for the design of real cooperative NOMA, achieving better outage performance.

The rest of the paper is structured as follows. Section ?? describes the system model and antenna selection policy. Section ?? studies the outage performance and the throughput performance of NOMA with two proposed schemes, while Section ?? investigates the numerical results of the NOMA system. In particular, the numerical results are presented and Monte Carlo simulations are applied to verify the accuracy of the proposed analysis. Finally, Section ?? concludes the paper

Notation: This paper needs some main notations to easy considerations on following analysis: the Euclidean norm of the vector is $\|.\|$, $E\{.\}$ shows expectation computation; $f_X(.)$, $F_X(.)$ denote the probability density function (PDF) and cumulative distribution function (CDF) of a random variable (RV) X, respectively. $P(.)$ is represented as probability operation. $E_n(.)$ stands for the exponential integrals function, $\Gamma(.)$ is the gamma function.

2. System Model

In this study, a cooperative cellular scenario deploying NOMA is considered with respect to performance evaluation for downlink as in Figure ??. Two transmission modes are introduced—direct and relay mode. In particular, the system model includes one base station (BS), which intends to transmit information to two NOMA mobile users with the help of a DF relay D_1 in relay mode to distant node D_2, but in direct mode D_2 can also receive the signal directly from the BS without assistance of relay. For robust and effective transmission, both source BS and mobile user D_2 are equipped with many antennas. However, in this case only a single antenna is equipped for the relay node D_1 due to some disadvantages, such as small size and limited power. Relay is only provided by wireless power transfer scheme from the BS. In practice, relay is often installed at intermediate positions (outdoor). As a result, the construction of a power grid is hard, and hence wireless-powered relay needs to be designed. In particular, relay can perform signal processing and harvest energy, and such an energy signal can be extracted from the same received signal. In the context of energy harvesting, such a signal is transferred from the BS via an RF signal transmission environment. It is noted that all the nodes operate in a half-duplex mode due to simple deployment. All the wireless links are assumed to exhibit frequency non-selective Rayleigh block fading, and additive white Gaussian noise (AWGN). For mathematical tractability, we restrict our attention primarily to transmit antenna selection (TAS) topology, in which such antenna selection criteria are required to enhance system performance with low cost of computations.

In the concerned system, two scenarios are examined in this paper:

- Scheme I: The BS intends to communicate with the far user D_2 under the assistance of the near user D_1. In this situation, D_1 is regarded as the relaying user and the DF protocol is employed to decode and forward information to D_2. A direct link does not exist between BS and D_2.
- Scheme II: Under the existence of a direct link between BS and D_2, a relay link is still employed to support D_2. As a result, a more complex process can be seen at the far NOMA user, as two signal streams are received. The question is of which scheme is suitable for application in such a NOMA network.

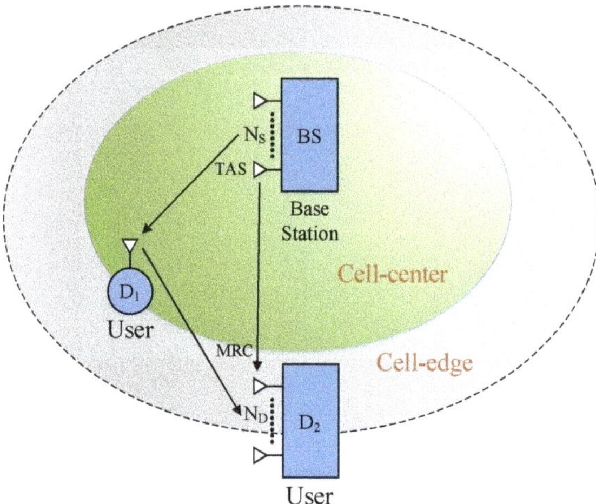

Figure 1. The proposed system model of NOMA facilitating multiple antennas at the BS and cell-edge user.

In this situation as Figure **??**, BS is equipped with N_S transmit antennas while D_2 has N_D received antennas. In the first phase, BS transmits mixed-signal x_S to D_1 and BS employs a single antenna selected from given N_S antennas. At BS, the transmit antenna, maximizing the instantaneous SNR at D_1, is selected for such transmission. Here, mixed-signal $x_S = \sqrt{a_1 P_S} x_1 + \sqrt{a_2 P_S} x_2$ is a summation of the coded modulation symbol of the two NOMA users, where a_i, $i \in \{1, 2\}$ are the power allocation coefficients to satisfy $a_1 + a_2 = 1$, P_S, P_R are transmit power at the BS, D_1, respectively. Without loss of generality, it is assumed that the effective channel gains are ordered to support the signal decoding operation of each NOMA user, and we assume that $a_2 > a_1$. It can be assumed that the $N_S \times 1$ channel vector between BS and D_1 is known as $\mathbf{h}_{SD_1} = \left[h_{1,1}, \dots, h_{1,i}, \dots h_{1,N_S} \right]$ while link BS and D_2 is modeled $\mathbf{h}_{SD_2} = \left[h_{0,1}, \dots, h_{0,j}, \dots, h_{0,N_D} \right]$, $i \in [1, N_S]$, $j \in [1, N_D]$. In this case, we assume that Rayleigh fading channels are deployed in such a model, i.e., $h_{1,i}$, $h_{0,i}$ are selected to perform signal transmission from the BS to D_1 in Scheme I and link $BS - D_2$, respectively; $\mathbf{h}_{D_2}$ characterizes for link from D_1 to D_2. We continue to denote the average channel power gains in links $BS - D_1, D_1 - D_2, BS - D_2$ are modeled as $\Omega_1, \Omega_2, \Omega_0$. Interestingly, only D_1 can furnish the capability of wireless energy transfer [? ?], due to the lower load of signal processing required here. The main reason for this is that we intend to design a single antenna at relay D_1 to reduce the complex computation at the intermediate device.

Following from the principle of wireless power transfer to relay node D_1 in such NOMA, the relay can harvest the amount of energy from the multiple-antenna BS and then the received power at the D_1 is given by [? ?]

$$P_R = \alpha \eta P_S |h_{1,i}|^2, \tag{1}$$

where α, η are the percentage of time for energy harvesting in the whole time of frame processing and energy efficiency factor (depending on how effective the circuit can operate), respectively.

The received signal can be given at D_1 as

$$y_{D_1} = \sqrt{(1 - \alpha) P_S} h_{1,i} \left(\sqrt{a_1 P_S} x_1 + \sqrt{a_2 P_S} x_2 \right) + n_{D_1}, \tag{2}$$

where n_{D_1} is denoted as AWGN noise with variance of σ^2 at node D_1. In the first computation, the received signal-to-interference-and-noise ratio (SINR) at D_1 to detect the D_2's message x_2 is given by

$$\gamma_{1i, D_1 x_2} = \frac{(1 - \alpha) P_S |h_{1,i}|^2 a_2}{(1 - \alpha) P_S |h_{1,i}|^2 a_1 + \sigma^2}. \tag{3}$$

In such NOMA, SIC will be carried out at the users. Therefore, the instantaneous SNR between the i-th antenna at BS and D_1 can be derived to evaluate outage performance. Applying the principle of NOMA for Scheme I, SIC is employed at D_1. After SIC, the received SNR at D_1 to detect its own message x_1 is given by

$$\gamma_{1i, D_1 x_1} = \frac{(1 - \alpha) P_S |h_{1,i}|^2 a_1}{\sigma^2}. \tag{4}$$

Regarding the TAS scheme, we deploy an interesting policy of maximal SNR to improve system performance. In particular, the optimal transmit antenna selection is employed at relay link as below, and the index of the antenna is selected with respect to x_2, x_1 respectively as

$$i^* = \arg \max_{1 \leq i \leq N_S} \left(\gamma_{1i, D_1 x_2} \right). \tag{5}$$

and

$$i^* = \arg \max_{1 \leq i \leq N_S} \left(\gamma_{1i, D_1 x_1} \right). \tag{6}$$

3. Exact Outage and Throughput in Delay-Limited Mode of Two Proposed Schemes

3.1. Scheme I: NOMA Network without Direct Link between BS and Far User D_2

The coexistence of near/far users and multiple/single antennas in NOMA systems is the main factor that results in varying performance for these users. For example, the performance of far users with poor channel conditions will be changed, and hence we improve the outage probability of them. It is worth noting that this multiple-antenna architecture at both the BS and D_2 is suitable for practical situations, wherein the far users are much farther away from the BS in comparison with near users and thus the poor channel conditions need to be considered. We then examine outage probability of D_1, D_2 as below.

According to the NOMA protocol, the complementary events of outage at D_1 can be explained as: D_1 can detect x_2 as well as its own message x_1. From the above description, the outage probability of D_1 with respect to threshold SNRs for NOMA user D_1, D_2 (denoting $\gamma_{th1}, \gamma_{th2}$, respectively) and such outage probability can be expressed as below

$$\mathcal{OP}_{1,x_1} = 1 - \Pr\left(\gamma_{1i,D_1x_2} > \gamma_{th2}, \gamma_{1i,D_1x_1} > \gamma_{th1}\right), \tag{7}$$

where $\gamma_{th1} = 2^{2R_1} - 1$ with R_1 being the target rate at D_1 to detect x_1 and $\gamma_{th2} = 2^{2R_2} - 1$ with R_2 being the target rate at D_2 to detect x_2. It is rewritten in following formula

$$
\begin{aligned}
\mathcal{OP}_{1,x_1} &= 1 - \Pr\left(\max_{1 \leq i \leq N_S}\left\{|h_{1,i}|^2\right\} > \xi\right) \\
&= \sum_{n=0}^{N_S} \binom{N_S}{n} (-1)^n \exp\left(-n\xi\right),
\end{aligned}
\tag{8}
$$

where $\xi = \max\left(\frac{\gamma_{th2}}{(1-\alpha)\bar{\gamma}_1(a_2-\gamma_{th2}a_1)}, \frac{\gamma_{th1}}{(1-\alpha)\bar{\gamma}_1a_1}\right)$, $\bar{\gamma}_1 = \frac{P_S\Omega_1}{\sigma^2}$. It is noted that (??) is derived on the condition of $a_2 > a_1\gamma_{th2}$. In the second phase, D_1 first decodes the received source signal, and then forwards the detected symbol by using the harvested energy. We then compute SNR for second-hop $D_1 - D_2$ transmission during the second phase and it is given by

$$\gamma_{2i} = \frac{P_R\left\|\mathbf{h}_{D_2}\right\|_F^2}{\sigma^2} = \alpha\eta\frac{P_S}{\sigma^2}|h_{1,i}|^2\left\|\mathbf{h}_{D_2}\right\|_F^2. \tag{9}$$

In this case, the DF protocol is deployed at user D_1. Thus, when the i-th antenna at BS is selected, the instantaneous SNR of end-to-end $BS - D_1 - D_2$ link can be expressed as

$$\gamma_{SD_1D_2} = \min\left(\gamma_{1i,D_1x_2}, \gamma_{2i}\right). \tag{10}$$

Using several expressions of SNR as previous calculations, i.e., $\gamma_{1i,D_1x_2} = \frac{(1-\alpha)P_S|h_{1,i}|^2 a_2}{(1-\alpha)P_S|h_{1,i}|^2 a_1 + \sigma^2}$, $\gamma_{2i} = \alpha\eta\frac{P_S}{\sigma^2}|h_{1,i}|^2\left\|\mathbf{h}_{D_2}\right\|_F^2$ we further compute an outage event related to D_2 to provide system performance analysis.

More specifically, the outage event of D_2 can be explained for two reasons. The first is that D_1 cannot detect x_2 . The second is that D_2 cannot detect its own message x_2 on the condition that D_1 can detect x_2 successfully. Based on this, the outage probability of D_2 can be expressed as below. In particular, the outage probability at the far user D_2 is given by

$$
\begin{aligned}
\mathcal{OP}_{2,x_2} =\ &\Pr\left(\gamma_{1i,D_1x_2} < \gamma_{th2}\right) \\
&+ \Pr\left(\min\left\{\gamma_{1i,D_1x_1}, \gamma_{2i}\right\} < \gamma_{th2}, \gamma_{1i,D_1x_2} > \gamma_{th2}\right).
\end{aligned}
\tag{11}
$$

Next, it can be re-expressed by

$$OP_{2,x_2} = \Pr\left(\max_{1 \le i \le N_S} \left(\frac{(1-\alpha)P_S|h_{1,i}|^2 a_2}{(1-\alpha)P_S|h_{1,i}|^2 a_1 + \sigma^2} \right) < \gamma_{th2} \right)$$
$$+ \Pr\left(\max_{1 \le i \le N_S} \min\left\{ \gamma_{1i,D_1 x_1}, \gamma_{2i} \right\} < \gamma_{th2}, \max_{1 \le i \le N_S} \left(\frac{(1-\alpha)P_S|h_{1,i}|^2 a_2}{(1-\alpha)P_S|h_{1,i}|^2 a_1 + \sigma^2} \right) > \gamma_{th2} \right). \tag{12}$$

To ease understanding, we divide the two components to compute independently as below J_1, J_2. It is worth noting that new denotation J_1 is given by

$$J_1 = \Pr\left(\frac{(1-\alpha)P_S|h_{1,i}|^2 a_2}{(1-\alpha)P_S|h_{1,i}|^2 a_1 + \sigma^2} < \gamma_{th2} \right)$$
$$= \sum_{n=0}^{N_S} \binom{N_S}{n} (-1)^n \exp\left(-\frac{n\gamma_{th2}}{(a_2 - a_1\gamma_{th2})\bar{\gamma}_1(1-\alpha)} \right). \tag{13}$$

Lemma 1. *It can be expressed the following outage event as*

$$J_2 = \Pr\left\{ \max_{1 \le i \le N_S} \min\left\{ \gamma_{1i,D_1 x_1}, \gamma_{2i} \right\} < \gamma_{th2} \right\}. \tag{14}$$

And it is shown in the closed-form expression as

$$J_2 = 1 - \frac{N_S \gamma_{th}}{\bar{\gamma}_1(1-\alpha)} \sum_{n=0}^{N_S-1} \binom{N_S-1}{n} \sum_{m=0}^{N_D-1} \sum_{k}^{\infty} \frac{(-1)^{k+n}}{m!k!}$$
$$\times \left(\frac{1-\alpha}{\alpha\mu\Omega_2} \right)^{m+k} E_{m+k}\left(\frac{\gamma_{th}(n+1)}{(1-\alpha)\bar{\gamma}_1} \right). \tag{15}$$

Proof. See in Appendix ??. $\square$

Finally, by applying Lemma ??, we obtain the outage probability for signal x_2 at D_2 as

$$\mathcal{OP}_{2,x_2} = J_1 + J_2(1 - J_1), \tag{16}$$

where J_1, J_2 are calculated as previous steps.

3.2. Scheme II: NOMA with Presence of Direct Link between BS and Far User D_2

In this scenario, extending signal gained at relaying link, the BS intends to serve directly D_2 based on selected antenna at index j, $h_{0,j}$. In particular, the observation on the received signal at D_2 for the direct link can be written as

$$\mathbf{y}_{D_2} = \mathbf{h}_{0,j}\left(\sqrt{a_1 P_S}x_1 + \sqrt{a_2 P_S}x_2 \right) + \mathbf{n}_{D_2}. \tag{17}$$

where $\mathbf{n}_{D_2}$ is noise matrix at D_2 following AWGN.

Similarly, the instantaneous SINR between transmitter on the j-th antenna at BS and D_2 during the first phase can be expressed as below. The received SINR at D_2 to detect x_2 is given by

$$\gamma_{0j,D_2 x_2} = \frac{P_S ||\mathbf{h}_{0,j}||_F^2 a_2}{P_S ||\mathbf{h}_{0,j}||_F^2 a_1 + \sigma^2}. \tag{18}$$

Due to existence of the relaying link, D_2 has two links to achieve a composed signal. More importantly, to reduce the implementation cost of the system, we adopt the TAS policy at BS and the selection-combining technique at D_2 to combine the direct signal and the relaying signal. Thus, the instantaneous end-to-end SNR of the system with the j-th antenna selected at BS is derived as

$$\gamma_{e,j^*} = \max(\gamma_{0j,D_2x_2}, \gamma_{SD_1D_2}). \tag{19}$$

Then, the general expression to examine outage performance at D_2 is given by

$$\mathcal{OP}_{2,x_2} = Pr(\gamma_{e,j^*} < \gamma_{th2}). \tag{20}$$

Based on obtained expressions of SNR, it can be achieved that

$$\mathcal{OP}_{2,x_2} = Pr\left(\underbrace{\max_{1 \leq i \leq N_S}}\ \max\left\{\gamma_{0i,D_2x_2}, \min\left\{\gamma_{1i,D_1x_1}, \gamma_{2i}\right\}\right\} < \gamma_{th2}\right). \tag{21}$$

Proposition 1. *The outage probability for D_2 in Scheme II can be expressed by*

$$\mathcal{OP}_{2,x_2} = \left[1 - \frac{1}{\Gamma(N_D)}\Gamma\left(N_D, \frac{\frac{\gamma_{th2}}{a_2 - a_1\gamma_{th2}}}{\bar{\gamma}_0}\right)\right]^{N_S}$$

$$\times \left[1 - \frac{N_S\gamma_{th2}}{(1-\alpha)a_1\bar{\gamma}_1}\sum_{n=0}^{N_S-1}\binom{N_S-1}{n}\sum_{m=0}^{N_D-1}\sum_{k}^{\infty}\frac{(-1)^{k+n}}{m!k!}\right. \tag{22}$$

$$\left.\times\left(\frac{(1-\alpha)a_1}{\alpha\mu\Omega_2}\right)^{m+k}\mathrm{E}_{m+k}\left(\frac{\gamma_{th2}(n+1)}{(1-\alpha)a_1\bar{\gamma}_1}\right)\right].$$

where $\bar{\gamma}_0 = \frac{P_S\Omega_0}{\sigma^2}$

Proof. See in Appendix ?? $\square$

Remark 1. *Such outage performance exhibits optimal value for consideration on varying time-splitting allocation for the function of energy harvesting. However, due to the complexity in the derived expressions of these outage probabilities, exact expressions of such optimal time to achieve lowest outage performance cannot be found. Fortunately, we can show the optimal time-splitting coefficient by exploiting numerical methodology. In addition, from such a derived formula, it is worth noting that the transmit SNR at the BS and the target rates are the main factors affecting outage performance. This can be verified in numerical results section.*

To further the comparison, we consider a special case where such a network only contains a direct link. It is based on obtained SNR $\gamma_{0j,D_2x_2} = \frac{P_S||\mathbf{h}_{0,j}||_F^2 a_2}{P_S||\mathbf{h}_{0,j}||_F^2 a_1 + \sigma^2}$, and by performing SIC at destination D_2, we have SNR to detect x_1.

$$\gamma_{0j,D_2x_1} = \frac{P_S||\mathbf{h}_{0,j}||_F^2 a_1}{\sigma^2}. \tag{23}$$

However, we only examine the outage probability for D_2 as detecting signal x_1

$$\mathcal{OP}_{3,x_1} = Pr\left(\gamma_{0j,D_2x_1} < \gamma_{th2}\right)$$

$$= \left[1 - \frac{1}{\Gamma(N_D)}\Gamma\left(N_D, \frac{\gamma_{th2}}{a_1\bar{\gamma}_0}\right)\right]^{N_S}. \tag{24}$$

3.3. Throughput Performance

For the delay-limited transmission mode, the source BS transmits at a constant rate $R_1 = \frac{1}{2}\log\left(1 + \gamma_{th1}\right)$ and $R_2 = \frac{1}{2}\log\left(1 + \gamma_{th2}\right)$ corresponding requirement of each signal x_1, x_2, which may be subjected to outage due to fading. Hence the average throughput can be expressed as

$$\tau_{k,x_l} = \left(1 - \mathcal{OP}_{n,x_l}\right) R_l, \tag{25}$$

where $n = \{1,2,3\}$ and $l = \{1,2\}$

Remark 2. *We firstly recall among many promising strategies to exploit multiple antennas in NOMA, TAS has widely been used in the traditional (MIMO) NOMA systems. In such a TAS scheme, there is excellent performance with full diversity, and it is simple to implement. Secondly, from the derived expressions here, the number of transmit and receive antennas contributes to a smaller outage value and exhibits outage performance with expected improvement. Furthermore, an extra advantage of TAS is that it requires only very low feedback signaling overhead.*

4. Numerical Results

In this section, it is assumed that the BS, the near NOMA user (relay), and the far NOMA user are located in approximate a straight line. We denote d_0 as the distance between BS and D_2, d_1 as the distance between BS and D_1, and d_2 as the distance between D_1 and the D_2. To ease computation, the distance between BS and the NOMA users is normalized with a factor m as $d_1 + d_2 = 5$ m. Furthermore, we can obtain $\Omega_0 = \left(1 + d_0^z\right)^{-1}$, $\Omega_1 = \left(1 + d_1^z\right)^{-1}$ and $\Omega_2 = \left(1 + d_2^z\right)^{-1}$, where z is denoted as the path-loss exponent. Here, $z = 2$. The time-switching factor for the energy-harvesting phase is $\alpha = 0.5$ and energy conversion efficiency $\eta = 1$. To satisfy requirements of QoS for each NOMA user, we set $R_1 = R_2 = 0.5$ bps/Hz.

In Figure **??**, we illustrate the outage performance versus transmit SNR at the BS. There is an excellent agreement between the exact analytical results and the simulations observed, and the performance gap of two users D_1, D_2 can be raised significantly across the whole range of SNR. Most important is that the system outage performance can be improved significantly by increasing the number of antennas at the BS while the relay node has a single antenna. Moreover, it reveals that the outage performance of user D_2 is worse than that of D_1. The main reason for this is that a lower amount of transmit power at the relay for signal processing second-hop $D_1 - D_2$ results in deceasing the outage performance of the second hop, and the total outage performance of two-hop transmission link falls as well. To show the impact of the number of received antennas at the far NOMA user, we perform a similar simulation to Figure **??**. In this case, we keep the number of transmit antennas at the BS as $N_S = 2$, then the outage performance will be enhanced with an increasing number of received antennas at D_2. It is shown that $N_S = 2, N_D = 3$ exhibits the best performance of D_2 among three illustrated cases. Please note that two NOMA users have different power allocation factors. Since NOMA allocates different power allocation factors for two NOMA users, then it is the existence of performance gap between the two NOMA users.

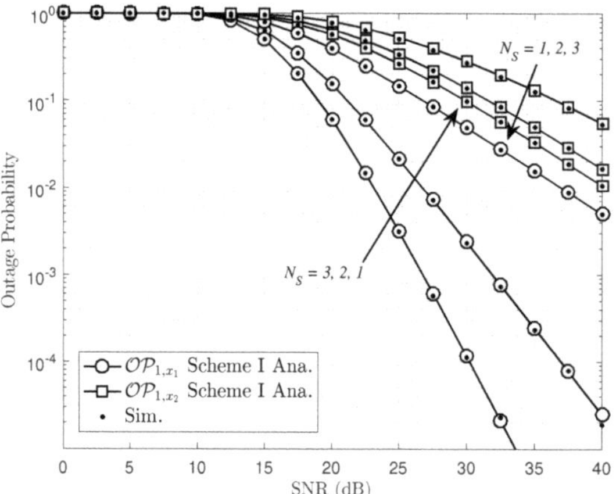

Figure 2. Outage probability of Scheme I with different numbers of transmit antennas at the *BS*, $N_D = 1$, $d_1 = 2$ m, $d_2 = 3$ m, $a_1 = 0.2$ and $a_2 = 0.8$.

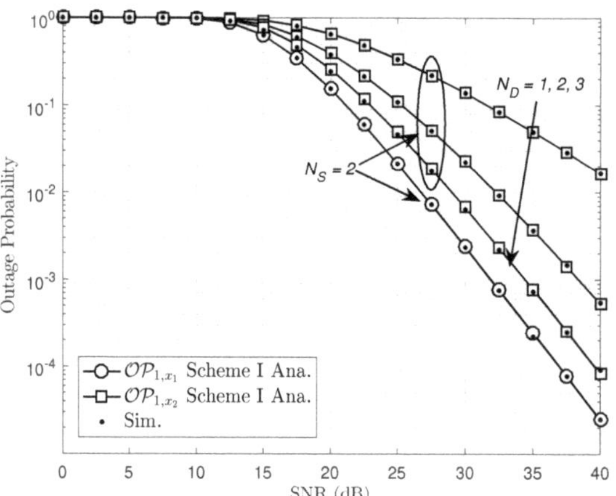

Figure 3. Outage probability of Scheme I with different N_D, $N_S = 2$, $d_1 = 2$ m, $d_2 = 3$ m, $a_1 = 0.2$ and $a_2 = 0.8$.

In Figure **??**, we present the outage performance by varying the number of transmit antennas at *BS*. We compare three cases of Scheme II as fixed single antenna at D_2, i.e., $N_D = 1$. It can be observed that with more transmit antennas at the *BS*, the outage performance will be improved. It confirms that the direct link in Scheme II has outage performance worse than the combined scheme, in which relay and direct link join to serve the far NOMA user. Furthermore, we compare these cases of such proposed NOMA relaying networks, and the worst performance can be seen with a conventional single-antenna NOMA, i.e., $N_S = N_D = 1$.

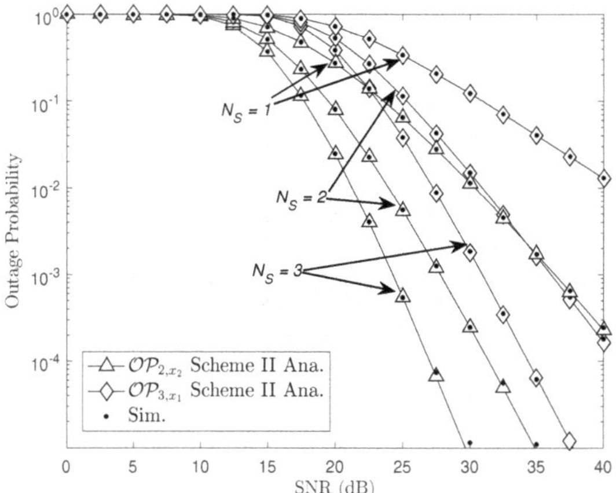

Figure 4. Outage probability of schemes II with different number of antennas N_S, $N_D = 1$, $d_1 = 2$ m, $d_2 = 3$ m, $a_1 = 0.2$ and $a_2 = 0.8$.

In Figure **??**, the impact number of received antennas at D_2 on outage performance can be illustrated by keeping a fixed number of transmit antennas at the BS, i.e., $N_S = 2$. It can be confirmed that the highest number of transmit/receive antennas at the BS, D_2 provide the best outage performance. As a result, it shows the advantage of multiple antennas assigned to proper users. Besides, energy harvesting is reasonable when facing the energy shortage situation at D_1. By employing the time-switching factor $\alpha = 0.5$ in the energy-harvesting approach, outage performance will be satisfied as multiple antennas are provided.

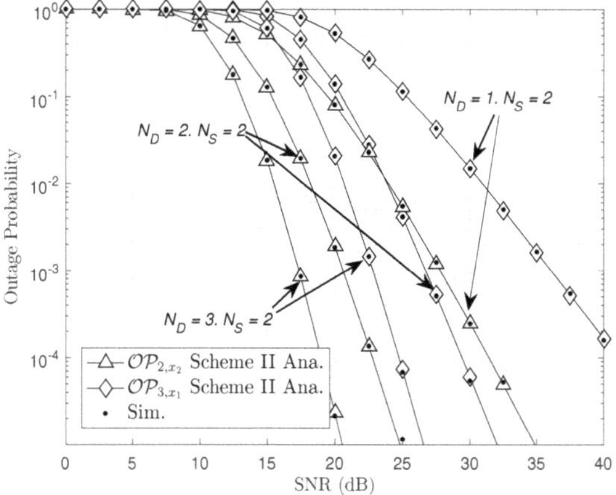

Figure 5. Outage probability of Scheme II with different N_D, $N_S = 2$, $d_1 = 2$ m, $d_2 = 3$ m, $a_1 = 0.2$ and $a_2 = 0.8$.

In Figure **??**, we compare two schemes in terms of outage performance. The first observation is that the best outage performance at D_1 in single-antenna NOMA. The worst performance can be raised at D_2 in Scheme I as only the relay link is implemented. It can be seen clearly that outage performance at D_2 in Scheme II is better than that in Scheme I. In the case of multiple antennas assigned at the BS, D_2, performance of D_2 can be improved significantly. In particular, outage probability at D_2 when

$N_S = N_D = 4$ is the best case among the four simulated curves. It reveals that multiple antennas at the far NOMA user contribute to outage improvement.

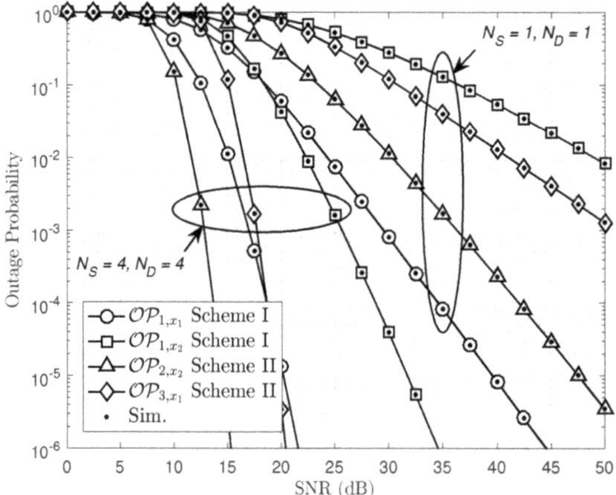

Figure 6. Comparison study on outage probability between Scheme I and Scheme II with $d_1 = 2$ m, $d_2 = 3$ m, $a_1 = 0.2$ and $a_2 = 0.8$.

In Figure **??**, we present the outage performance by varying the location of the relay D_1. It is noted that the curve showing outage performance of only the direct link in Scheme II is not affected by varying such distance d_1. Three users have different optimal relay locations, while outage performance of D_2 is slightly changed in the considered range of distance d_1. The most important observation is that the outage performance of D_1 falls significantly at locating D_1 far from the BS. Since NOMA allocates less transmit power to the users with better channel conditions, the optimal relay location for the user with better channel conditions should be nearer to BS to achieve its high SNR. It confirms that optimal location arrangement plays an important role in the remaining outage performance at an acceptable level.

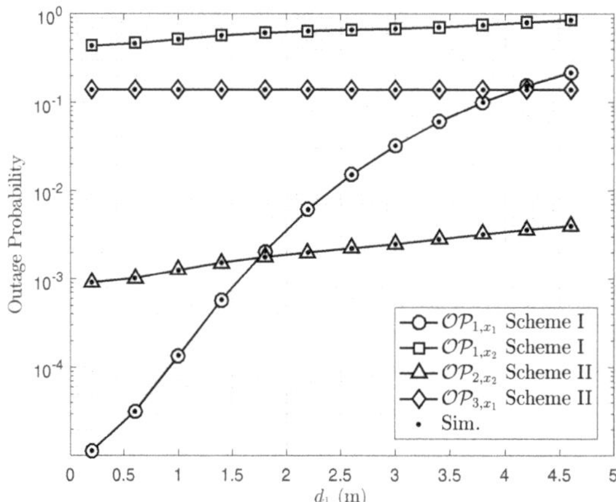

Figure 7. The impact of relay location on outage probability with $N_S = 2$, $N_D = 2$, $P_S = 20$ dB, $a_1 = 0.2$ and $a_2 = 0.8$.

Next, we further examine how the time-switching factor of energy-harvesting policy affects outage performance, as in Figure **??**. In this case, the special case of Scheme II, the direct link between the BS and D_2 is not supported by wireless power charge, and hence outage performance here is constant. Interestingly, optimal outage performance of D_2 can be seen at approximately $\alpha = 0.65$ for both schemes, while when changing time-switching factor in energy harvesting $\alpha = 0.1$ to $\alpha = 0.9$ we obtain higher power at D_1, which can harvest wireless energy. However, the remaining time for signal processing is inversely proportion to the time allocated for energy harvesting. The main reason is that a small amount of time for energy harvesting corresponding to a larger amount of time for signal processing at the first hop $BS - D_1$ causes better outage performance of D_1, and the best case occurs at around $\alpha = 0.1$.

In Figure **??**, we present the throughput performance by varying the threshold rate $\gamma_{th1} = \gamma_{th2} = \gamma_{th}$. It is noted that the highest curve showing throughput performance of D_2 is in Scheme II at set values $N_S = N_D = 4$. However, throughput will be decreased at a high threshold rate. A higher number of transmit antennas at the BS and receive antennas at D_2 leads to improved throughput performance, and they can be affordable with higher threshold data rates. Once again, we proved that multiple antennas design leads to system performance improvement.

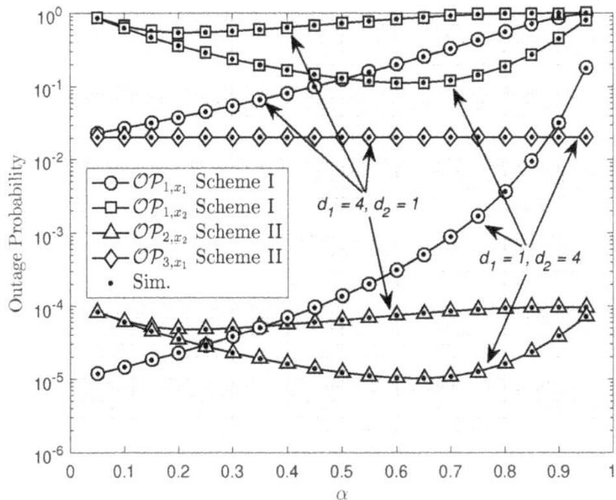

Figure 8. Outage probability versus the power-splitting ratio α with $N_S = 2$, $N_D = 3$, $P_S = 20$ dB, $a_1 = 0.2$ and $a_2 = 0.8$.

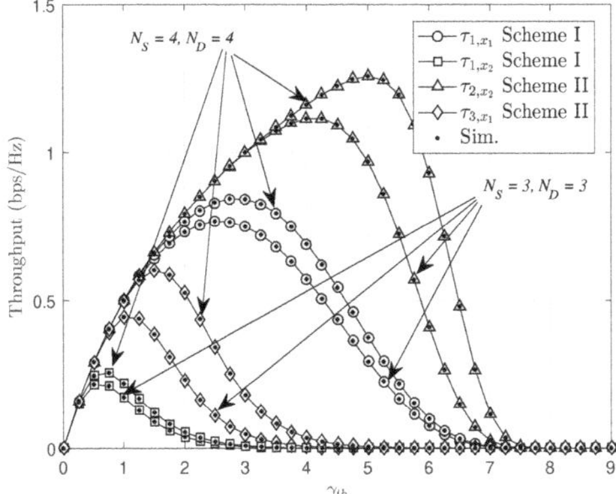

Figure 9. Impact of outage threshold on the throughput with transmit SNR at source $SNR = 20dB$, $d_0 = 5$ m, $d_1 = 3$ m, $d_2 = 2$ m, $a_1 = 0.1$ and $a_2 = 0.9$.

5. Conclusions

In this paper, we studied the impact of the number of transmit antennas at *BS* on outage performance in a multiple-antenna NOMA. In this paper, we compare two schemes related to the existence or non-existence of direct link between the *BS* and the far NOMA user. It is confirmed that optimal TAS policy deployed in this paper helps to improve performance in such a NOMA since multiple antennas lead to higher diversity order compared to single-antenna NOMA system. Simulation results revealed that the considered multiple-antenna NOMA system employing energy harvesting and a TAS scheme can achieve significantly higher performance than many situations of such systems. Furthermore, our results confirmed the robustness of the proposed scheme can be seen with respect to varying distance of these nodes, time-switching factors for energy harvesting, number of antennas at both the *BS*, and the far NOMA user. There are, intuitively, results can improve the performance of the far NOMA user, when considering and selecting various appropriate system parameters to achieve performance improvement.

Author Contributions: D.-T.D. provided idea of system, performed the theoretical analysis and wrote the manuscript. C.-B.L. implemented the simulation. M.V. contributed to the manuscript preparation.

Funding: This research was funded by VSB Technical University of Ostrava grant number SGS reg. No. SP2019/41 and the APC was funded by VSB Technical University of Ostrava.

Acknowledgments: This research received support from the grant SGS reg. No. SP2019/41 conducted at VSB Technical University of Ostrava, Czech Republic.

Appendix A. Proof of Lemma 1

To perform following outage probability

$$J_2 = \Pr\left\{\min\left\{\gamma_{1i,D_1x_1}, \gamma_{2i}\right\} < \gamma_{th2}\right\} \tag{A1}$$

We first recall these SNR as $\gamma_{1i,D_1x_1} = \frac{(1-\alpha)P_S|h_{1,i}|^2 a_1}{\sigma^2}$, $\gamma_{2i} = \alpha\eta\frac{P_S}{\sigma^2}|h_{1,i}|^2\|\mathbf{h}_{D_2}\|_F^2$. To derive the outage probability of optimal TAS scheme, we first present the CDF of RV as $Z = \min\left\{(1-\alpha), \alpha\eta\|\mathbf{h}_{D_2}\|_F^2\right\}$ as follows

$$F_Z(z) = \begin{cases} 1, & z > 1 - \alpha \\ \Pr\left(\|\mathbf{h}_{D_2}\|_F^2 < \frac{z}{\alpha\eta}\right), & z < 1 - \alpha \end{cases} \tag{A2}$$

It is noted that $\|\mathbf{h}_{D_2}\|_F^2$ follows Chi-square distribution, we have

$$F_Z(z) = \begin{cases} 1, & z > 1 - \alpha \\ 1 - \frac{1}{\Gamma(N_D)}\Gamma\left(N_D, \frac{z}{\alpha\eta\Omega_2}\right), & z < 1 - \alpha \end{cases} \tag{A3}$$

Then, we have expectation operation as below

$$\begin{aligned} E_Z\left[\left(\Pr\left(\gamma_{SD_1D_2,i} < \gamma_{th}\right)\right)^{N_S}\right] =1 &- \frac{N_S}{\Omega_1}\sum_{n=0}^{N_S-1}\binom{N_S-1}{n}(-1)^n\sum_{m=0}^{N_D-1}\frac{1}{m!}\left(\frac{\gamma_{th}}{\alpha\eta\Omega_2\bar{\gamma}}\right)^m \\ &\times \int_{\frac{\gamma_{th}}{(1-\alpha)\bar{\gamma}}}^{\infty}\frac{1}{x^m}\exp\left(-\frac{\gamma_{th}}{\alpha\eta\Omega_2\bar{\gamma}}\right)\exp\left(-\frac{n+1}{\Omega_1}x\right)dx \end{aligned} \tag{A4}$$

Using $\exp\left(-\frac{u}{x}\right) = \sum_{k=0}^{\infty}\frac{(-1)^k}{k!}\left(\frac{u}{x}\right)^k$, it can be achieved that

$$E_Z\left[\left(\Pr\left(\gamma_{SD_1D_2,i} < \gamma_{th}\right)\right)^{N_S}\right] = 1 - \frac{N_S\gamma_{th}}{\bar{\gamma}_1(1-\alpha)} \sum_{n=0}^{N_S-1}\binom{N_S-1}{n}\sum_{m=0}^{N_D-1}\sum_{k}^{\infty}\frac{(-1)^{k+n}}{m!k!}$$
$$\times\left(\frac{1-\alpha}{\alpha\mu\Omega_2}\right)^{m+k}E_{m+k}\left(\frac{\gamma_{th}(n+1)}{(1-\alpha)\bar{\gamma}_1}\right) \tag{A5}$$

This is end of the proof.

Appendix B. Proof of Proposition 1

It first recalls outage event at D_2 as below

$$OP_{2,x_2} = \Pr\left(\underbrace{\max}_{1\leq i\leq N_S}\max\left\{\gamma_{0j,D_2x_2}, \min\left\{\gamma_{1i,D_1x_1}, \gamma_{2i}\right\}\right\} < \gamma_{th2}\right) \tag{A6}$$

Then, it is re-expressed by

$$OP_{2,x_2} = E_Z\left[\Pr\left(\max\left\{\gamma_{0i^*,D_2x_2}, \gamma_{SD_1D_2,j^*}\right\} < \gamma_{th2}\right)\right]$$
$$= E_Z\left[\left(\Pr\left(\gamma_{0j^*,D_2x_2} < \gamma_{th2}\right)\Pr\left(\gamma_{SD_1D_2,j^*} < \gamma_{th2}\right)\right)^{N_S}\right]$$
$$= \left[\Pr\left(\gamma_{0j^*,D_2x_2} < \gamma_{th2}\right)\right]^{N_S}E_Z\left[\left(\Pr\left(\gamma_{SD_1D_2,i^*} < \gamma_{th2}\right)\right)^{N_S}\right] \tag{A7}$$

where $E_Z[.]$ denotes the expectation operator with respect to the RV Z. Considering on SNR $\gamma_{0j,D_2x_2} = \frac{P_S\|\mathbf{h}_{0,j}\|_F^2 a_2}{P_S\|\mathbf{h}_{0,j}\|_F^2 a_1 + \sigma^2}$ follows Chi-square distribution, we have

$$\Pr\left(\gamma_{0,i} < \frac{\gamma_{th2}}{a_2 - a_1\gamma_{th2}}\right) = 1 - \frac{1}{\Gamma(N_D)}\Gamma\left(N_D, \frac{\frac{\gamma_{th2}}{a_2-a_1\gamma_{th2}}}{\bar{\gamma}_0}\right) \tag{A8}$$

where $\bar{\gamma}_0 = \frac{P_S\Omega_0}{\sigma^2}$. Then, the second term can be formulated by

$$\Pr\left(\gamma_{SD_1D_2,i} < \gamma_{th2}\right) = \Pr\left(\bar{\gamma}|h_{1,i}|^2 z < \gamma_{th2}\right)$$
$$= 1 - \exp\left(-\frac{\gamma_{th2}}{z\bar{\gamma}_1}\right) \tag{A9}$$

where $\bar{\gamma}_1 = \bar{\gamma}\Omega_1$. It is noted that recalling $\gamma_{2i} = \alpha\eta\frac{P_S}{\sigma^2}|h_{1,i}|^2\|\mathbf{h}_{D_2}\|_F^2$, then it can be obtained outage probability as

$$\Pr\left(\alpha\eta\frac{P_S}{\sigma^2}|h_{1,i}|^2\|\mathbf{h_{D2}}\|_F^2 < \gamma_{th2}\right) = \Pr\left(\bar{\gamma}|h_{1,i}|^2 z < \gamma_{th2}\right)$$
$$= 1 - \exp\left(-\frac{\gamma_{th2}}{z\bar{\gamma}_1}\right) \tag{A10}$$

Therefore, it can be shown such outage probability as

$$E_Z\left[\left(\Pr\left(\gamma_{SD_1D_2,i} < \gamma_{th2}\right)\right)^{N_S}\right] = 1 - \frac{N_S}{\Omega_1}\sum_{n=0}^{N_S-1}\binom{N_S-1}{n}(-1)^n\sum_{m=0}^{N_D-1}\frac{1}{m!}\left(\frac{\gamma_{th2}}{\alpha\eta\Omega_2\bar{\gamma}}\right)^m$$
$$\times\int_{\frac{\gamma_{th2}}{(1-\alpha)\bar{\gamma}}}^{\infty}\frac{1}{x^m}\exp\left(-\frac{\gamma_{th2}}{\alpha\eta\Omega_2\bar{\gamma}}\right)\exp\left(-\frac{n+1}{\Omega_1}x\right)dx \tag{A11}$$

Using $\exp\left(-\frac{u}{x}\right) = \sum_{k=0}^{\infty}\frac{(-1)^k}{k!}\left(\frac{u}{x}\right)^k$, we have

$$E_Z\left[\left(\Pr\left(\gamma_{SR,i} < \gamma_{th2}\right)\right)^{N_S}\right] = 1 - \frac{N_S \gamma_{th2}}{\bar{\gamma}_1 (1-\alpha)} \sum_{n=0}^{N_S-1} \binom{N_S - 1}{n} \sum_{m=0}^{N_d-1} \sum_{k}^{\infty} \frac{(-1)^{k+n}}{m!k!}$$
$$\times \left(\frac{1-\alpha}{\alpha\mu\Omega_2}\right)^{m+k} E_{m+k}\left(\frac{\gamma_{th2}(n+1)}{(1-\alpha)\bar{\gamma}_1}\right) \tag{A12}$$

References

1. Islam, S.M.R.; Zeng, M.; Dobre, O.A.; Kwak, K. Performance analysis of cooperative NOMA schemes in spatially random relaying networks. *IEEE Wirel. Commun.* **2018**, *25*, 40–47. [CrossRef]
2. Islam, S.M.R.; Avazov, N.; Dobre, O.A.; Kwak, K.-S. Power-Domain Non-Orthogonal Multiple Access (NOMA) in 5G Systems: Potentials and Challenges. *IEEE Commun. Surv. Tutor.* **2017**, *19*, 721–742. [CrossRef]
3. Wan, D.; Wen, M.; Ji, F.; Yu, H.; Chen, F. Non-Orthogonal Multiple Access for Cooperative Communications: Challenges, Opportunities, and Trends. *IEEE Wirel. Commun.* **2018**, *25*, 109–117. [CrossRef]
4. Ding, Z.; Liu, Y.; Choi, J.; Sun, Q.; Elkashlan, M.; Poor, H.V. Application of non-orthogonal multiple access in LTE and 5G networks. *IEEE Commun. Mag. Technol.* **2017**, *55*, 185–191. [CrossRef]
5. Do, D.-T.; Nguyen, H.-S.; Voznak, M.; Nguyen, T.-S. Wireless powered relaying networks under imperfect channel state information: System performance and optimal policy for instantaneous rate. *Radioengineering* **2017**, *26*, 869–877. [CrossRef]
6. Nguyen, X.-X.; Do, D.-T. Optimal power allocation and throughput performance of full-duplex DF relaying networks with wireless power transfer-aware channel. *EURASIP J. Wirel. Commun. Netw.* **2017**, *2017*, 152. [CrossRef]
7. Nguyen, X.-X.; Do, D.-T. Maximum Harvested Energy Policy in Full-Duplex Relaying Networks with SWIPT. *Int. J. Commun. Syst.* **2017**, *30*, e3359. [CrossRef]
8. Nguyen, K.-T.; Do, D.; Nguyen, X.-X.; Nguyen, N.-T.; Ha, D.-H. Wireless information and power transfer for full duplex relaying networks: Performance analysis. In *Recent Advances in Electrical Engineering and Related Sciences (AETA 2015)*; Springer: Berlin/Heidelberg, Germany, 2015; pp. 53–62.
9. Zhang, Z.; Ma, Z.; Xiao, M.; Ding, Z.; Fan, P. Full-duplex device-to-device aided cooperative non-orthogonal multiple access. *IEEE Trans. Veh. Technol.* **2017**, *66*, 4467–4471. [CrossRef]
10. Liu, H.; Ding, Z.; Kim, K.J.; Kwak, K.S.; Poor, H.V. Decode-and-Forward Relaying for Cooperative NOMA Systems With Direct Links. *IEEE Trans. Wirel. Commun.* **2018**, *17*, 8077–8093. [CrossRef]
11. Nguyen, T.-L.; Do, D.-T. Exploiting Impacts of Intercell Interference on SWIPT-assisted Non-orthogonal Multiple Access. *Wirel. Commun. Mobile Comput.* **2018**, *2018*, 2525492. [CrossRef]
12. Do, D.-T.; Nguyen Van, M.-S.; Hoang, T.-A.; Voznak, M. NOMA-Assisted Multiple Access Scheme for IoT Deployment: Relay Selection Model and Secrecy Performance Improvement. *Sensors* **2019**, *19*, 736. [CrossRef]
13. Kader, M.F.; Uddin, M.B.; Islam, S.M.R.; Shin, S.Y. Capacity and outage analysis of a dual-hop decode-and-forward relay-aided NOMA scheme. *Dig. Signal Process.* **2019**, *88*, 138–148. [CrossRef]
14. Deng, D.; Fan, L.; Lei, X.; Tan, W.; Xie, D. Joint user and relay selection for cooperative NOMA networks. *IEEE Access* **2017**, *5*, 20220–20227. [CrossRef]
15. Han, T.; Gong, J.; Liu, X.; Islam, S.M.R.; Li, Q.; Bai, Z.; Kwak, K.S. On Downlink NOMA in Heterogeneous Networks with Non-Uniform Small Cell Deployment. *IEEE Access* **2018**, *6*, 31099–31109. [CrossRef]
16. Liu, Y.; Qin, Z.; Elkashlan, M.; Gao, Y.; Hanzo, L. Enhancing the physical layer security of non-orthogonal multiple access in large-scale networks. *IEEE Trans. Wirel. Commun.* **2017**, *16*, 1656–1672. [CrossRef]
17. Do, D.-T.; Le, C. Application of NOMA in Wireless System with Wireless Power Transfer Scheme: Outage and Ergodic Capacity Performance Analysis. *Sensors* **2018**, *18*, 3501. [CrossRef]
18. Nguyen, T.; Do, D. Power Allocation Schemes for Wireless Powered NOMA Systems with Imperfect CSI: System model and performance analysis. *Int. J. Commun. Syst.* **2018**, *31*, e3789. [CrossRef]
19. Zaidi, S.K.; Hasan, S.F.; Gui, X. Evaluating the Ergodic Rate in SWIPT-Aided Hybrid NOMA. *IEEE Commun. Lett.* **2018**, *22*, 1870–1873. [CrossRef]

20. Pei, L.; Yang, Z.; Pan, C.; Huang, W.; Chen, M.; Elkashlan, M.; Nallanathan, A. Energy-Efficient D2D Communications Underlaying NOMA-Based Networks With Energy Harvesting. *IEEE Commun. Lett.* **2018**, *22*, 914–917. [CrossRef]
21. Wang, Y.; Wu, Y.; Zhou, F.; Chu, Z.; Wu, Y.; Yuan, F. Multi-Objective Resource Allocation in a NOMA Cognitive Radio Network With a Practical Non-Linear Energy Harvesting Model. *IEEE Trans. Commun.* **2017**, *65*, 1077–1091. [CrossRef]
22. Hedayati, M.; Kim, I. On the Performance of NOMA in the Two-User SWIPT System. *IEEE Trans. Veh. Technol.* **2018**, *67*, 11258–11263. [CrossRef]
23. Ding, Z.; Adachi, F.; Poor, H.V. The application of MIMO to non-orthogonal multiple access. *IEEE Trans. Wirel. Commun.* **2016**, *15*, 537–552. [CrossRef]
24. Al-Abbasi, Z.Q.; So, D.K.C.; Tang, J. Resource allocation for MU-MIMO non-orthogonal multiple access (NOMA) system with interference alignment. In Proceedings of the 2017 IEEE International Conference on Communications (ICC), Paris, France, 21–25 May 2017; pp. 1–6.
25. Wang, H.; Zhang, R.; Song, R.; Leung, S.H. A novel power minimization precoding scheme for MIMO-NOMA uplink systems. *IEEE Commun. Lett.* **2018**, *22*, 1106–1109. [CrossRef]
26. Ding, Z.; Schober, R.; Poor, H.V. A general MIMO framework for NOMA downlink and uplink transmission based on signal alignment. *IEEE Trans. Wirel. Commun.* **2016**, *15*, 4438–4454. [CrossRef]
27. Chen, Z.; Ding, Z.; Dai, X.; Karagiannidis, G.K. On the Application of Quasi-Degradation to MISO-NOMA Downlink. *IEEE Trans. Signal Process.* **2016**, *64*, 6174–6189. [CrossRef]
28. Chen, X.; Zhang, Z.; Zhong, C.; Ng, D.W.K. Exploiting MultipleAntenna Techniques for Non-Orthogonal Multiple Access. *IEEE J. Sel. Areas Commun.* **2017**, *35*, 2207–2220. [CrossRef]
29. Zhang, X.; Chen, F.; Wang, W. Outage Probability Study of Multiuser Diversity in MIMO Transmit Antenna Selection Systems. *IEEE Signal Process Lett.* **2007**, *14*, 161–164. [CrossRef]
30. Lei, L.; Yuan, D.; Ho, C.K.; Sun, S. Power and Channel Allocation for Non-Orthogonal Multiple Access in 5G Systems: Tractability and Computation. *IEEE Trans. Wirel. Commun.* **2016**, *15*, 8580–8594. [CrossRef]
31. Wei, Z.; Ng, D.W.K.; Yuan, J.; Wang, H.M. Optimal Resource Allocation for Power-Efficient MC-NOMA with Imperfect Channel State Information. *IEEE Trans. Commun.* **2017**, *65*, 3944–3961. [CrossRef]
32. Sun, Y.; Ng, D.W.K.; Ding, Z.; Schober, R. Optimal Joint Power and Subcarrier Allocation for MC-NOMA Systems. In Proceedings of the 2016 IEEE Global Communications Conference (GLOBECOM), Washington, DC, USA, 4–8 December 2016; pp. 1–6.
33. Cui, J.; Ding, Z.; Fan, P. Outage Probability Constrained MIMO-NOMA Designs Under Imperfect CSI. *IEEE Trans. Wirel. Commun.* **2018**, *17*, 8239–8255. [CrossRef]

 *electronics* 

Article

Frequency and Pattern Reconfigurable Antenna for Emerging Wireless Communication Systems

Amjad Iqbal [1], **Amor Smida** [2,3], **Nazih Khaddaj Mallat** [4], **Ridha Ghayoula** [3,5], **Issa Elfergani** [6,*], **Jonathan Rodriguez** [6,7] **and Sunghwan Kim** [8,*]

[1]   Centre for Wireless Technology (CWT), Faculty of Engineering, Multimedia University, Cyberjaya 63100, Malaysia; amjad730@gmail.com
[2]   Department of Medical Equipment Technology, College of Applied Medical Sciences, Majmaah University, 11952 AlMajmaah, Saudi Arabia; a.smida@mu.edu.sa
[3]   Unit of Research in High Frequency Electronic Circuits and Systems, Faculty of Mathematical, Physical and Natural Sciences of Tunis, Tunis El Manar University, Tunis 2092, Tunisia
[4]   College of Engineering, Al Ain University of Science and Technology, Al Ain 64141, UAE; nazih.mallat@aau.ac.ae
[5]   Department of Electrical and Computer Engineering, Laval University, Quebec City, QC G1V0A6, Canada; ridha.ghayoula.1@ulaval.ca
[6]   Instituto de Telecomunicações, Campus Universitário de Santiago, 3810-193 Aveiro, Portugal; jonathan@av.it.pt
[7]   Faculty of Computing, Engineering and Science, University of South Wales, Pontypridd CF37 1DL, UK
[8]   School of Electrical Engineering, University of Ulsan, Ulsan 44610, Korea
[*]   Correspondence: i.t.e.elfergani@av.it.pt (I.E.); sungkim@ulsan.ac.kr (S.K.); Tel.: +82-52-259-1401 (S.K.)

Received: 13 March 2019; Accepted: 3 April 2019; Published: 7 April 2019

Abstract: A printed and minimal size antenna having the functionality of frequency shifting as well as pattern reconfigurability is presented in this work. The antenna proposed in this work consists of three switches. Switch 1 is a lumped switch that controls the operating bands of the antenna. Switch 2 and Switch 3 controls the beam switching of the antenna. When the Switch 1 is ON, the proposed antenna operates at 3.1 GHz and 6.8 GHz, covering the 2.5–4.2 GHz and 6.2–7.4 GHz bands, respectively. When Switch 1 is OFF, the antenna operates only at 3.1 GHz covering the 2.5–4.2 GHz band. The desired beam from the antenna can be obtained by adjusting the ON and OFF states of Switches 2 and 3. Unique beams can be obtained by different combination of ON and OFF states of the Switches 2 and 3. A gain greater than 3.7 dBi is obtained for all four cases.

Keywords: pattern reconfigurable; patch antenna; s-parameters; frequency reconfigurable

---

1. Introduction

With the development of communication technology, there is considerable interest in reconfigurable antennas, mainly for their applications in various wireless communication systems. Their properties can be adjusted so as to achieve the desired frequency band, radiation direction or polarization. Tunable antennas have many advantages over wide band antennas, such as smaller size, comparable radiation patterns among all multiple frequency bands, productive utilization of electromagnetic range, and frequency discernment, which is helpful for decreasing the antagonistic impacts of co-channel interference and jamming [1,2]. Most of the work in the literature has focused on tuning a single antenna property, rather than multiple properties. Microfluidic controlled polypropylene tubes, inserted between the main radiators and the ground plane, have been used for the purpose of frequency reconfigurablity [3]. A pin diode has been used to vary the antenna's bandwidth from narrow to wide, while varactor diodes have been used to continuously change the resonant frequency in the narrow band [4], making it more competitive for cognitive radio applications.

A PIN diode-based multiband reconfigurable printed monopole antenna for WLAN/WiMAX was presented in [5]. A wide range of frequency tunability was achieved in a probe-feed patch antenna using voltage-controlled varactor diodes [6]. A hooked-shaped, stub-loaded printed antenna with reconfigurable frequency for multiple applications was presented in [7]. Frequency tunability was achieved in [8,9] by using lumped switches. Frequency tunability in [10] was achieved by operation of PIN and varactor diodes. Frequency tunability in Vivaldi antenna was studied in [11]. High impedance surfaces [12], loops with diodes [13], multiple lumped switches [14] and multiple diodes in irregular manner [15], are used for reshaping the beam direction. A dual-polarized antenna with different frequency choices using metamaterial was presented in [16]. A dual-band antenna with independently controllable bands using varactor diodes is presented in [17].

The rapid development of communication systems demands configurations where the frequency and pattern of the system can be independently tuned. Combining all these configurations in a single antenna is a major challenge. In recent years, some antennas have been designed successfully for frequency and pattern tunability. A slot antenna having the characteristics of frequency and the pattern reconfigurability using PIN diodes is reported in [18]. The antenna uses two switches on the slot to produce multiple resonant bands, while the four slits along with the switches produce pattern reconfigurability. The antenna presented in [18] does not cover the whole elevation or azimuthal plane. The aforementioned antenna uses numerous RF PIN diodes, thus increasing the complexity of the system as well as increasing the insertion losses. In [19], multiple pin diodes are used for frequency and pattern reconfigurability. The antenna performance at each standard is good enough however its big size limits its application in modern communication. In [20], matching stubs are used to shift the resonant frequency, while the PIN diodes in the annular slot are inserted to direct the main lobe and the null in the desired direction. Liquid crystal technology is used in [21] to produce the frequency and pattern reconfigurability in the antenna. In [22], frequency and pattern reconfigurability are achieved using slits connected through PIN diodes with the main radiating part. The designed antenna in [22] can only tilt the radiation pattern at $30°$, $0°$, $-30°$. A flexible antenna for 1.9 G and 2.4 G is proposed in [23] with frequency tuning and beam switching characteristics. However, the antenna is not suitable for modern wireless communication architecture because of its complex structure (more RF diodes) and large dimensions. Stub and varactor diodes loaded array antenna capable of beam steering and frequency shifting is presented in [24], however the antenna has less beam shifting capability $(-30°$ to $+30°)$ as well as large size. A high gain High Impedance Structure (HIS) based array antenna is reported in [25]. The antenna uses four pin diodes for full frequency reconfiguration and partial pattern reconfiguration $(+/-11°$ phase shift). A frequency switchable antenna with three beam choices is reported in [26]. Couple of PIN diodes are inserted in the slot etched on the upper face of the antenna to tune the resonant bands while the two PIN diodes are added in the feeding network for beam shifting.

This paper provides a best solution for rapid development in communication systems configurations where the frequency and pattern of the system can be independently tuned. The proposed antenna works in either single or dual frequency mode according to the state of the lumped switch (Switch 1) while two lumped switches (Switch 2 and Switch 3) are deployed within the ground to produce pattern reconfigurability.

2. Antenna's Design Methodology

The proposed antenna's configuration is illustrated in Figure 1. The patch and ground of the proposed antenna is printed on 1.6 mm thicker substrate of FR-4, having the dielectric constant of 4.4. The proposed antenna has a small geometry of 23 mm × 31 mm × 1.6 mm. The microstrip transmission line is 3 mm wide resulting in 50 Ω characteristic impedance. Values of various parameters used in the proposed antenna are listed in Table 1.

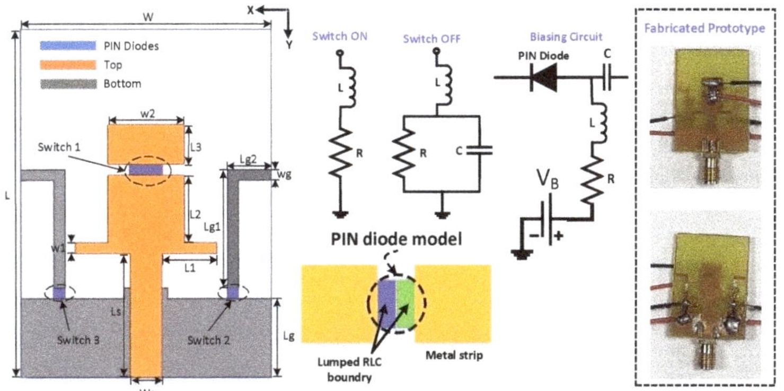

Figure 1. Proposed antenna diagram, biasing circuit for PIN diode and Fabricated antenna.

Table 1. Different Parameter and values of the antenna.

Parameter	Value (mm)	Parameter	Value (mm)	Parameter	Value (mm)
W	23	L1	5.05	Lg2	4
L	31	L2	6	Lg1	10.5
L3	3.5	W1	1	Wg	1
W2	7	Ws	3	Lg	7
Ls	9.5				

Equation (1) is used to calculate the length of the radiating element [27].

$$L_{f_r} = \frac{c}{4 f_r \sqrt{\epsilon_{eff}}} \tag{1}$$

$$\epsilon_{eff} \approx \frac{\epsilon_r + 1}{2} + \frac{\epsilon_r - 1}{2}\left(1 + 12\left(\frac{w}{h}\right)\right)^{-0.5} \tag{2}$$

In the above two equations, "c" is the speed of light in a vacuum , "λ_g" is the guided wavelength, "ϵ_{eff}" is the effective dielectric constant, "w" is the width of the substrate and "h" is the thickness of the substrate. Length of the monopole is calculated as 11.91 mm while using (1) and (2), however the optimized monopole length (L1 + L2) is noted as 11.05 mm.

A different shape of stubs are connected with the ground plane for observing the antenna's reflection coefficient response for all switching states. By connecting I-shaped stub with the ground (Figure 2a), the antenna resonates at 3.5 GHz and 6.7 GHz for case 1 with good impedance match. The antenna resonates at 3.9 GHz and 6.7 GHz for case 2 and case 3. An impedance mismatch is seen at the lower frequency band for case 4, and the antenna resonates at higher frequency band only. In the second step, H-shaped stub is loaded in the ground plane as shown in the Figure 2b. The antenna resonate at 3.5 GHz and 7.3 GHz when the switch 1 is ON and the rest of the switches are OFF. The antenna resonate at 3.5 GHz and 7.3 GHz for case 2 and case 3. A reasonable impedance matching at higher frequency band and mismatch at lower frequency band is seen in case 4. Lastly, an L-shaped stub is introduced in the ground plane and its impact during the all four cases is noted at Figure 2c. It is evident that the antenna performance in term of reflection coefficient is good for all four cases.

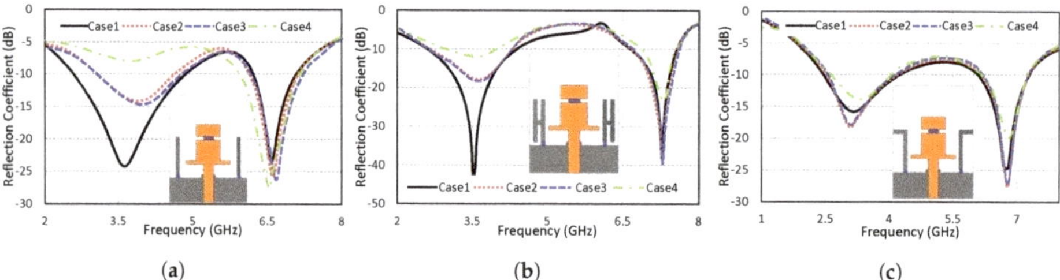

Figure 2. Reflection coefficient analysis for all cases with (**a**) I-shaped stub, (**b**) H-shaped stub, (**c**) L-shaped stub.

2.1. Switching Techniques

Usually the PIN diode behaves as a variable resistor in the RF frequency range, however, the ON and OFF states have more complex circuitry. Equivalent circuits of the ON and OFF states of the PIN diode consist of an inductor (L) and resistor (R). Forward biasing for the diode is obtained when the inductor and resistor (R) are connected in series. In case of the OFF state, the inductor is connected with a parallel-connected resistor (R) and capacitor (C). The ON and OFF behavior of the PIN diode are studied as RL and RLC circuits, respectively [28]. The lower value of the R in the RL circuit allows current flow between the radiating parts. The higher value of RC in the RLC circuit blocks the current from flowing between the radiating elements. Thus for the sake of simplicity we modeled our PIN diode as an RL circuit in the simulation. The value of the inductor (L) is kept constant. A resistor (R) value is kept low as 1 Ω for ON state of the diode and high as 5 MΩ for OFF state of the diode. A biasing voltage (V_B) of 3 V and 0 V is applied to the circuitry for switch ON and switch OFF condition, respectively.

2.2. Antenna's Parametric Analysis

High Frequency Structure Simulator (HFSS 13.0) is used for designing of the proposed antenna. The reflection coefficient (S_{11}) of the designed antenna under ON and OFF states of the switch 1 is illustrated in Figure 3. It is clear from Figure 3 that the proposed antenna has a single frequency band in the OFF state of Switch 1. Dual band antenna performance can be achieved by changing the state of Switch 1 to the ON state, as shown in Figure 3.

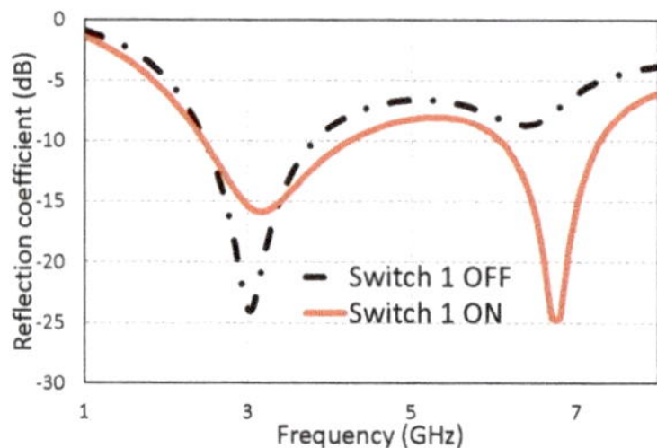

Figure 3. Reflection coefficient of the proposed antenna at ON and OFF switch.

Parametric examination of various parameters of the proposed antenna is performed so as to assess the impact of different parameters on the antenna performance. The proposed antenna is assessed using the parameters *w1* and *L2*. It can be seen from the graph of varying parameter *w1* that *w1* has a large effect on the band centered at 6.8 GHz. Changing the parameter *w1* dramatically changes the resonant frequency of the 6.8 GHz band. By increasing the value of the parameter *w1*, the resonant frequency of the second band shifts towards the lower frequency, whereas the effect on the first resonant band is negligible. Thus, it is concluded from the parametric analysis that the second resonant band can be controlled by the parameter *w1*. Figure 4a demonstrates the simulated S_{11} of the proposed antenna versus frequencies for varied *w1*.

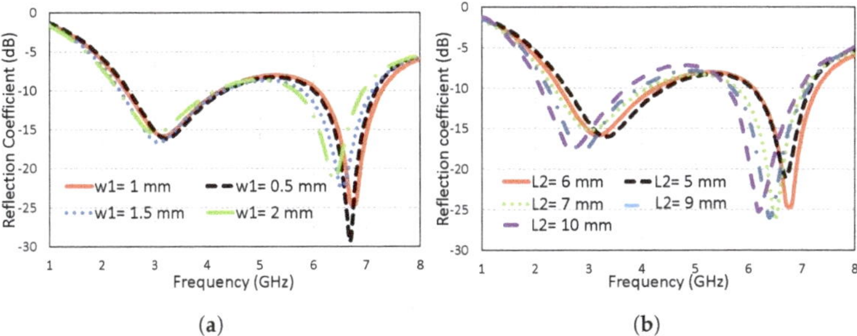

Figure 4. Reflection coefficient against frequencies for varied parameter (**a**) *w1* (**b**) *L2*.

Figure 4b presents the simulated S_{11} of the proposed antenna versus frequencies for a range of values of the parameter *L2*. It is clear from the figure that the parameter *L2* is effective in shifting of both bands. By increasing the value of *L2*, both bands shift toward lower frequency. From [27], it is clear that the frequency is inversely related to length of the monopole. The physical length of the radiating element increases, which corresponds to a lower operating frequency (1).

3. Results and Discussions

Figure 1 shows the top and bottom of the fabricated antenna. A low-capacitance PIN diode, MPP4203 (Microsemi) is used in the fabricated antenna. The DC path is completely isolated from the feeding path. The capacitors have the ability to block DC and pass RF signals while the RF choke (RFC) blocks RF and passes DC. The fabricated antenna uses 125 nH inductors, and 470 pF capacitors. The fabricated prototype is measured for reflection coefficient, and gain using Agilent Vector Network Analyzer (VNA). The calibration of the VNA is performed using SOLT (short-open-load-through) technique. After calibrating the VNA, fabricated prototype is connected with VNA and reflection coefficient is obtained. A 6dB log periodic antenna is placed at a distance of 8 m from the proposed antenna for far field pattern measurements. The fabricated antenna is placed on turn-table and the motor is allowed to rotate the antenna with 10° steps in both principle planes.

The proposed antenna has a resonant frequency of 3.1 GHz covering a 10 dB bandwidth of 1.7 GHz, as shown in Figure 5, when the lumped switch (Switch 1) is OFF. The antenna covers two bands when the lumped switch (Switch 1) is ON. The antenna has resonant frequencies of 3.1 GHz and 6.8 GHz, covering 10 dB bandwidths of 1.7 GHz and 1.2 GHz, respectively, when the lumped switch of the proposed antenna is ON.

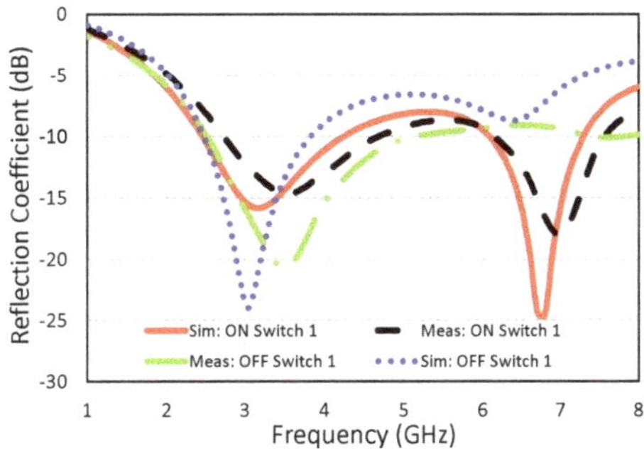

Figure 5. Simulated and measured reflection coefficient of the proposed antenna.

Figure 5 compares the simulated and measured reflection coefficient (S_{11}) results in both ON and OFF state of the switch (Switch 1). The simulated and measured results are well-matched. The slight mismatch in the curves for S_{11} is caused by the introduction of the three lumped switches in the fabricated antenna, SMA connector losses and cable losses.

To further characterize the behavior of the proposed antenna, the surface current distributions at 3.1 GHz and 6.8 GHz are shown in Figure 6. It is clear from Figure 6 that the current density is higher at the lower side of the radiating part with respect to the remaining portion of the radiator for 3.1 GHz. It is clear from Figure 6 that the lower region is responsible for resonance at 3.1 GHz. The surface current density of the proposed antenna is higher at lower as well as upper portions of the antenna at 6.8 GHz indicating that the 6.8 GHz resonance is affected by lower as well as upper parts of the antenna. The simulated gain and efficiency of the antenna under all switching cases is illustrated in Figure 7.

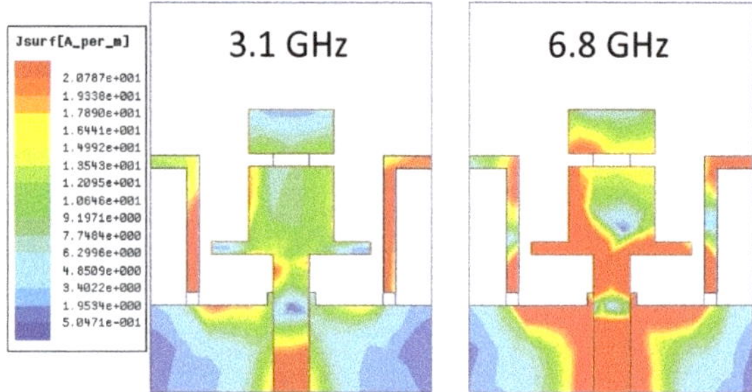

Figure 6. Surface current density of the proposed antenna at ON switch condition.

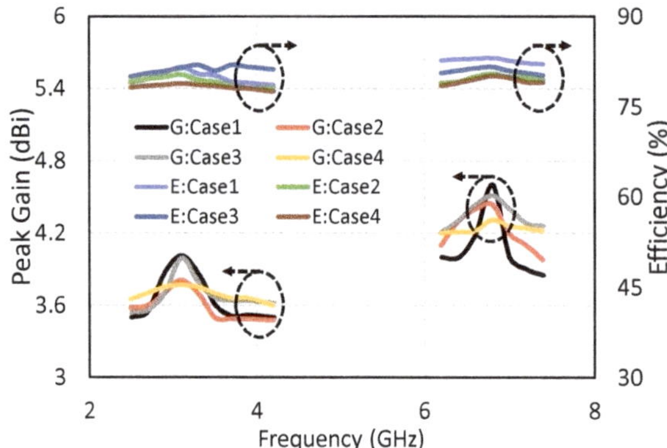

Figure 7. Simulated peak gains and efficiencies for all switching cases (G = Peak gain and E = efficiency).

Figures 8–11 present the radiation patterns for Case 1, Case 2, Case 3, and Case 4, respectively. The radiation patterns for both Phi = 0° and Phi = 90° are simulated and measured. Figures for all cases compare the simulated and measured radiation patterns of the proposed antenna at 3.1 and 6.8 GHz. Due to time varying current on the ground plane, size and structure of the ground plane is important factor in determining the impedance and radiation properties of the antenna. The introduction of L-shaped stub on the ground has very little impact on the impedance, however a huge impact on the radiation of the antenna is observed. Unique current distribution is observed in case of all switching condition which leads to different beam tilting phenomenon (Figure 12).

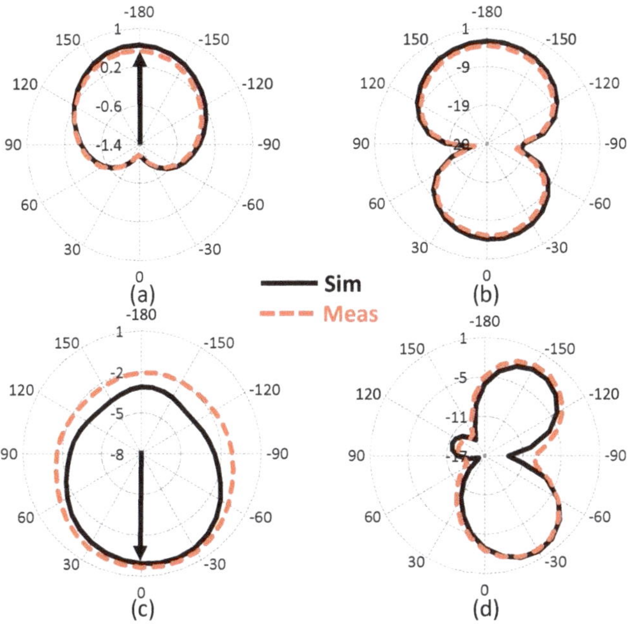

Figure 8. Case 1: (**a**) Phi = 0° at 3.1 GHz (**b**) Phi = 90° at 3.1 GHz (**c**) Phi = 0° at 6.8 GHz (**d**) Phi = 90° at 6.8 GHz.

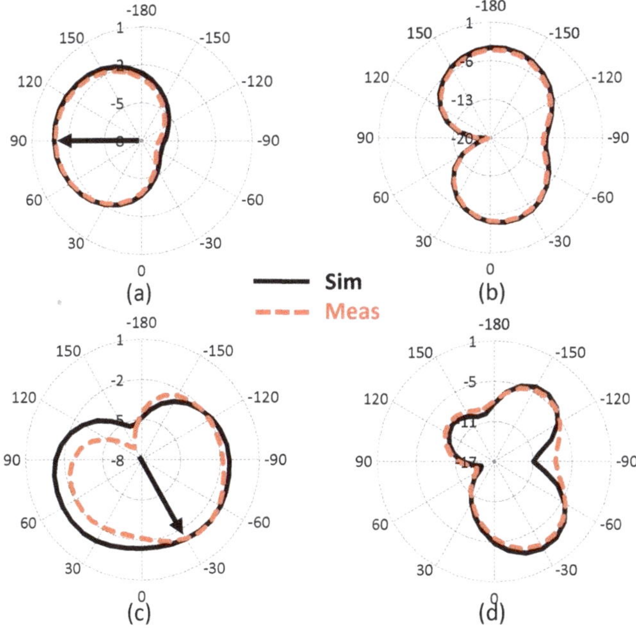

Figure 9. Case 2: (**a**) Phi = 0° at 3.1 GHz (**b**) Phi = 90° at 3.1 GHz (**c**) Phi = 0° at 6.8 GHz (**d**) Phi = 90° at 6.8 GHz.

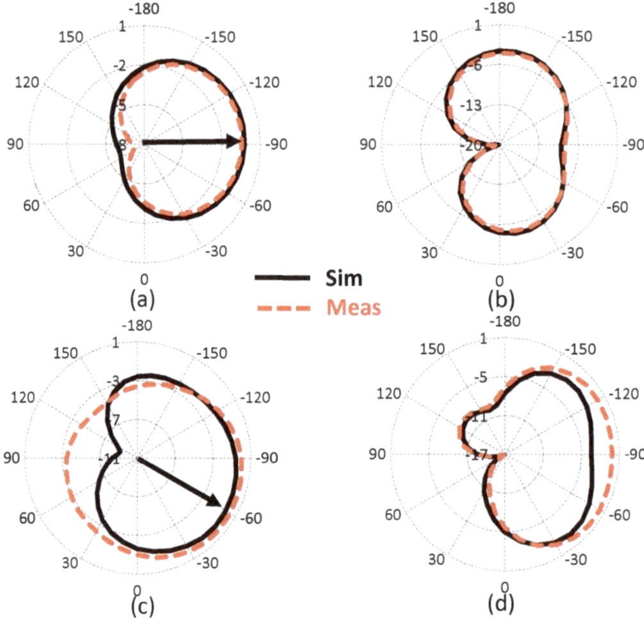

Figure 10. Case 3: (**a**) Phi = 0° at 3.1 GHz (**b**) Phi = 90° at 3.1 GHz (**c**) Phi = 0° at 6.8 GHz (**d**) Phi = 90° at 6.8 GHz.

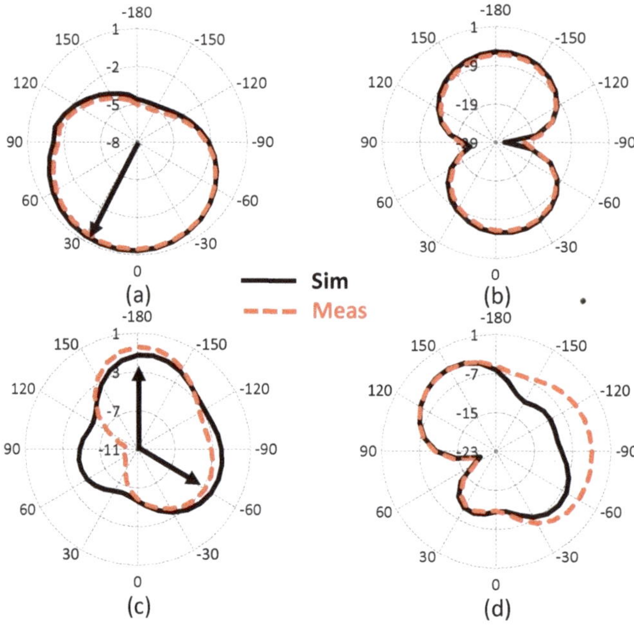

Figure 11. Case 4: (**a**) Phi = 0° at 3.1 GHz (**b**) Phi = 90° at 3.1 GHz (**c**) Phi = 0° at 6.8 GHz (**d**) Phi = 90° at 6.8 GHz.

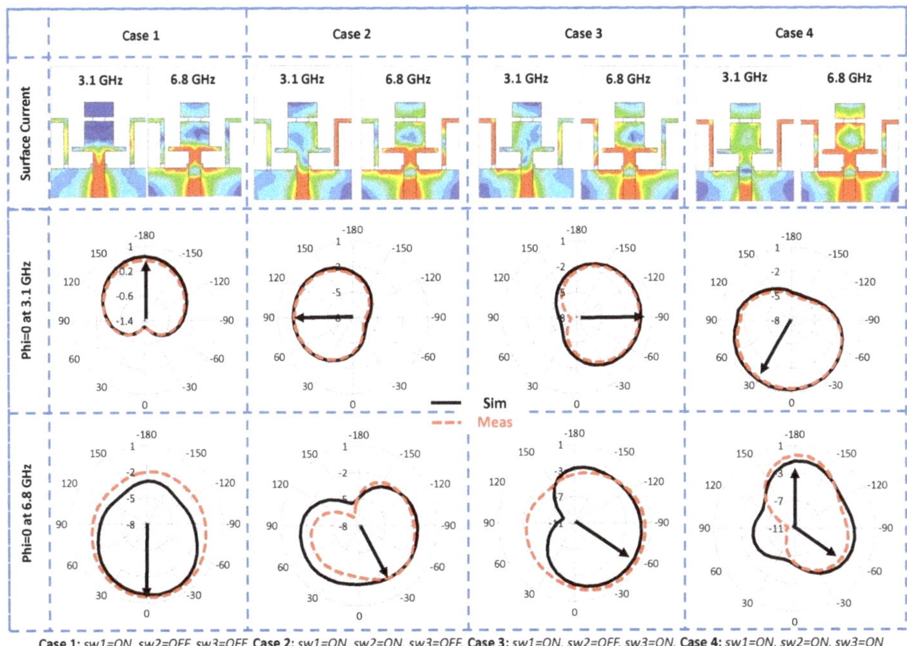

Figure 12. Summary of the beam switching using switch 2 and switch 3.

Figure 8 shows the radiation pattern for Case 1 where Switch 1 is ON and the rest of the switches are OFF. It can be seen from Figure 8 that the main lobe at Phi = 0° is directed towards 180° while a low backward lobe is observed as compared to the main lobe at 3.1 GHz. In the Phi = 0° plane, the main lobe is directed towards 0° and a low backward lobe is observed at 6.8 GHz. At lower frequency band the designed antenna radiates predominantly as a "Figure of Eight" in Phi = 90° plane while a little distortion is observed at high frequency band.

Figure 9 shows the radiation pattern for Case 2 where Switch 1 and Switch 2 are ON, and Switch 3 is OFF. In this case, the current flows along the L-shaped stub connected with ground plane as shown in Figure 12. The major current contents are present on the right sided L-shaped stub, resulting in enhancing the beam in one direction (90° at 3.1 GHz and −30° at 6.8 GHz) while suppressing the beam in opposite direction. It is quite clear from Figure 9 that the main lobe of the proposed antenna is directed towards 90° for Phi = 0° at 3.1 GHz, while the main lobe is directed towards −30° for 6.8 GHz. At lower frequency band the designed antenna radiates predominantly as a distorted "Figure of Eight" in Phi = 90° plane, while much more distortion is observed at high frequency band.

Figure 10 shows the radiation pattern for Case 3 where Switch 1 and Switch 3 are ON, and Switch 2 is OFF. Most of the current on the ground plane is concentrated on left sided L-shaped stub (Figure 12) which act as a director and tilt the beam towards −90° at 3.1 GHz and −60° at 6.8 GHz. It is clear from Figure 10 that the main lobe of the proposed antenna is directed towards 270° for Phi = 0° at 3.1 GHz while main lobe is directed towards −60° at 6.8 GHz.

Figure 11 shows the radiation pattern for Case 4 when Switch 1 and Switch 2 and Switch 3 are ON. Current in the ground plane is concentrated on both L-shaped stubs which results in suppressing the power in the opposite direction (resulting in directing beam at 30° for 3.1 GHz and −60°/180° for 6.8 GHz). It is clear from Figure 11 that the main lobe of the proposed antenna is directed towards 30° for Phi = 0° at 3.1 GHz, while the main lobe is directed towards −60° and 180° at 6.8 GHz. The proposed antenna exhibits stable radiation pattern in all four cases. Also, front-to-back-ratios (FBR) are satisfactory in the directed beams.

Table 2 summarizes the pattern reconfigurability of the proposed antenna achieved by adjusting different switching states. Table 3 compares the simulated gain, measured gain and simulated efficiency of the antenna for four possible cases. It can be concluded from the Table 3 that the antenna has good simulated and measured gain and satisfactory simulated efficiency at four possible cases. Table 4 compares the proposed antenna with the antennas reported in the literature. The proposed antenna novelty lies in its simple shape and its ability to efficiently reconfigure the frequency and beam direction. It can be well noted that the proposed antenna has smaller dimensions than the antennas reported in the literature. Also, the proposed antenna uses minimum number of switches which in turn reduces the losses occurred by the switches and also reduce the complexity level of the design. The antenna presented in this work has wider bandwidth than the existing antennas along with higher peak gain.

Table 2. Summary of the state of switches and corresponding main lobe direction.

Case	Switch 1	Switch 2	Switch 3	Main Lobe Direction at Phi = 0° for 3.1 GHz	Main Lobe Direction at Phi = 0° for 6.8 GHz
1	ON	OFF	OFF	180°	0°
2	ON	ON	OFF	90°	−30°
3	ON	OFF	ON	−90°	−60°
4	ON	ON	ON	30°	−60° and 180°

Table 3. Summary of the different cases and its corresponding gains and efficiencies at both bands.

Case	Gain (dBi)		Efficiency (%)	Gain (dBi)		Efficiency (%)
	Sim (3.1 GHz)	Meas (3.1 GHz)		Sim (6.8 GHz)	Meas (6.8 GHz)	
1	4.01	3.93	81.6	4.60	4.58	83.1
2	3.81	3.78	80.3	4.44	4.35	80.5
3	3.99	3.89	81.4	4.51	4.46	81.6
4	3.77	3.70	78.8	4.31	4.21	80.1

Table 4. Performance comparison with previously published work.

Ref.	Size (mm²)	Reconfiguration	Actuators	Bandwidth (MHz)	Peak Gain (dBi)
[18]	130 × 160	Frequency and Pattern	11 PIN diodes	200/150/150	5.6/4.6/3.3
[19]	80 × 45.8	Frequency and Pattern	5 PIN diodes	580/290	2.1/4.8
[20]	50 × 50	Frequency and Pattern	2 PIN diodes	100/70	4/5.6
[22]	50 × 50	Frequency and Pattern	4 PIN diodes	180/200/180/200	4/3.8/4.4/5
[23]	42 × 44	Frequency and Pattern	8 PIN diodes	160/220	NG
[24]	160.9 × 151.5	Frequency and Pattern	2 Varactor diodes	230	9
[25]	70 × 70	Frequency and Pattern	4 PIN diodes	NG	0.521/7.833
[26]	40 × 30	Frequency and Pattern	4 PIN diodes	400/500	2.24–2.76
This Work	23 × 31	Frequency and Pattern	3 PIN diodes	1700/1200	4.01/4.60

4. Conclusions

A compact printed antenna having the functionality of frequency shifting as well as pattern reconfigurability is successfully designed and experimentally validated in this paper. The proposed antenna is switch-dependent and has dissimilar performance for ON and OFF states of Switch 1. The designed antenna can be operated at two unique frequencies, depending on the state of Switch 1. When Switch 1 is OFF, the antenna operates at 3.1 GHz. Changing the state of Switch 1 causes the antenna to operate at 3.1 GHz (UWB health monitoring band for pets) and 6.8 GHz (indoor UWB band) with good gain (4.01 dBi and 4.6 dBi, respectively). The pattern reconfigurability is obtained by changing the states of Switch 2 and Switch 3 introduced in the ground plane. Four different beam directions may be obtained by adjusting the states of Switch 2 and Switch 3 while the maintaining high gain and efficiency.

Author Contributions: A.I. and A.S. provided the idea, performed the experiments, and managed the paper. N.K.M., R.G., I.E., J.R., and S.K. assisted in the idea development and paper writing.

Funding: This research was supported by the Research Program through the National Research Foundation of Korea (NRF-2016R1D1A1B03934653, NRF-2019R1A2C1005920). The authors also extend their appreciation to the Deanship of Scientific Research at Majmaah University for supporting this work through research project No.1440-80.

Conflicts of Interest: The authors declare no conflict of interest.

References

1. Augustin, G.; Aanandan, C.; Mohanan, P.; Vasudevan, K. A reconfigurable dual-frequency slot-loaded microstrip antenna controlled by pin diodes. *Microw. Opt. Technol. Lett.* **2005**, *44*, 374–376.
2. Ojaroudi Parchin, N.; Jahanbakhsh Basherlou, H.; Al-Yasir, Y.I.; Abd-Alhameed, R.A.; Abdulkhaleq, A.M.; Noras, J.M. Recent Developments of Reconfigurable Antennas for Current and Future Wireless Communication Systems. *Electronics* **2019**, *8*, 128. [CrossRef]
3. Tang, H.; Chen, J.X. Microfluidically frequency-reconfigurable microstrip patch antenna and array. *IEEE Access* **2017**, *5*, 20470–20476. [CrossRef]
4. Tang, M.C.; Wen, Z.; Wang, H.; Li, M.; Ziolkowski, R.W. Compact, Frequency-Reconfigurable Filtenna with Sharply Defined Wideband and Continuously Tunable Narrowband States. *IEEE Trans. Antennas Propag.* **2017**, *65*, 5026–5034. [CrossRef]
5. Abdulraheem, Y.I.; Oguntala, G.A.; Abdullah, A.S.; Mohammed, H.J.; Ali, R.A.; Abd-Alhameed, R.A.; Noras, J.M. Design of frequency reconfigurable multiband compact antenna using two PIN diodes for WLAN/WiMAX applications. *IET Microw. Antennas Propag.* **2017**, *11*, 1098–1105. [CrossRef]
6. Rouissi, I.; Floc'h, J.M.; Rmili, H.; Trabelsi, H. Design of a frequency reconfigurable patch antenna using capacitive loading and varactor diode. In Proceedings of the 2015 9th European Conference on Antennas and Propagation (EuCAP), Lisbon, Portugal, 13–17 April 2015; pp. 1–4.
7. Sharma, N.; Yadav, M.; Kumar, A. Design of quad-band microstrip-fed stubs-loaded frequency reconfigurable antenna for multiband operation. In Proceedings of the 2017 4th International Conference on Signal Processing and Integrated Networks (SPIN), Noida, India, 2–3 February 2017; pp. 275–279.
8. Iqbal, A.; Ullah, S.; Naeem, U.; Basir, A.; Ali, U. Design, fabrication and measurement of a compact, frequency reconfigurable, modified T-shape planar antenna for portable applications. *J. Electr. Eng. Technol.* **2017**, *12*, 1611–1618.
9. Iqbal, A.; Saraereh, O.A. A Compact Frequency Reconfigurable Monopole Antenna for Wi-Fi/WLAN Applications. *Prog. Electromagn. Res.* **2017**, *68*, 79–84.
10. Li, H.; Xiong, J.; Yu, Y.; He, S. A simple compact reconfigurable slot antenna with a very wide tuning range. *IEEE Trans. Antennas Propag.* **2010**, *58*, 3725–3728. [CrossRef]
11. Hamid, M.R.; Gardner, P.; Hall, P.S.; Ghanem, F. Reconfigurable vivaldi antenna. *Microw. Opt. Technol. Lett.* **2010**, *52*, 785–787. [CrossRef]
12. Deo, P.; Mehta, A.; Mirshekar-Syahkal, D.; Nakano, H. An HIS-based spiral antenna for pattern reconfigurable applications. *IEEE Antennas Wirel. Propag. Lett.* **2009**, *8*, 196–199. [CrossRef]
13. Kang, W.; Lee, S.; Kim, K. Design of symmetric beam pattern reconfigurable antenna. *Electron. Lett.* **2010**, *46*, 1536–1537. [CrossRef]
14. Zhang, S.; Huff, G.; Feng, J.; Bernhard, J. A pattern reconfigurable microstrip parasitic array. *IEEE Trans. Antennas Propag.* **2004**, *52*, 2773–2776. [CrossRef]
15. Cai, X.; Wang, A.; Chen, W. A circular disc-shaped antenna with frequency and pattern reconfigurable characteristics. In Proceedings of the 2011 China-Japan Joint Microwave Conference Proceedings (CJMW), Hangzhou, China, 20–22 April 2011; pp. 1–4.
16. Chen, X.; Zhao, Y. Dual-band Polarization and Frequency Reconfigurable Antenna Using Double Layer Metasurface. *AEU-Int. J. Electron. Commun.* **2018**, *95*, 82–87. [CrossRef]
17. Nguyen-Trong, N.; Hall, L.; Fumeaux, C. A dual-band dual-pattern frequency-reconfigurable antenna. *Microw. Opt. Technol. Lett.* **2017**, *59*, 2710–2715. [CrossRef]
18. Majid, H.A.; Rahim, M.K.A.; Hamid, M.R.; Ismail, M.F. Frequency and Pattern Reconfigurable Slot Antenna. *IEEE Trans. Antennas Propag.* **2014**, *62*, 5339–5343. [CrossRef]

19. Li, P.K.; Shao, Z.H.; Wang, Q.; Cheng, Y.J. Frequency and Pattern Reconfigurable Antenna for Multi-Standard Wireless Applications. *Group* **2015**, *1*, D3.

20. Nikolaou, S.; Bairavasubramanian, R.; Lugo, C.; Carrasquillo, I.; Thompson, D.C.; Ponchak, G.E.; Papapolymerou, J.; Tentzeris, M.M. Pattern and frequency reconfigurable annular slot antenna using PIN diodes. *IEEE Trans. Antennas Propag.* **2006**, *54*, 439–448. [CrossRef]

21. Zhao, Y.; Huang, C.; Qing, A.; Luo, X. A Frequency and Pattern Reconfigurable Antenna Array Based on Liquid Crystal Technology. *IEEE Photonics J.* **2017**, *9*, 1–7. [CrossRef]

22. Selvam, Y.P.; Kanagasabai, M.; Alsath, M.G.N.; Velan, S.; Kingsly, S.; Subbaraj, S.; Rao, Y.V.R.; Srinivasan, R.; Varadhan, A.K.; Karuppiah, M. A Low-Profile Frequency- and Pattern-Reconfigurable Antenna. *IEEE Antennas Wirel. Propag. Lett.* **2017**, *16*, 3047–3050. [CrossRef]

23. Zhu, Z.; Wang, P.; You, S.; Gao, P. A Flexible Frequency and Pattern Reconfigurable Antenna for Wireless Systems. *Prog. Electromagn. Res.* **2018**, *76*, 63–70.

24. Zainarry, S.N.M.; Nguyen-Trong, N.; Fumeaux, C. A Frequency-and Pattern-Reconfigurable Two-Element Array Antenna. *IEEE Antennas Wirel. Propag. Lett.* **2018**, *17*, 617–620. [CrossRef]

25. Dewan, R.; Abd Rahim, M.K.; Hamid, M.R.; Himdi, M.; Majid, H.B.A.; Samsuri, N.A. HIS-EBG Unit Cells for Pattern and Frequency Reconfigurable Dual Band Array Antenna. *Prog. Electromagn. Res.* **2018**, *76*, 123–132. [CrossRef]

26. Han, L.; Wang, C.; Zhang, W.; Ma, R.; Zeng, Q. Design of Frequency-and Pattern-Reconfigurable Wideband Slot Antenna. *Int. J. Antennas Propag.* **2018**, *2018*. [CrossRef]

27. Stutzman, W.L.; Thiele, G.A. *Antenna Theory and Design*; John Wiley & Sons: Hoboken, NJ, USA, 2013.

28. Yeom, I.; Choi, J.; Kwoun, S.s.; Lee, B.; Jung, C. Analysis of RF front-end performance of reconfigurable antennas with RF switches in the far field. *Int. J. Antennas Propag.* **2014**, *2014*, 385730. [CrossRef]

Article

Maximum Transmit Power for UE in an LTE Small Cell Uplink

Amir Haider and Seung-Hoon Hwang *

Division of Electronics and Electrical Engineering, Dongguk University-Seoul, Seoul 04620, Korea
* Correspondence: shwang@dongguk.edu; Tel.: +82-2-2260-3994

Received: 19 May 2019; Accepted: 15 July 2019; Published: 16 July 2019

Abstract: To furnish the network with small cells, it is vital to consider parameters like cell size, interference in the network, and deployment strategies to maximize the network's performance gains expected from small cells. With a small cell network, it is critical to analyze the impact of the uplink power control parameters on the network's performance. In particular, the maximum transmit power (P_{max}) for user equipment (UE) needs to be revisited for small cells, since it is a major contributor towards interference. In this work, the network performance was evaluated for different P_{max} values for the small cell uplink. Various deployment scenarios for furnishing the existing macro layer in LTE networks with small cells were considered. The P_{max} limit for a small cell uplink was evaluated for both homogenous small cell and heterogeneous networks (HetNet). The numerical results showed that it would be appropriate to adopt P_{max} = 18 dBm in uniformly distributed small cells rather than P_{max} = 23 dBm, as in macro environments. The choice of P_{max} = 18 dBm was further validated for three HetNet deployment scenarios. A decrease of 0.52 dBm and an increase of 0.03 dBm and 3.29 dBm in the proposed P_{max} = 18 dBm were observed for the three HetNet deployments, respectively. Furthermore, we propose that the fractional power control mode can be employed instead of the full compensation mode in small cell uplinks.

Keywords: small cell; maximum transmit power; UE; open-loop power control; interference

---

1. Introduction

Mobile broadband traffic has exponentially increased with the advent of modern smart phones, tablets, and communication devices, resulting in an ever-increasing data rate demand by end users. This requires an increase in network capacity and available data rates, power efficient connectivity, and decreased latency in the network. To increase capacity, network densification is necessary for radio access networks [1]. By deploying additional network nodes per unit area, the distance between the eNodeB and the user equipment (UE) can be minimized, which allows accommodating more UE per unit area, resulting in a link–budget improvement [2]. While addressing the capacity demands in areas with a high density of UE, such as urban and metropolitan areas (often referred to as hotspots), it is vital to choose the appropriate deployment option as well as the right configuration regarding the network's parameters to ensure acceptable data rates, minimum interference generation, and improved service quality for users. To configure network settings with newly furnished small cells, it is necessary to analyze the impact of different parameters such as cell size, distance between UE and the serving eNodeB, inter-site distance among neighboring cells, and the power control parameters particularly in the uplink direction [3].

To ensure power efficient connectivity, mitigate interference, and improve service quality, the 3rd generation partnership project (3GPP) standards introduced transmit power control (TPC) which enables power control for both the downlink and uplink directions in long term evolution (LTE) [4]. A precise control mechanism for the UE uplink transmission power is emphasized to mitigate the adverse impact of the interference on the network performance. There are two methods for TPC in the LTE uplink: open loop (OL) and closed loop (CL). The OP/TPC is the capability of the UE to set its uplink transmit power to a specified value suitable for the receiver. The UE can set its output power level by estimating the pathloss between the UE and the serving eNodeB based on the reference symbol received power (RSRP) and the OL/TPC parameters specified in Equation (1). The CL/TPC is the capability of the UE to adjust the uplink transmit power in accordance with the closed-loop correction value, also known as TPC commands. The eNodeB transmits the TPC commands towards the UE, based on the closed-loop signal-to-interference and noise ratio (SINR) target and measured received SINR. In CL/TPC, the uplink receiver at the eNodeB estimates the SINR of the received signal and compares it with the desired SINR target value. When the received SINR is below the SINR target, a TPC command is transmitted to the UE to request an increase in the transmitter power. Otherwise, the TPC command will request a decrease in transmitter power. The closed-loop power control operates around an open-loop point of operation. The initial transmit power for the UE is set using the open-loop power control. The initial power is further adjusted using a closed-loop correction value via downlink signaling of TPC commands to the UE. In summary, OL/TPC determines the initial power settings of the network while CL/TPC aims to correct errors in OL/TPC configurations. The 3GPP specification defines the UE transmit power P_{Tx} (dBm) for the LTE uplink as [5]:

$$P_{\mathrm{Tx}} = \min\left\{P_{\max}, 10\log_{10} \mathrm{M} + P_\mathrm{o} + \alpha.\mathrm{PL} + \Delta_{mcs} + f(i)\right\} \tag{1}$$

- $P_{\max}$ is the maximum allowable transmit power for the UE.
- M is the number of physical resource blocks (PRBs). For simplicity, M is set to 1 in this paper.
- P_o is the cell/UE specific parameter. P_o is assumed to be UE specific in this paper.
- α is the pathloss compensation factor.
- PL is the pathloss between the UE and serving eNodeB.
- Δ_{mcs} and $f(i)$ are the CL/TPC parameters defining the modulation and coding scheme and closed-loop correction, respectively, as defined in the 3GPP standards. To focus our analysis on the OL/TPC, Δ_{mcs} and $f(i)$ are assumed to be zero.

At this point, it is important to specify the role of the pathloss compensation factor α in terms of the mode of TPC being used. The modes of TPC based on the α value are:

- $\alpha = 0$ implies "no power control" mode.
- $\alpha > 0$ designates the "fractional power control" mode. In practice, the values from 0.7 to 0.9 are adopted.
- $\alpha = 1$ indicates a "full compensation' mode for the uplink pathloss.

To accomplish the task of network densification and optimize the network performance, it is crucial to investigate the uplink power control mechanism for small cells, particularly the maximum transmit power limit $P_{\max}$ for the case of the small cell uplink. In general, the 3GPP standards define a $P_{\max}$ limit of 23 dBm regardless of cell size. However, in the case of small cells, the distance between the eNodeB and UE is shorter, as compared to the macro cell, resulting in different transmit power levels for UE in small cells. An increase in the transmit power improves the received signal strength, and consequently, the received signal-to-interference ratio (SIR). However, in the case of small cells, transmitting at a higher power might result in the degradation of the received SIR due to the significant increase in the inter-cell interference [6]. Similarly, choosing a lower value of $P_{\max}$ can decrease the interference, but might affect the SIR performance as the received signal strength might go below the receiver threshold in the case of choosing too low a value for $P_{\max}$. Also, it is important to consider the deployment strategy to furnish the existing macro cell layer with small cells, as the choice of TPC parameters is highly dependent on the location of small cells. This issue implies a need to carefully reconsider the $P_{\max}$ for small cell deployment.

In general, the fractional power control (FPC) mode is adopted for macro cell deployment, which ensures better service quality for the cell's average users at the cost of a slight degradation in the cell edge performance [7]. In References [8–10], the impact of α and P_o on the transmit power was investigated in detail; however, the consideration of variation in the $P_{\max}$ limit was not considered. In Reference [11], a resource allocation-based uplink transmission energy reduction scheme was presented; however, it is limited only to a macro cell environment and considers the conventional 23 dBm limit for the $P_{\max}$. Also, for a small cell deployment, no separate description is available in the literature regarding the TPC mode. In Reference [12], the authors presented a detailed study on the downlink transmit power and bandwidth allocation problem for a small cell eNodeB in HetNet deployment. By considering the downlink scenario, the results indicated that for HetNet deployment, the energy efficiency for small cell users as well as the overall system capacity can be improved by increasing the number of small cells and the number of users per small cell. However, the analysis was limited only for the downlink transmit power of the small cell eNodeB and the inter-cell interference analysis was also left for future consideration. In our previous work [13], we proposed a full compensation mode for small cells in homogenous as well as in heterogeneous network (HetNet) deployment scenarios assuming the standard $P_{\max}$ limit, i.e., 23 dBm. However, we can expect that, with a lower value of $P_{\max}$, the situation might become different. To provide better insight regarding the impact of the $P_{\max}$, it is essential to analyze its impact on small and macro cell performances and the TPC mode by adopting OL/TPC. Furthermore, the deployment strategy for small cells in the existing macro layer needs to be considered carefully, as the performance gains offered by small cells are highly dependent on the location of small cells within a macro cell. To optimize the energy efficient resource allocation problem in a HetNet environment, a novel subchannel and power allocation algorithm was presented in Reference [14], by taking the user requirements and inter-cell interference into account. However, the system model considered the heterogenous approach by the co-existence of a software-defined visible light communication (VLC) and radio frequency (RF) small cell network rather than the conventional macro/small cell case. In Reference [15], the authors presented a novel learning-based resource allocation framework for optimizing the use of uplink–downlink decoupling, by operating the small cell in a HetNet network in the LTE unsilenced band (LTE-U). By adopting such an approach, a significant gain in terms of data rates and load balancing can be achieved, However, most existing network deployments consider simple association mechanisms where the small cells operate in the licensed band. Therefore, in this work, we considered the conventional association mechanism for the small cells in the licensed band. Performance analysis of various small cell deployment scenarios was presented in Reference [16,17]; however, the analysis was limited to downlink received SIR and system capacity only, and no details regarding the uplink side were presented. An LTE UE power consumption model was presented in Reference [18], suggesting that the uplink transmit power and the downlink data rate greatly affect the UE power consumption. Therefore, to improve battery life of UE, it is critical to study the impact of $P_{\max}$ limit for the LTE uplink. A study carried out in Reference [19] concerning the radio frequency electromagnetic field (RF-EMF) exposure from UEs in an LTE network indicated that knowledge on the realistic power levels of UE is important for accurate assessment of RF-EMF exposure, and UE transmit power should be kept lower in order to reduce the emissions. Keeping in view the abovementioned issues, it is vital to investigate the $P_{\max}$ limit for the LTE uplink specially in the case of small cell networks, as future networks are heterogeneous in nature.

As discussed earlier, small cell deployment in an existing macro layer is a promising solution to meet the capacity demands for future cellular networks, and the energy efficiency for such an ultra-dense small cell network is a challenging issue. In particular, the maximum transmit power limit $P_{\max}$ for small cell UE needs to be optimized to minimize the battery power consumption and to enable the efficient use of the available power while limiting the interference to neighboring small cells. The main motivation for the work presented in this manuscript was to specify the upper bound for the transmit power of the UE $P_{\max}$, not the actual transmit power, as the 3GPP specifications only define $P_{\max} = 23$ dBm for the macro cell case and do not explicitly define the $P_{\max}$ limit in the case of small cells.

To the best of our knowledge, this consideration has not been taken into account until now, but it proves to be vital while shifting toward small cells from the current macro cell deployment. Therefore, in this paper, the network performance was evaluated in order to show the need for a different P_{max} limit for a small cell uplink compared to a macro cell uplink. The distinctive contributions and innovations of this paper are summarized as follow:

1. Outlining four small cell deployment scenarios within the existing macro layer which were:

    ○ Homogenous small cell deployment;
    ○ HetNet deployment with uniformly distributed small cells;
    ○ HetNet deployment with small cells near macro cell eNodeB;
    ○ HetNet deployment with small cells near macro cell edge.

2. Derivation for interference, received SIR and spectral efficiency expressions for homogenous and HetNet deployment scenarios for small cells.
3. Numerical computation of reduced P_{max} limit for each scenario for the small cell uplink.
4. Ensuring similar interference levels as the macro cell deployment at $P_{max} = 23$ dBm, with reduced P_{max} limit in each small cell deployment scenario, while achieving similar received SIR performance as in the macro cell deployment.

The rest of the paper is organized in the following way: Section 2 presents the overview of the small cell base station deployment topology and related work. Section 3 presents the network model under consideration. Performance analysis for the proposed model is depicted in Section 4. Finally, the conclusion of this work is presented.

2. Cell Topology and Related Work

Recently, energy-efficient communication in cellular networks has earned tremendous attention [20]. The goals of energy-efficient communication can be categorized to maximize the energy efficiency which is defined as the amount of delivered data per energy consumption and minimizing the total network energy consumption while satisfying the required data rate and other quality of service (QoS) requirements of the users [21]. While deploying small cells in the existing macro layer, one of the major issues under consideration for ultra-dense networks is to minimize the energy consumption of the network. The cell topology adopted under deployment of small cell plays a key role in this respect. Concerning the small cell deployment strategy, there are two approaches to deploying small cell base stations within the macro layer to offload the macro traffic:

- A static small-base station deployment, where once deployed, the small cell base stations cannot be moved to new locations.
- A dynamic small cell base station deployment strategy, where the mobility aided small cell base stations (aerial drones or vehicular mounted) can be moved to new locations depending on various criteria such as high traffic load, disaster management or some mega event, to improve the performance of the network.

Several recent works adopted the static small cell base station deployment strategy due to the fact of its simple implementation as the base stations are fixed and the system modeling can be done by considering the link–budget analysis at the time of deployment by network operators. Most of the downlink parameters and inter-base station communication (termed as backhaul link) are fixed. The performance optimization is mainly considered by optimizing the network parameters based on the uplink measurements from the UE. However, with such static deployment, an important problem that arises in energy-efficient cellular networks is that of meeting a set of specified SIR targets using the minimum possible network power [22,23]. A practical policy introduced by the 3GPP specifications for achieving energy efficiency in such networks is to appropriately turn the small cells on/off. In the literature, several works address the small cell on/off approach to improve energy-efficient communications in dense small cell networks. An interference-aware small cell on/off mechanism was presented in Reference [24], where the small cells can be turned on based on the generated interference in the network for the downlink case. By considering the dormant state of small cells, that is, when no UE is associated to the small cell, a small cell on/off scheme was outlined in Reference [25], where the small cell base stations sense and store the reference signal received power (RSRP) of the neighboring base stations and UEs to determine their activation and deactivation. To reduce the signaling involved in the on/off mechanism of small cells, the authors in Reference [26] presented an interference contribution rate-based algorithm, resulting in a lower inter-cell interference in the downlink. In Reference [27], a dynamic small cell on/off mechanism along with dual connectivity was elaborated, which takes the load balancing and downlink outage probability of the users into account while turning the small cell on or off. Additionally, the dual-connectivity-based seamless handover procedure to transfer the UE from small cell to macro cell was discussed. Although the abovementioned schemes can efficiently mitigate the inter-cell interference and reduce the network energy consumption, the focus was to improve the energy efficiency in the downlink and the energy efficiency improvement for the UE in uplink direction was not considered. In summary, a static base station deployment proves to be simple in terms of implementation, and thus, preferred by the network operators while initializing the network deployment. In terms of uplink power control for the UE, the optimization of the network performance is fundamentally dependent on the uplink measurements. Therefore, from the user's perspective, energy consumption at the UE needs to be addressed for static small cell deployment. In particular, the maximum transmit power limit P_{max} needs to be revisited, as it is the major contributor towards interference.

Similarly, by focusing on reduction of the downlink energy consumption, a dynamic small cell deployment strategy can be considered, where the base station can be deployed in areas with higher UE density to improve the service quality in the case of mega events or disaster management—that is, in the case of network failure. By considering approaches such as stochastic geometry, several recent works presented dynamic base station deployment schemes by considering mobile base stations (aerial drone or vehicular mounted), which can roam in the network coverage area to address capacity issues. In Reference [28], a game theory-based mobility control scheme for flying drone base stations was presented, which can guide the movement of the drones in macro cell hotspots. The results designate that the proposed scheme can efficiently optimize the spectral efficiency by minimizing the number of drones in the network, while maintaining the safety perspectives to avoid the risk of the collision of drones and maintain an optimal altitude. However, the focus was on downlink performance enhancement and the issue of frequent handoffs on the UE's end was not considered while discussing the user association. Similarly, by considering the recovery of the network infrastructure to sustain user communication for disaster management, the authors in Reference [29] studied the efficient utilization of autonomous mobile base stations (aerial or vehicular) to improve network reachability in emergency situations. A mobile base station fit mechanism was proposed which employs mobile base stations to automatically construct new routes and assess and adaptively respond to the fitness of pause positions, such that mobile base stations can provide links to recover the disconnected network infrastructure based on dynamic and local interaction with the stationary macro and small cells. In Reference [30], an extensive analysis of cell coverage, capacity, and inter-cell interference for an aerial base station was presented. From the coverage results, the optimal altitude for an aerial base station was reported and the coverage dependency for power transmitted by a base station was shown. However, the focus was only on the downlink and the impact on UE association was not considered. Due to the free movement of mobile base stations over the entire service area, UE may frequently find different base stations available for communication, resulting in increased power consumption due to the frequent handoffs. More importantly, the negative impact of base station height relative to the UE discussed in Reference [31] was neglected, which presented a new and significant theoretical discovery, i.e., the serious problem of area spectral efficiency crash. It showed that if the absolute height difference between base station antenna and UE antenna is larger than zero, then the area spectral efficiency performance will continuously decrease with network densification for an ultra-dense small cell network. Thus, the network capacity suffers from severer degradation compared to fixed-cell topology as the drone base station flies higher. Additionally, during the travel period, a mobile small cell is unable to provide service due to the serious channel fading, and thus, a loss in the service time occurs, as discussed in Reference [32]. Due to the mobility of the base station, the availability of reliable wireless backhaul links and the related resource allocation are principal issues that should be considered while adopting the dynamic small cell base station deployment strategy [33]. In summary, dynamic small cell base station deployment can be beneficial to reduce the downlink energy consumption. However, with mobile base stations, the issues of base station location, height, and mobility management will be more complex and will have a great impact on the quality of service in future cellular networks, as these networks will have to handle a large number of UE and their frequent handoffs due to the very dense short-footage small cells.

Conventionally, a static small cell base station deployment strategy is considered while optimizing the UE energy consumption in the uplink direction. In terms of downlink energy efficiency, a small cell on/off scheme can be adopted along with the static small cell base station deployment. Also, the abovementioned downlink-centric design considerations of dynamic small cell base station deployment can be adopted. As the focus of this work was to minimize the UE power consumption by finding the upper bound for the transmit power of UE in the uplink direction only, we adopted the conventional static small cell deployment strategy to model the network.

3. Network and Propagation Model

This paper considered four network configurations to analyze the P_{max} in an LTE uplink. As a reference case, the homogenous macro cell deployment was considered, assuming $P_{max} = 23$ dBm in the uplink direction. For the small cell uplink, a homogenous network deployment scenario with uniformly distributed small cells was considered at first and analyzed for different P_{max} limits, starting with the current standard limit of 23 dBm. Secondly, a HetNet deployment scenario was considered with uniformly distributed small cells within the homogenous macro cell environment mentioned above. Thirdly, the HetNet deployment scenario with small cells deployed near the macro cell eNodeB was considered, that is, the central area of the macro cell. Finally, the deployment scenario with small cell deployed away from the macro cell eNodeB was considered, that is, on the macro cell edge region.

The uniform macro cell deployment and first scenario for small cell deployment are described in the following subsection together as the only difference among them was the cell size. This is followed by the subsections describing the three HetNet scenarios for small cell deployment separately. A fluid model in Reference [34] was employed to conduct the analysis, and was further extended for each network deployment, as the model was only available for a macro cell deployment scenario. The interference offered to the central small cell was evaluated in each scenario and the power control algorithm was optimized by selecting the optimal P_{max} limit. It was obvious that the transmit power for small cell UE must be reduced, since the cell size significantly decreased in the cases with small cells relative to the macro cell deployment. However, the P_{max} limit varied from one deployment scenario to the other, as the location of small cells within the macro cell plays a key role in the configuration of the network's parameters. This is described in detail in the following subsection.

3.1. Homogenous Macro/Small Cell Deployment

This section describes the network model considered for the homogenous deployment of macro and small cells. To determine the P_{max} in small cells, which ensures the same performance when compared to the macro cell deployment, both network deployments were analyzed with similar conditions. In Figure 1, a homogenous distribution of both the macro and small cells is depicted. It is important to mention here that both the macro and small cells were considered independently, that is, the macro cells were considered first, followed by the uniformly distributed small cells. Two rings of neighboring cells were considered around the central cell. At eNodeB, omni-directional antennas were modeled to provide network coverage in each cell. In Figure 1, the central cell has a radius R and is surrounded by two rings of cells termed "interfering rings", at distance $2nR$ ($n = 1, 2$). The network size is given as $R_{nt} = 5R$. In this case, $R = 200$ m and 1000 m specify the cell radius, while $R_{nt} = 1000$ m and 5000 m denote the network range for uniformly distributed macro and small cell networks, respectively.

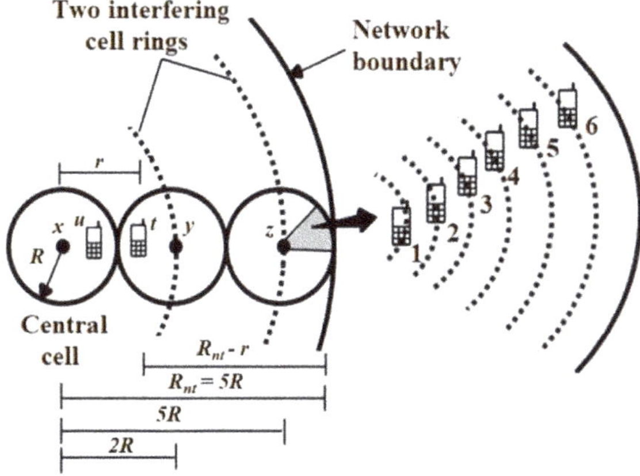

Figure 1. Fluid model for uniformly distributed cells. Zoomed view shows a uniform UE distribution in each cell.

3.1.1. Transmit and Receive Power

For the above described network model, the UE's transmit power P_{Tx} in dBm is given as:

$$P_{Tx} = P_o + \alpha.PL \tag{2}$$

and the received power P_{Rx} in dBm at the eNodeB is:

$$P_{Rx} = P_{Tx} - PL = (P_o + \alpha.PL) - PL = P_o + (\alpha - 1).PL \tag{3}$$

As the network was analyzed for FPC mode as well as the full pathloss compensation mode, the α value varied from 0.7 to 1. To model the wireless channel, a free space model was considered for the pathloss estimation depending on the distance of the UE from the serving eNodeB.

A study carried out in Reference [35] clearly showed that for OL/TPC, the signal degradation due to the distance between UE and the serving eNodeB and the interference generated due to the improper power control settings were dominant. In this work, we investigated P_{max} for the uplink using OL/TPC. Therefore, only the pathloss model was considered. The simulation parameters for each deployment scenario are summarized in Table 1. The mathematical expression for the free space model is given as:

$$PL = 32.45 + 20\log_{10}(r) + 20\log_{10}(f_{MHz}) \tag{4}$$

where r is the distance of the UE from the serving eNodeB in meters, and f_{MHz} is the operating frequency. Equation (4) can be rewritten to calculate the distance from the UE to its serving eNodeB, when the corresponding pathloss and the frequency values are known:

$$r = \log 10^{-1}(PL - 20\log_{10}(f_{MHz}) - 32.45)/20 \tag{5}$$

Table 1. System and simulation parameters for each small cell deployment scenario.

Parameters	Homogenous Macro Cells	Homogenous Small Cells	HetNet with Uniformly Distributed Small Cells	HetNet with Small Cells Deployed near Macro Cell eNodeB	HetNet with Small Cells Deployed near Macro Cell Edge
Number of interfering rings (N_i)-cell type	2-macro	2-small	2-small, 2-macro	1-small, 2-macro	1-small, 2-macro
Number of cells	19 macro cells	19 small cells	19 macro, 19 small cells	19 macro, 7 small cells	19 macro, 13 small cells
Cell radius	R_{macro} = 1000 m	R_{small} = 200 m	R_{small} = 200 m, R_{macro} = 1000 m		
UE distribution	Uniform distribution				
Pathloss model	Free space model PL = 32.45 + 20log$_{10}$(r) + 20log$_{10}$(f_{MHz})				
Antenna type at eNodeB	Omni-directional antenna				
Pathloss compensation factor "α" range	α_{macro} = 0.7, 0.8, 0.9, 1	α_{small} = 0.7, 0.8, 0.9, 1	α_{small} = 0.7, 0.8, 0.9, 1 α_{macro} = 0.7, 0.8, 0.9, 1		

3.1.2. Interference

The central cell is prone to the interference offered by the cells in the two interfering rings. In each cell, the UEs are uniformly distributed. Following the outage probability in the fluid model presented in Reference [35], 20 UEs can be scheduled by each eNodeB with a 5% outage probability. To quantify the system performance, the analysis was conducted by considering six positions of the UE within a cell. The position of the UE was such that UE with a lower index was near the eNodeB, while a large index indicated an increase in the distance between the UE and the serving eNodeB.

Let us consider UE u_i (where i = 1~6 indicates the UE index) scheduled by the central cell eNodeB x. As the frequency reuse factor of unity is assumed, the user u_i suffers external interference due only to one UE per PRB from each of the network's cells. From Equation (3), the received power at the eNodeB x from UE u_i can be written as:

$$p_{rx,u_i} = p_0 \cdot pl_{u_i,x}^{\alpha-1} \tag{6}$$

where P_0 = 10log (p_0) (P_0 in dBm and p_0 in mW) is the target received power and $pl_{u_i,x}$ is the pathloss between the UE u_i and eNodeB x. By this approach, the mathematical expression for the external interference offered by the interfering rings to the central cell eNodeB x is approximated from Reference [34] as:

$$I = \int_R^{R_{nt}} \int_0^{2\pi} \rho_{ue} p_0 pl_{u_i,x}^{\alpha} pl_{u_i,x}^{-1} r dr d\theta \tag{7}$$

where ρ_{ue} designates the uniform distribution function of the UE scheduled by eNodeB x. Employing a uniform UE distribution provides a better insight to model and analyze the network as the non-uniform distribution assumption complicates the analysis, especially when the locations of the UE and eNodeBs are dependent. The pathloss pl can be expressed as path gain pg, where $pg = 1/pl$. The assumption here is that path loss only depends on the distance between the UE and the serving eNodeB. The expression for path gain is given as:

$$pg_{u_i,x}(r) = Ar^{-\eta} \tag{8}$$

where η = 3.5 is the pathloss coefficient and A is a constant. Consider user t scheduled by eNodeB y, located on the first interfering ring of cells around the central cell. The distance range of user t from eNodeB x is $r \in (R; 2R)$. As depicted in Figure 1, the user t is at a distance of 2R-r from eNodeB y. The interference offered by the area $2\pi r dr$ around UE t is $\rho_{ue} p_0 A^{-\alpha}(2R-r)^{\alpha\eta} Ar^{-\eta} 2\pi r dr$. Similarly, if UE t is located in the distance range $r \in (2R; 3R)$ from eNodeB x, it is at a distance $r - 2R$ from eNodeB y. Hence, the interference offered now by the area $2\pi r dr$ around UE t can be estimated to be

$\rho_{ue} p_o A^{-\alpha}(-2R + r)^{\alpha\eta} A r^{-\eta} 2\pi r dr$. Therefore, the generalized expression for the interference offered by the nth interfering ring to the central cell using Equation (7) can be given as:

$$
\begin{aligned}
I_n = {} & 2\pi \int_{(2n-1)R}^{2nR} \rho_{ue} p_o A^{-\alpha}(2nR - r)^{\alpha\eta} A r^{-\eta} r dr \\
& + 2\pi \int_{2nR}^{(2n+1)R} \rho_{ue} p_o A^{-\alpha}(-2nR + r)^{\alpha\eta} A r^{-\eta} r dr
\end{aligned}
\tag{9}
$$

Hence, the total interference offered to the central cell from the interfering rings can be approximated as:

$$
I_{Ext} = \sum_{n=1}^{N_i} I_n
\tag{10}
$$

where N_i represents the number of interfering rings around the central cell. $N_i = 2$ was considered for this work.

3.1.3. Received SIR

The received SIR for user u_i can be obtained by dividing Equation (6) by Equation (10) as:

$$
SIR_{rx,u_i} = \frac{p_{rx,u_i}}{I_{Ext}}
\tag{11}
$$

The impact of the noise and shadowing were not considered in this analysis. If the impact of shadowing and noise need to be included, the variation can be expected as in Reference [36]. The subscript "small" and "macro" are used for parameters in all the above equations, using them for small and macro cells, respectively.

3.1.4. Spectral Efficiency

To see the impact of variation in the $P_{\max}$ limit on the system performance, the cell (or average) spectral efficiency $\overline{SE}$ (in bps/Hz/cell) for the proposed system can be obtained using the classical Shannon formula and Equation (11) as:

$$
\overline{SE} = \frac{2}{R^2 - r_o^2} \int_{r_o}^{R} \log_2(1 + SIR_{Rx}(r)) r dr
\tag{12}
$$

where, $r_o \neq 0$ represents the minimum distance between a UE and its serving eNodeB. Equation (12) can be employed for both a small and macro cell network, where $r_o = 5$ m and 50 m are employed for small and macro cell cases, respectively. To evaluate the spectral efficiency for a UE at a certain distance from the serving eNodeB, Equation (12) can be rewritten as:

$$
SE(r) = \log_2(1 + SIR_{Rx}(r))
\tag{13}
$$

3.2. HetNet Deployment with Uniformly Distributed Small Cells

In this section, the HetNet deployment with uniformly distributed small cells within a macro cell was considered. The deployment was considered to be a HetNet environment as the macro and small cells co-exist. The small cells were deployed in the central macro cell, mentioned as the reference case in Section 3.1. The central small cell was placed in the center of the macro cell and was surrounded by a two interfering rings of small cells. Figure 2 depicts the deployment scenario. The focus was on the small cell uplink TPC parameters, in particular the $P_{\max}$ limit.

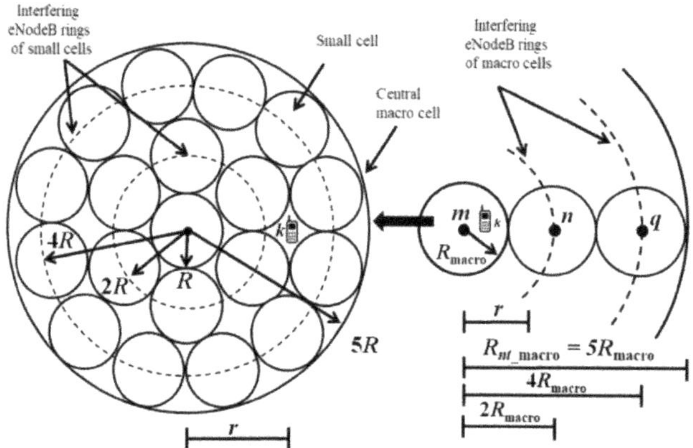

Figure 2. HetNet scenario for uniformly distributed small cells within a macro cell.

For such deployment, the central small cell is prone to interference offered by the small cells in the interfering rings around the central small cell as well as the macro cell interference. It is important to mention here that the impact of interference from the macro cell was considered for the macro cell containing the small cells, as well as the neighboring macro cells. The interference from the neighboring macro cells was expected to be lower, as the distance between the central small cell eNodeB and neighboring macro cell boundary was 5R.

3.2.1. Transmit and Receive Power

The expressions for the transmit and receive power of the macro and macro cells were taken from Equations (2) and (3).

3.2.2. Interference

At first, let us consider the small cell interference offered to the central small cell. The interference I_n offered by one interfering ring to the central small cell eNodeB x can be given using Equation (9) as:

$$
\begin{aligned}
I_n = {} & 2\pi \int_{(2n-1)R}^{2nR} p_{ue} p_{o\_small} A^{-\alpha_{small}} (2nR - r)^{\alpha_{small}\eta} Ar^{-\eta} r\, dr \\
& + 2\pi \int_{2nR}^{(2n+1)R} p_{ue} p_{o\_small} A^{-\alpha_{small}} (-2nR + r)^{\alpha_{small}\eta} Ar^{-\eta} r\, dr
\end{aligned}
\tag{14}
$$

As there are two interfering rings of small cells at a distance of 2R and 4R, therefore the total interference offered by the interfering rings of small cells to the central small cell eNodeB x can be given using Equations (10) and (14) as:

$$
I_{small} = \sum_{n=1}^{2} I_n
\tag{15}
$$

To compute the macro cell interference, the macro cell containing small cells as well as the neighboring macro cells were also considered. At first, the interference from the macro cell containing the small cells was computed, followed by the interference from the two rings of macro cells around the central macro cell. From Figure 2, it is clear that the central small cell was in the middle of the macro cell. Consider user k scheduled by macro cell eNodeB m, co-located with the central small cell eNodeB x. The distance range of user k from eNodeB x is $r \in (0; R_{macro})$ as depicted in Figure 2. The interference offered by the area $2\pi r dr$ around UE k is $p_{ue\_macro} p_{o\_macro} A^{-\alpha_{macro}} (R_{macro} - r)^{\alpha_{macro}\eta} Ar^{-\eta} 2\pi r dr$. Therefore, the expression for the interference offered by the central macro cell to the central small cell eNodeB x can be given using Equation (7) as:

$$I_{c\_macro} = 2\pi \int_0^{R_{macro}} p_{ue\_macro} p_{o\_macro} A^{-\alpha_{macro}} r^{\alpha_{macro}\eta} A r^{-\eta} r \, dr \tag{16}$$

To calculate the interference offered by the interfering ring of macro cells, consider user k scheduled by eNodeB n, located on the first interfering ring of cells around the central cell. The distance range of user k from eNodeB x is $r \in (R_{macro}; 2 R_{macro})$ as depicted in Figure 2. The user k is at a distance of $2R_{macro}$-r from eNodeB n. The interference offered by the area $2\pi r dr$ around UE k is $p_{ue\_macro} p_{o\_macro} A^{-\alpha_{macro}} (2R_{macro} - r)^{\alpha_{macro}\eta} A r^{-\eta} 2\pi r dr$. Similarly, if UE k is located in the distance range $r \in (2R_{macro}; 3R_{macro})$ from eNodeB x, it is at a distance $r - 2R_{macro}$ from eNodeB n. Hence, the interference offered now by the area $2\pi r dr$ around UE k can be estimated to be $p_{ue\_macro} p_{o\_macro} A^{-\alpha_{macro}} (-2R_{macro} + r)^{\alpha_{macro}\eta} A r^{-\eta} 2\pi r dr$. The interference offered by the nth interfering ring of macro cells around the central small cell can be formulated from Equation (9) as:

$$\begin{aligned}
I_n = {} & 2\pi \int_{(2n-1)R_{macro}}^{2nR_{macro}} p_{ue_macro} p_{o_macro} A^{-\alpha_{macro}} (2nR_{macro} - r)^{\alpha_{macro}\eta} A r^{-\eta} r \, dr \\
& + 2\pi \int_{2nR_{macro}}^{(2n+1)R_{macro}} p_{ue_macro} p_{o_macro} A^{-\alpha_{macro}} (-2nR_{macro} + r)^{\alpha_{macro}\eta} A r^{-\eta} r \, dr
\end{aligned} \tag{17}$$

Hence, the total macro cell interference offered to the central small cell can be approximated as:

$$I_{macro} = I_{c\_macro} + \sum_{n=1}^{N_i} I_n \tag{18}$$

where N_i represents the number of macro cell interfering rings around the central small cell. $N_i = 2$ was considered for this section as well. The total interferences in the HetNet environment can then be given by using Equations (15) and (18) as:

$$I_{HetNet} = I_{macro} + I_{small} \tag{19}$$

3.2.3. Received SIR

The received SIR for user u_i can be obtained by dividing Equation (6) for small cells by Equation (19) as:

$$SIR_{rx,u_i} = \frac{p_{rx\_small,u_i}}{I_{HetNet}} \tag{20}$$

3.2.4. Spectral Efficiency

To see the impact of the variation in the P_{max} limit on the system performance, the spectral efficiency for the cell average and at the specific position of the UE in the cell was computed from Equations (12) and (13) by using the received SIR given in Equation (20).

3.3. HetNet Deployment with Small Cells near Macro Cell eNodeB

In this section, the HetNet deployment scenario with small cell deployed near macro cell eNodeB inside a macro cell is considered. One interfering ring of small cells is considered around the central small cell. Figure 3 depicts the deployment scenario. The central small cell is prone to interference offered by the small cells in the interfering ring as well as the macro cell in which the small cells are deployed. The macro cell interference is considered from the macro cell containing the small cells as well as the neighboring macro cells.

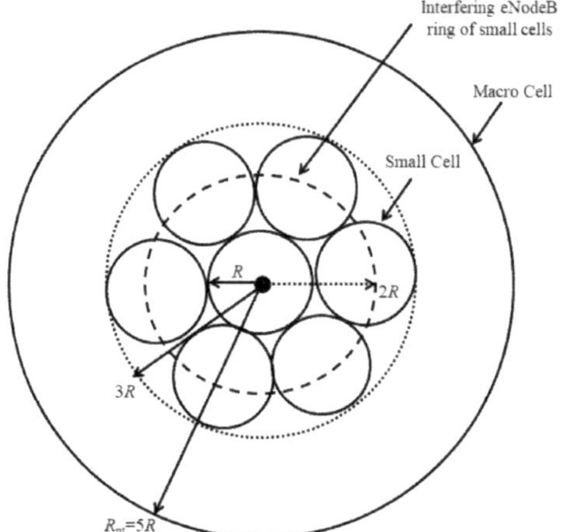

Figure 3. HetNet scenario for deployment of small cells near the macro cell eNodeB.

3.3.1. Transmit and Receive Power

The expressions for transmit and receive power of macro and small cells are taken from Equations (2) and (3).

3.3.2. Interference

The interference I_{small} offered by one interfering ring at a distance of $2R$ from the central small cell eNodeB x can be given using Equation (9) as:

$$I_{\text{small}} = 2\pi \int_R^{2R} P_{ue} P_{\text{o\_small}} A^{-\alpha_{\text{small}}} (2R - r)^{\alpha_{\text{small}}\eta} A r^{-\eta} r \, dr$$
$$+ 2\pi \int_{2R}^{3R} P_{ue} P_{\text{o\_small}} A^{-\alpha_{\text{small}}} (-2R + r)^{\alpha_{\text{small}}\eta} A r^{-\eta} r \, dr \tag{21}$$

The macro cell interference is similar to that considered in Section 3.2. The total HetNet interference using Equations (18) and (21) is given as:

$$I_{\text{HetNet}} = I_{\text{macro}} + I_{\text{small}} \tag{22}$$

Compared to homogenous small cell deployment scenario, the interference is expected to be lower as there is only one interfering ring of small cells in this case.

3.3.3. Received SIR

The received SIR for user u_i can be obtained by dividing Equation (6) for small cells by Equation (22) as:

$$SIR_{\text{rx},u_i} = \frac{P_{\text{rx\_small},u_i}}{I_{\text{HetNet}}} \tag{23}$$

3.3.4. Spectral Efficiency

To see the impact of the variation in the $P_{\max}$ limit on the system performance, the spectral efficiency for the cell average and at the specific position of the UE in the cell was computed from Equations (12) and (13) using the received SIR given in Equation (23).

3.4. HetNet Deployment with Small Cells near Macro Cell Edge

In this section, the HetNet environment with the small cells deployed at the edge of the macro cell was considered, as shown in Figure 4. In practice, this kind of deployment is most effective at decreasing the interference among macro cells, as the UE in the macro cell edge area associated to the macro cell can be scheduled by the small cells. This traffic offloading from the macro layer will result in reduced distance between UE and the serving eNodeB, resulting in a decrease of the transmit power and interference of the UE, hence improving the received SIR. Furthermore, the interfering ring is separated from the central small cell eNodeB x by a distance of $4R$, which results in a decreased interference level offered to the central small cell.

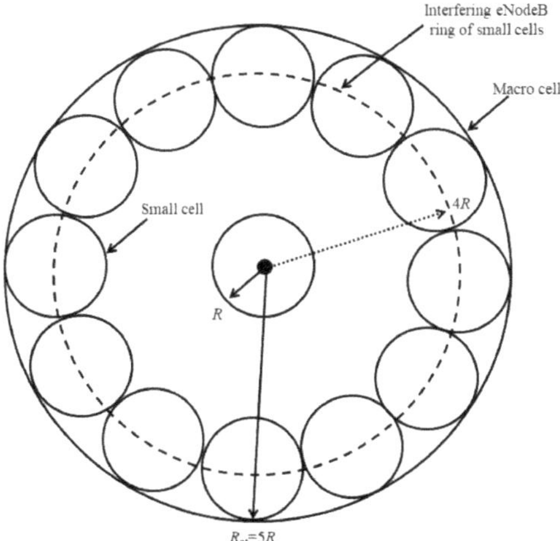

Figure 4. HetNet scenario for deployment of macro cells near the edge of the macro cell.

3.4.1. Transmit and Receive Power

The expressions for the transmit and receive power of the macro and small cells were taken from Equations (2) and (3).

3.4.2. Interference

The interference I_{small} offered by one interfering ring at a distance of $4R$ from the central small cell eNodeB x can be given using Equation (9) as:

$$I_{small} = 2\pi \int_{3R}^{4R} p_{ue}p_{o\_small}A^{-\alpha_{small}}\left(4R - r\right)^{\alpha_{small}\eta}Ar^{-\eta}rdr$$
$$+2\pi \int_{4R}^{5R} p_{ue}p_{o\_small}A^{-\alpha_{small}}\left(-4R + r\right)^{\alpha_{small}\eta}Ar^{-\eta}rdr \tag{24}$$

Using Equations (18) and (24), the total interference in the HetNet environment can then be given as:

$$I_{HetNet} = I_{macro} + I_{small} \tag{25}$$

3.4.3. Received SIR

The received SIR for user u_i can be obtained by dividing Equation (6) for small cells by Equation (25) as:

$$SIR_{rx,u_i} = \frac{p_{rx\_small,u_i}}{I_{HetNet}} \tag{26}$$

3.4.4. Spectral Efficiency

To see the impact of the variation in the P_{max} limit on the system performance, the spectral efficiency for the cell average and at the specific position of the UE in the cell was computed from Equations (12) and (13) using the received SIR given in Equation (26).

4. Performance Evaluation

This section presents the analysis of P_{max} for both macro and small cell deployments using OL/TPC. The performance of the system was quantified using the interference generated in the network and the received SIR level at the serving eNodeB. The analysis started with the uniform distribution of macro/small cells followed by the HetNet deployment scenarios which were uniformly distributed small cells within a macro cell, small cells near the macro cell eNodeB, and small cells on the cell edge within a macro cell. Let us consider each deployment scenario one by one.

4.1. Homogenous Macro/Small Cell Deployment

Firstly, the uniform macro cell deployment was considered as the reference case. The P_{max} limit was set to be 23 dBm. The P_0 value for both macro and small cells was computed using Equation (2). The pathloss compensation factor α_{macro} for the macro cell varied from 0.7 to 1, and the received SIR was obtained using Equation (11), as presented in Figure 5. The trend with an increase in α_{macro} can be observed clearly for the UE near the eNodeB (i.e., UE with the index 1), where the received SIR decreased from 10.7 dB to 4.2 dB. Meanwhile, for the UE near the cell edge (i.e., UE with the index 6), the SIR improved from 3.2 dB to 4.2 dB. For lower value of α_{macro}, the UE near the eNodeB had better SIR, while the cell edge UE suffered from degradation in the SIR and vice versa. This is one of the key objectives, to use the FPC to improve the average cell performance at the cost of the degradation in the cell edge UE's performance. On the other hand, the convergence of all SIR values for $\alpha_{macro} = 1$ specifies the advantages of full pathloss compensation, where all users have the same SIR regardless of their position in the cell.

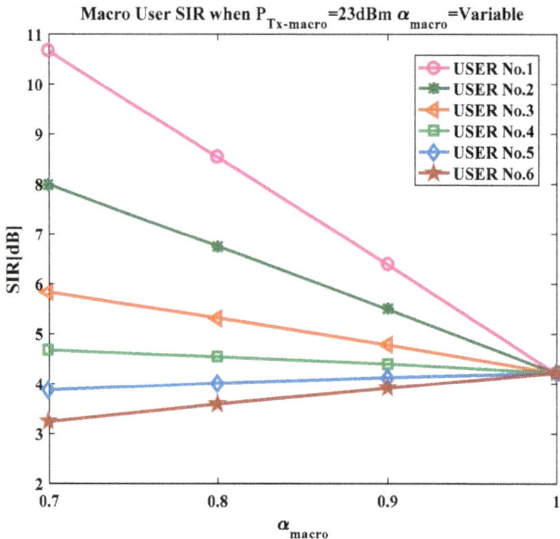

Figure 5. SIR as a function of α_{macro} in the homogenous macro cell deployment.

The coverage of a base station is defined as the area where the received SIR by the UE reaches a certain value for a given outage probability. The generic impact of shadowing and fading on outage probability for the macro cell users with a pathloss coefficient $\eta = 3.5$ was studied by following the recommendations in Reference [35], as this was not the focus of our work. To see the impact of shadowing and fading on the system performance, the outage probability for the system using a fluid model is depicted in Figure 6. The outage probability was plotted as a function of received SIR at the cell edge UE in the case of the macro cell only i-e at $r = R_{\text{macro}}$. Figure 6 shows that without shadowing and fading, the received SIR of the cell edge UE reaches -3 dB. As there was no random variation in the signal propagation, this value was deterministic, that is, the outage was 0% up to -3 dB and 100% for higher values of received SIR. In presence of shadowing ($\sigma = 3$ dB) only, the received SIR reached -9 dB with an outage probability of 10%. However, when the impact of fading was also considered along with the shadowing, the received SIR reached only -15 dB for the same outage probability. As the received SIR highly depends on the position and number of the UE in the cell and the environment, that is, the shadowing, power control parameters, cell dimensions, and the type of environment (e.g., urban or rural), such results are helpful for the network operators while admitting or rejecting new connections according to the specific environment for the new entering UE. As in this work our focus was to find the upper bound for the transmit power P_{max} for the UE rather than the actual transmit power, therefore, the impact of shadowing and fading was not taken into consideration. However, the analysis of the impact of shadowing and fading on the system performance while setting the actual uplink transmit power for the small cell uplink will be considered in detail in our future work.

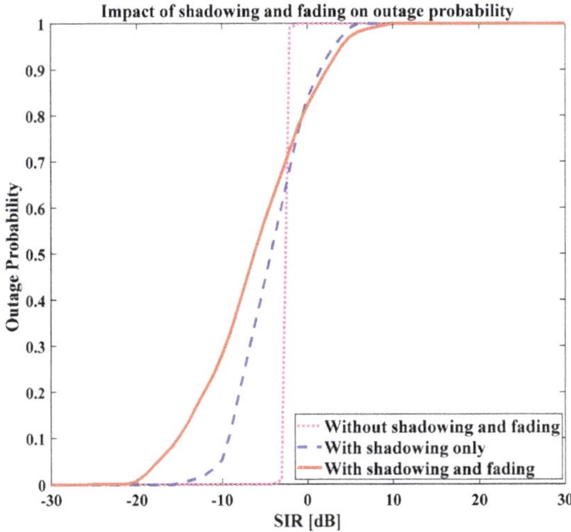

Figure 6. Outage probability as a function of SIR for the UE at the cell edge.

Secondly, a uniform small cell deployment was considered by adopting similar conditions as in the macro cell case. The cell size reduction resulted in an increase in the interference level, as the interfering UE come closer to each other. It can be seen in Figure 7 that the performance for all UE was degraded compared to Figure 5. This indicates that the network interference was more dominant. For lower value of α_{small}, the UE near the eNodeB could not achieve enough SIR gain while the cell edge UE suffered from a significant degradation in the SIR. Therefore, it was proposed not to employ the FPC, but the full pathloss compensation in the small cell environment instead [25].

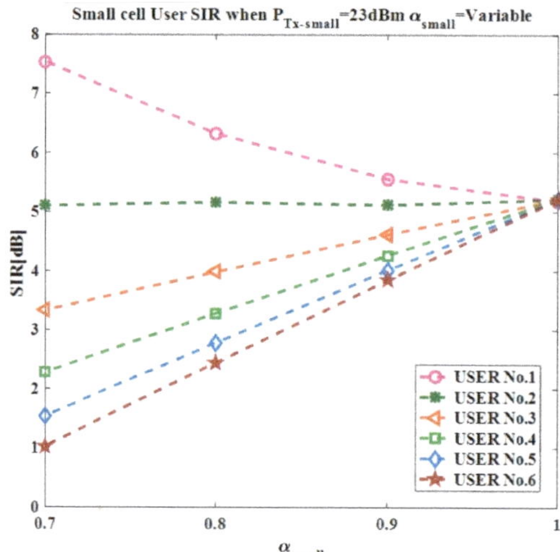

Figure 7. SIR as a function of α_{small} in the homogenous small cell deployment.

To see the impact of P_{max} on the interference generation, Equation (10) was plotted in Figure 8 for both macro and small cell cases. For the small cell case, P_{max} = 23 dBm resulted in a very significant increase in interference compared to the macro cell case. The target was to find the P_{max} limit which resulted in almost similar received SIR performance as compared to the uniform macro cell deployment with P_{max} = 23 dBm. The maximum transmit power for the small cell was varied so that the same interference level as the macro cell could be achieved. Figure 8 shows that a decrease in P_{max} resulted in a decrease in the interference. At P_{max} = 18 dBm, the received SIR for the small cell was almost similar to that achieved in the macro cell case, as depicted in Figure 9. This justifies our intention of reconsidering the P_{max} for the small cell case and not adopting the same value as that adopted in the macro cell case.

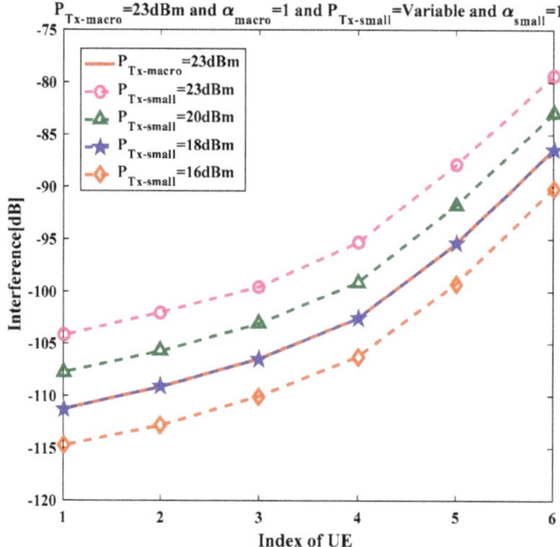

Figure 8. Interference comparison as a function of P_{max} in the uniform macro and small cell deployments.

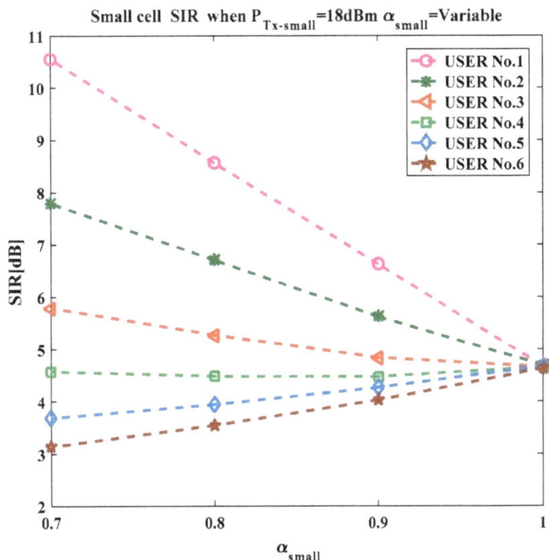

Figure 9. SIR as a function of α_{small} in the homogenous small cell deployment.

Also, it is interesting to note from comparing Figures 5 and 9 that with the FPC, the SIR performance in the case of small cells with P_{max} = 18 dBm showed a similar pattern with that in the macro cell case in Figure 5, which suggests that the FPC can be adopted in small cells as well. A comparison between Figures 7 and 9 clearly supports our argument that the FPC mode with P_{max} = 18 dBm showed much better performance than the case in which P_{max} = 23 dBm, where the cell edge as well as the cell average received SIR showed a clear difference in performance.

In Figure 10, the spectral efficiency for the cell average as well as for the user near the eNodeB and cell edge computed using Equations (12) and (13) is presented as a function of the pathloss compensation factor α_{small}. It can be seen in Figure 10, that for the users near the serving eNodeB, the spectral efficiency dropped as the pathloss compensation factor increased and for cell edge users there was an increase in the spectral efficiency. This trend was attributed to the FPC, where the cell edge user performance can be improved at the cost of a decrease in performance for users near the cell center. The average spectral efficiency showed a slight decrease as the α_{small} value approached from 0 to 1. From this trend, we can infer that the FPC mode can be adopted in small cells with a slight trade-off between the cell-average and cell-edge UE performance. The above analysis suggests that for the case with the small cells, the P_{max} level must be kept to 18 dBm instead of 23 dBm, and the FPC mode can be employed instead of the full compensation mode since it shows a significant improvement in the network performance.

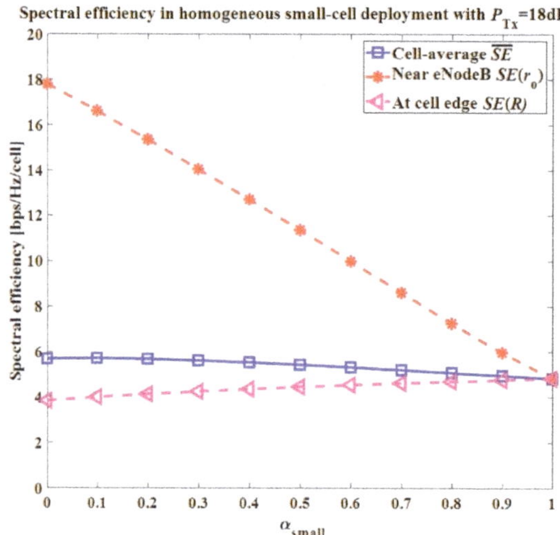

Figure 10. Spectral efficiency for small cells as a function of distance and α_{small} in homogenous small cell deployment.

4.2. HetNet Deployment with Uniformly Distributed Small Cells

In this section, the performance analysis for the first HetNet deployment scenario is presented. Uniform macro/small cell deployments in above section were considered as reference cases. From Figure 2, it is clear that there were two interfering rings of small cells around the central small cell. The P_{max} limit was varied for the small cell uplink from 16 dBm to 23 dBm. The target was to find the P_{max} value which resulted in the same SIR value as compared to the uniform macro cell deployment with P_{max} = 23 dBm and uniform small cell deployment with P_{max} = 18 dBm.

To see the result of interference offered to the central small cell from the interfering rings as well as the macro cells, Equation (19) was plotted in Figure 11. It can be observed that the interference was slightly increased as compared to Figure 8 due to the addition of macro cell interference. However, the interference from the small cells was dominant as compared to the macro cell interference. The macro cell interference was lower, as the major interference was only from the macro cell containing the small cells. The neighboring macro cells were at $5R$ from the central small cell eNodeB; therefore, their interference had a minor impact on the central small cell. In Figure 11, it can be clearly seen that for small cells, P_{max} = 17.48 dBm corresponded to the same interference level as compared to the reference cases with P_{max} = 23 dBm and P_{max} = 18 dBm in uniformly distributed macro and homogenous small cell deployments, respectively.

To see the impact of increased interference and decreased P_{max} value, received SIR was obtained in Figure 12 using Equation (20). The trend with an increase in α_{small} can be observed clearly for the UE near eNodeB (i.e., UE with the index 1); the received SIR decreased from 10.53 dB to 4.5 dB. Compared to Figure 5, the received SIR showed a loss of 0.17 dB and a gain of 0.3 dB for α_{small} = 0.7 and α_{small} = 1, respectively.

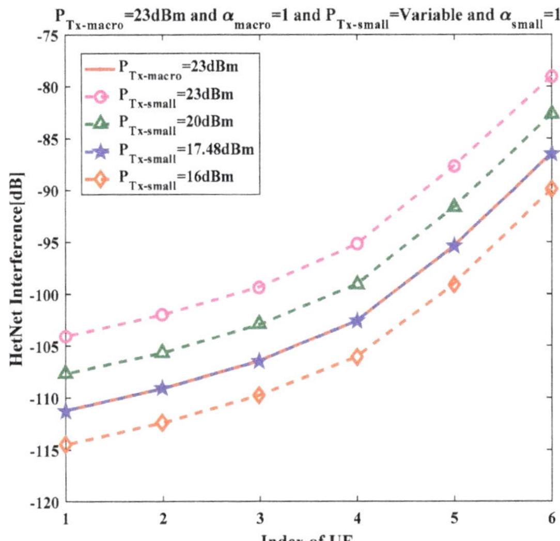

Figure 11. Interference comparison as a function of P_{max} in the uniform macro cell and HetNet deployments.

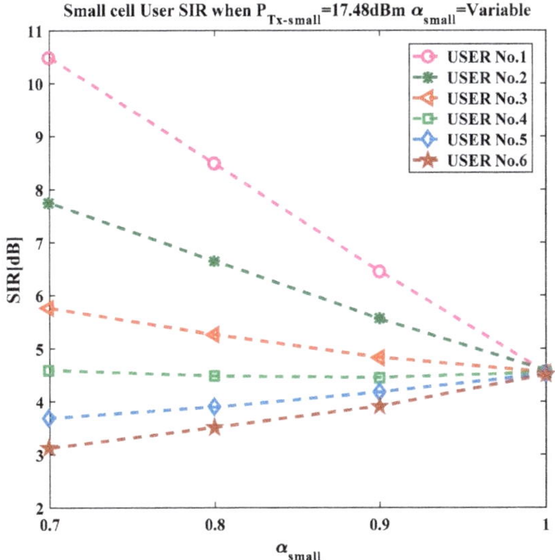

Figure 12. SIR as a function of α_{small} in the HetNet deployment.

Similarly, compared to Figure 9, the received SIR loss was 0.03 dB and 0.15 dB for α_{small} = 0.7 and α_{small} = 1, respectively. Meanwhile, for the UE near the cell edge (i.e., UE with the index 6), the received SIR improved from 3.15 dB to 4.5 dB. The received SIR compared to Figure 5 showed a loss of 0.05 dB and a gain of 0.3 dB, for α_{small} = 0.7 and α_{small} = 1, respectively. While the loss compared to Figure 9 was 0.05 dB and 0.15 dB. Like the previous section, the lower value of α_{small} resulted in a better SIR for the UE near the eNodeB, while the UE near the cell edge suffered from degradation in received SIR and vice versa. Also, the results support our argument of adopting FPC instead of full compensation in the small cell uplink in a HetNet environment with a reduced P_{max} value, as the comparison of Figure 7 with Figure 12 showed a clear gain in received SIR by adopting FPC. This trend can be validated further by comparing the spectral efficiency in Figure 10 with Figure 13, where the reduced P_{max} value in small cells had a negligible impact on the small cell UE's performance.

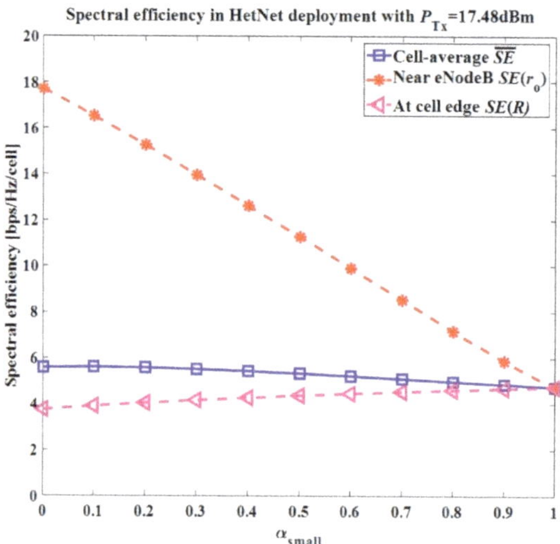

Figure 13. Spectral efficiency for the small cell as a function of distance and α_{small} in the HetNet deployment with uniformly distributed small cells.

4.3. HetNet Deployment with Small Cells near the Macro Cell eNodeB

In this section, the performance analysis for the second HetNet deployment scenario is presented. The uniform macro/small cell deployment described in Section 4.1 was considered as the reference case. Figure 3 shows that there was only one interfering ring of small cells around the central small cell. The P_{max} value varied for the small cell uplink from 16 dBm to 23 dBm.

To see the results of the interference offered to the central small cell from the interfering ring as well as the macro cell, Equation (22) was plotted in Figure 14. It is observed that due to the absence of one interfering ring as compared to the previous scenario, the interference offered to the central small cell decreased. This was because the second interfering ring, which was at a distance of $4R$ from the central cell eNodeB, was not present. The macro cell interference was present, but the dominant interference was the one offered by the first interfering ring of small cells. By comparing Figures 8 and 14, it can be observed that the interference level decreased by 0.9 dB on average. Due to the decrease in interference, the transmit power in the small cells could be increased. In Figure 14, it can clearly be seen that for the small cells, P_{max} = 18.03 dBm corresponded to the same interference level as compared to the reference cases with P_{max} = 23 dBm and P_{max} = 18 dBm in uniformly distributed macro and homogenous small cell deployments, respectively.

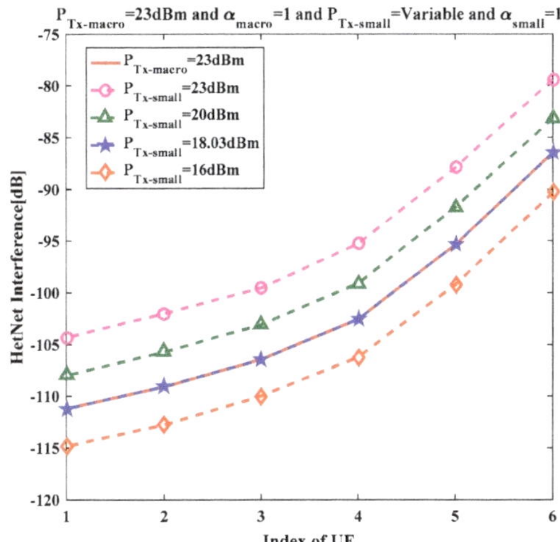

Figure 14. Interference comparison as a function of $P_{\max}$ in the uniform macro cell and HetNet deployments.

To see the results of the decreased interference and increased $P_{\max}$ value, received SIR was obtained in Figure 15 using Equation (23). The trend with an increase in α_{small} can be observed clearly for the UE near the eNodeB (i.e., UE with the index 1); the received SIR decreased from 10.75 dB to 4.8 dB. Compared to Figure 5, the received SIR gain was 0.05 dB and 0.6 dB for $\alpha_{small} = 0.7$ and $\alpha_{small} = 1$, respectively. Similarly, compared to Figure 9, the received SIR gain was 0.19 dB and 0.15 dB for $\alpha_{small} = 0.7$ and $\alpha_{small} = 1$, respectively. Meanwhile, for the UE near the cell edge (i.e., UE with the index 6), the received SIR improved from 3.3 dB to 4.8 dB. The received SIR gains as compared to Figure 5 were 0.1 dB and 0.6 dB, while the gains compared to Figure 9 were 0.2 dB and 0.15 dB. A slight gain in the spectral efficiency can be observed with the increased $P_{\max}$ value for the HetNet deployment by comparing Figure 16 with Figures 10 and 13. This gain was attributed to the fact that the received SIR was improved due to the lowered interference level, as only one interfering ring of the small cells was present in this deployment scenario.

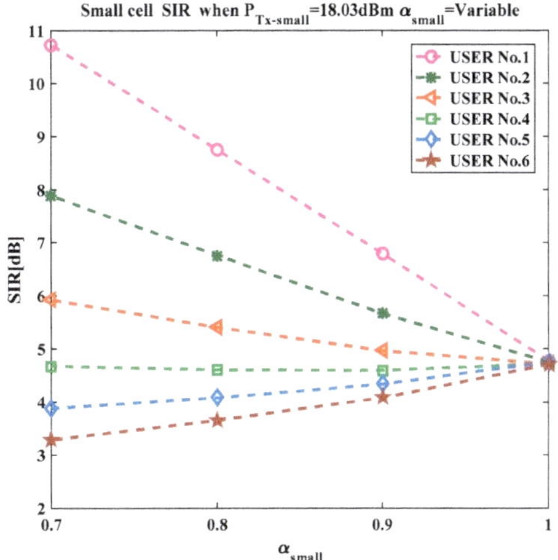

Figure 15. SIR as a function of α_{small} in the HetNet deployment.

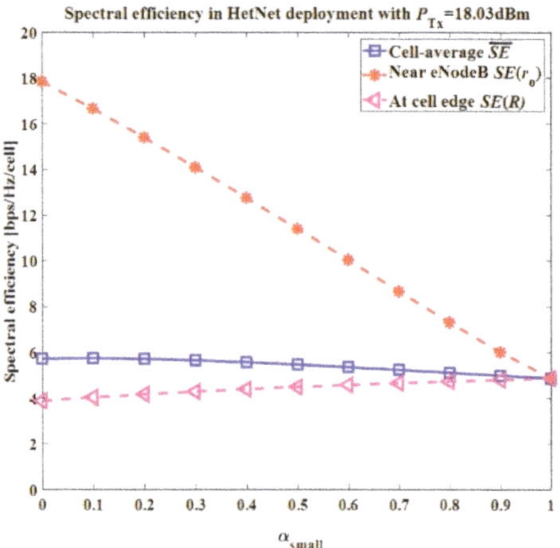

Figure 16. Spectral efficiency for the small cell as a function of distance and α_{small} in the HetNet deployment with small cells near the macro cell eNodeB.

4.4. HetNet Deployment with Small Cells near Macro Cell Edge

In this section, the performance analysis for the third HetNet deployment scenario is presented. The uniform macro/small cell deployment was considered as the reference case here as well. From Figure 4, there was only one interfering ring of small cells around the central small cell. However, the interfering ring was at the macro cell edge and at a distance $4R$ from the central small cell eNodeB. The P_{max} value varied for the small cell uplink from 16 dBm to 23 dBm. The target was to find the P_{max} value which resulted in the same SIR value as compared to the uniform macro cell deployment with $P_{\text{max}} = 23$ dBm and a uniform small cell deployment with $P_{\text{max}} = 18$ dBm.

The interference offered to the central small cell is presented in Figure 17 using Equation (25). Compared to Figure 8, Figure 11, and Figure 14, the interference level was lower by 6.5 dB, 6.6 dB, and 5.1 dB, respectively. In Figure 17, it can be seen clearly that for the small cells, $P_{\text{max}} = 21.29$ dBm corresponded to the same interference level as compared to the reference cases with $P_{\text{max}} = 23$ dBm and $P_{\text{max}} = 18$ dBm in the uniformly distributed macro and small cell deployments, respectively.

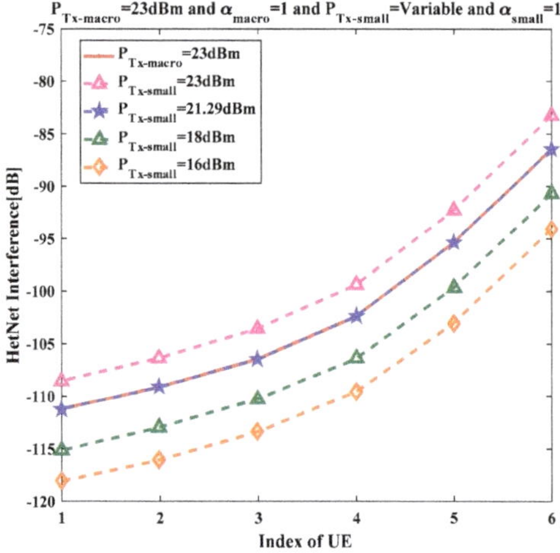

Figure 17. Interference comparison as a function of P_{max} in the uniform macro cell and HetNet deployments.

The impact of increased P_{max} and decreased interference was observed by plotting the received SIR in Figure 18 using Equation (26). The trend with an increase in α_{small} can clearly be observed for the UE near the eNodeB (i.e., UE with the index 1); the received SIR decreased from 11.71 dB to 5.4 dB. Compared to Figure 5, the received SIR gain was 1.01 dB and 1.2 dB for $\alpha_{\text{small}} = 0.7$ and $\alpha_{\text{small}} = 1$, respectively. Similarly, compared to Figure 9, the received SIR gain was 1.21 dB and 0.8 dB for $\alpha_{\text{small}} = 0.7$ and $\alpha_{\text{small}} = 1$, respectively. Meanwhile, for the UE near the cell edge (i.e., UE with the index 6), the received SIR improved from 3.7 dB to 5.4 dB. The received SIR gains as compared to Figure 5 were 0.5 dB and 1.2 dB, while the gain compared to Figure 9 were 0.6 dB and 0.8 dB. The results for the spectral efficiency presented in Figure 19 further validate our argument, where a significant gain in the system performance can be observed by revisiting the P_{max} limit. This HetNet deployment scenario shows significant gain in system performance, as the interfering ring of the small cell was far away from the central small cell, and the increased P_{max} value resulted in higher received SIR for all the UE. The P_{max} limits for the considered network deployments are summarized in Table 2.

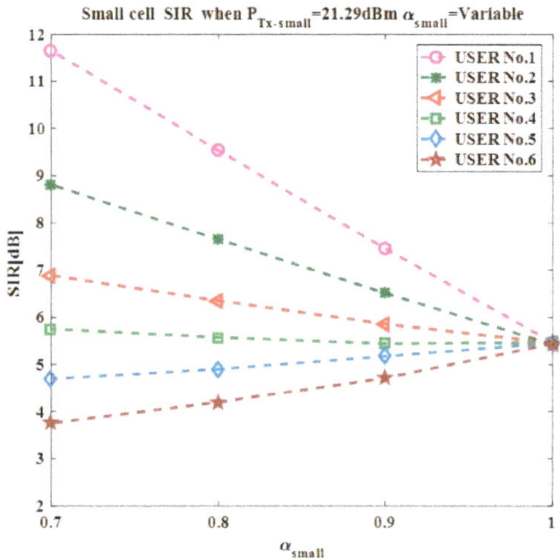

Figure 18. SIR as a function of α_{small} in the HetNet deployment.

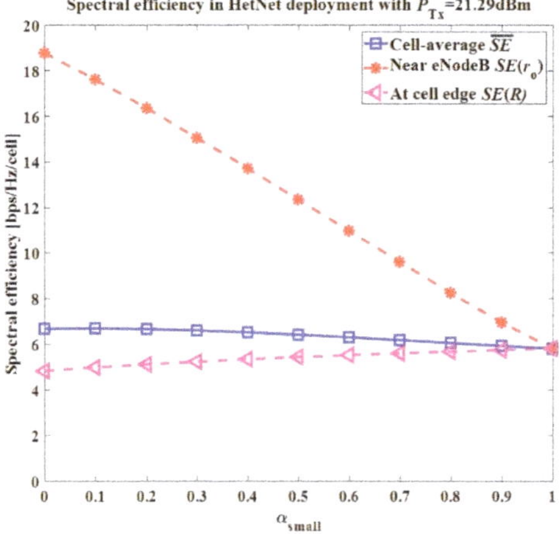

Figure 19. Spectral efficiency for the small cell as a function of distance and α_{small} in the HetNet deployment with small cells near the macro cell edge.

Table 2. $P_{\max}$ limit for various network deployments.

Deployment Scenario	$P_{\max}$
Homogenous small cell deployment	18 dBm
HetNet deployment with uniformly distributed small cells	17.48 dBm
HetNet deployment with small cells near the macro cell eNodeB	18.03 dBm
HetNet deployment with small cells near the macro cell edge	21.29 dBm

As expected earlier, such a HetNet deployment with $P_{\max}$ = 21.29 dBm offers advantages in terms of received SIR by adopting FPC instead of full compensation. The further optimization of the $P_{\max}$ value by considering various energy efficient deployment strategies such as small cell on/off, non-uniform, and mobile small cell deployment in the macro cell, and other cell-specific features will be continued in future work.

5. Conclusions

In this paper, the maximum transmit power $P_{\max}$ for a small cell uplink was investigated for OL/TPC in various deployment scenarios. We showed that, to achieve better network performance, $P_{\max}$ = 18 dBm is the appropriate choice in a uniform small cell deployment, which provides room for FPC operation to compensate for the pathloss. Furthermore, the HetNet environment was investigated with the three most efficient small cell deployments within the existing macro layer. For the HetNet environment, we observed a deviation of $P_{\max}$ from 18 dBm by −0.52 dBm in the first, +0.03 dBm in the second, and +3.29 dBm in the third HetNet deployment due to the variations in interference. However, significant SIR gain was achieved with an increase in $P_{\max}$ for the HetNet deployment as compared to the uniform small cell deployment. Use of FPC mode for the HetNet environment was also validated by our analysis. Further research will be conducted to consider the joint optimization of OL and CL/TPC schemes by employing system-level simulations, along with downlink energy efficiency enhancement schemes and event-specific HetNet deployment.

Author Contributions: A.H. and S.-H.H. contributed to the main idea of this research work. A.H. performed the simulations. A.H. and S.-H.H. contributed to the writing of this article. This research activity was planned and executed under the supervision of S.-H.H.

Funding: This work was supported by a grant from the Institute for Information and Communications Technology Promotion (IITP) of the Ministry of Science, ICT, and Future Planning (MSIP), Rep. of Korea (R0101-15-224, Development of 5G Mobile Communication Technologies for Hyper-Connected Smart Services).

Conflicts of Interest: The authors declare no conflicts of interest regarding the publication of this article.

References

1. Ge, X.; Tu, S.; Mao, G.; Wang, C.X.; Han, T. 5G Ultra-Dense Cellular Networks. *IEEE Wirel. Commun.* **2016**, *23*, 72–79. [CrossRef]
2. Ge, X.; Cheng, H.; Guizani, M.; Han, T. 5G wireless backhaul networks: Challenges and research advances. *IEEE Netw.* **2014**, *28*, 6–11. [CrossRef]
3. Coupechoux, M.; Kelif, J.M. How to set the fractional power control compensation factor in LTE? In Proceedings of the 34th IEEE Sarnoff Symposium (SARNOFF), Princeton, NJ, USA, 3–4 May 2011; pp. 1–5.
4. Simonsson, A.; Furuskar, A. Uplink power control in LTE—Overview and performance: Principles and benefits of utilizing rather than compensating for SINR variations. In Proceedings of the 68th IEEE Vehicular Technology Conference (VTC Fall), Calgary, BC, Canada, 21–24 September 2008; pp. 1–5.
5. Evolved Universal Terrestrial Radio Access (E-UTRA). *Physical Layer Procedures 'TR 36.213'*; 3rd Generation Partnership Project (3GPP): Phoenix, AZ, USA, January 2015; Available online: https://portal.3gpp.org/desktopmodules/Specifications/SpecificationDetails.aspx?specificationId=2427 (accessed on 1 May 2019).

6. Haider, A.; Seong-Hee, L.; Seung-Hoon, H.; Kim, D.I.; Jee, H.N. Uplink open loop power control for LTE HetNet. In Proceedings of the URSI Asia-Pacific Radio Science Conference (URSI AP-RASC), Seoul, Korea, 21–25 August 2016; pp. 83–85.

7. Li, J. Uplink power control for heterogeneous networks. In Proceedings of the IEEE Communications and Networking Conference (WCNC), Shanghai, China, 7–10 April 2013; pp. 773–777.

8. Cîrstea, E.; Ciochină, S. LTE Uplink Power Control and its Impact on System Performance. In Proceedings of the 3rd International Conference on Adaptive and Self-Adaptive Systems and Applications, Rome, Italy, 11–14 September 2011; pp. 57–61.

9. Koleva, P.; Poulkov, V.; Asenov, O.; Semov, P. Improved open loop power control for LTE uplink. In Proceedings of the 38th International Conference on Telecommunication and Signal Processing, Prague, Czech Republic, 9–11 July 2015; pp. 1–5.

10. Vallejo-Mora, A.B.; Toril, M.; Luna-Ramirez, S.; Mendo, A.; Pedraza, S. Congestion Relief in Subway Areas by Tuning Uplink Power Control in LTE. *IEEE Trans. Veh. Technol.* **2017**, *66*, 6489–6497. [CrossRef]

11. Lauridsen, M.; Jensen, A.R.; Mogensen, P. Reducing LTE uplink transmission energy by allocating resources. In Proceedings of the IEEE Vehicular Technology Conference (VTC Fall), San Francisco, CA, USA, 5–8 September 2011; pp. 1–5.

12. Zhang, H.; Liu, H.; Cheng, J.; Leung, V.C.M. Downlink Energy Efficiency of Power Allocation and Wireless Backhaul Bandwidth Allocation in Heterogeneous Small Cell Networks. *IEEE Trans. Commun.* **2018**, *66*, 1705–1716. [CrossRef]

13. Haider, A.; Sinha, R.S.; Hwang, S.H. Investigation of Open-Loop Transmit Power Control Parameters for Homogeneous and Heterogeneous Small-Cell Uplinks. *ETRI J.* **2018**, *40*, 51–60. [CrossRef]

14. Zhang, H.; Liu, N.; Long, K.; Cheng, J.; Leung, V.C.M.; Hanzo, L. Energy Efficient Subchannel and Power Allocation for Software-defined Heterogeneous VLC and RF Networks. *IEEE J. Sel. Areas Commun.* **2018**, *36*, 658–670. [CrossRef]

15. Chen, M.; Saad, W.; Yin, C. Echo State Networks for Self-Organizing Resource Allocation in LTE-U with Uplink-Downlink Decoupling. *IEEE Trans. Wirel. Commun.* **2017**, *16*, 3–16. [CrossRef]

16. Pak, Y.; Min, K.; Choi, S. Performance evaluation of various small-cell deployment scenarios in small-cell networks. In Proceedings of the 18th IEEE Symposium on Consumer Electronics (ISCE), Jeju Island, Korea, 22–25 June 2014; pp. 1–2.

17. Jung, Y.; Kim, H.; Lee, S.; Hong, D.; Lim, J. Deployment of small cells with biased density in heterogeneous networks. In Proceedings of the 22nd Asia-Pacific Conference on Communications (APCC), Yogyakarta, Indonesia, 25–27 August 2016; pp. 541–544.

18. Jensen, A.R.; Lauridsen, M.; Mogensen, P.; Sorensen, T.B.; Jensen, P. LTE UE power consumption model: For system level energy and performance optimization. In Proceedings of the IEEE Vehicular Technology Conference (VTC Fall), Quebec City, QC, Canada, 3–6 September 2012; pp. 1–5.

19. Joshi, P.; Colombi, D.; Thors, B.; Larsson, L.E.; Törnevik, C. Output Power Levels of 4G User Equipment and Implications on Realistic RF EMF Exposure Assessments. *IEEE Access* **2017**, *5*, 4545–4550. [CrossRef]

20. Mahdavi-Doost, H.; Prasad, N.; Rangarajan, S. Optimizing energy efficiency over energy-harvesting LTE cellular networks. In Proceedings of the 2016 IEEE International Symposium on Information Theory (ISIT), Barcelona, Spain, 10–15 July 2016.

21. Bhardwaj, A.; Agnihotri, S. Energy- and Spectral-Efficiency Trade-Off for D2D-Multicasts in Underlay Cellular Networks. *IEEE Wirel. Commun. Lett.* **2018**, *7*, 546–549. [CrossRef]

22. Madi, N.K.M.; Hanapi, Z.B.M.; Othman, M.; Subramaniam, S. Link Adaptive Power Control and Allocation for Energy-Efficient Downlink Transmissions in LTE Systems. *IEEE Access* **2018**, *6*, 18469–18483. [CrossRef]

23. Triantafyllopoulou, D.; Moessner, K. QoS and energy efficient resource allocation in downlink OFDMA systems. In Proceedings of the 2015 IEEE International Conference on Communications (ICC), London, UK, 8–12 June 2015; pp. 5967–5972.

24. Tung, L.P.; Wang, L.C.; Chen, K.S. An interference-aware small cell on/off mechanism in hyper dense small cell networks. In Proceedings of the International Conference on Computing, Networking and Communications (ICNC), Santa Clara, CA, USA, 26–29 January 2017; pp. 767–771.

25. Networks, U.S.C.; Lee, K.; Park, Y.; Heo, J.; Na, M.; Wang, H.; Hong, D. Dual Layer Small Cell On/Off Control for Ultra-Dense Small Cell Networks. In Proceedings of the 85th Vehicular Technology Conference (VTC-Spring), Sydney, NSW, Australia, 4–7 June 2017; pp. 1–5.

26. Shen, B.; Lei, Z.; Huang, X.; Chen, Q. An Interference Contribution Rate Based Small Cells On/Off Switching Algorithm for 5G Dense Heterogeneous Networks. *IEEE Access* **2018**, *6*, 29757–29769. [CrossRef]
27. Huang, X.; Tang, S.; Zheng, Q.; Zhang, D.; Chen, Q. Dynamic Femtocell gNB On/Off Strategies and Seamless Dual Connectivity in 5G Heterogeneous Cellular Networks. *IEEE Access* **2018**, *6*, 21359–21368. [CrossRef]
28. Fotouhi, A.; Ding, M.; Hassan, M. Flying Drone Base Stations for Macro Hotspots. *IEEE Access* **2018**, *6*, 19530–19539. [CrossRef]
29. Teng, R.; Li, H.B.; Miura, R. Dynamic Recovery of Wireless Multi-Hop Infrastructure with the Autonomous Mobile Base Station. *IEEE Access* **2016**, *4*, 627–638. [CrossRef]
30. Cileo, D.G.; Sharma, N.; Magarini, M. Coverage, capacity and interference analysis for an aerial base station in different environments. In Proceedings of the International Symposium on Wireless Communicasion Systems (ISWCS), Bologna, Italy, 28–31 August 2017; pp. 281–286.
31. Ding, M.; Perez, D.L. Please lower small cell antenna heights in 5G. In Proceedings of the IEEE Global Communication Conference (GLOBECOM), Washinton, DC, USA, 4–8 December 2016; pp. 1–6.
32. Chou, S.F.; Yu, Y.J.; Pang, A.C. Mobile Small Cell Deployment for Service Time Maximization over Next-Generation Cellular Networks. *IEEE Trans. Veh. Technol.* **2017**, *66*, 5398–5408. [CrossRef]
33. Kalantari, E.; Bor-Yaliniz, I.; Yongacoglu, A.; Yanikomeroglu, H. User association and bandwidth allocation for terrestrial and aerial base stations with backhaul considerations. In Proceedings of the 28th IEEE Annual International Symposium on Personal, Indoor and Mobile Radio Communications (PIMRC), Montreal, QC, Canada, 8–13 October 2017; pp. 1–6.
34. Kelif, J.M.; Coupechoux, M.; Godlewski, P. A fluid model for performance analysis in cellular networks. *EURASIP J. Wirel. Commun. Netw.* **2010**, *2010*, 435189. [CrossRef]
35. Kelif, J.-M.; Coupechoux, M. Joint Impact of Pathloss Shadowing and Fast Fading—An Outage Formula for Wireless Networks. *arXiv* **2010**, arXiv:1001.1110.
36. Kelif, J.M.; Coupechoux, M. Impact of topology and shadowing on the outage probability of cellular networks. In Proceedings of the IEEE International Conference on Communications, Dresden, Germany, 14–18 June 2009; pp. 1–6.

Article

Enabling Non-Linear Energy Harvesting in Power Domain Based Multiple Access in Relaying Networks: Outage and Ergodic Capacity Performance Analysis

Thanh-Luan Nguyen [1], Minh-Sang Van Nguyen [2], Dinh-Thuan Do[3,*], Miroslav Voznak [4]

[1] Faculty of Electrical and Electronics Engineering, Bach Khoa University, Ho Chi Minh City, Vietnam
[2] Faculty of Electronics Technology, Industrial University of Ho Chi Minh City (IUH),
Ho Chi Minh City, 700000, Vietnam
[3] Wireless Communications Research Group, Faculty of Electrical and Electronics Engineering, Ton Duc
Thang University, Ho Chi Minh City, 700000, Vietnam
[4] Faculty of Electrical Engineering and Computer Science, VSB-Technical University of Ostrava,
17. listopadu 2172/15, 708 00 Ostrava, Czech Republic
* Correspondence: dodinhthuan@tdtu.edu.vn

Received: 17 June 2019; Accepted: 19 July 2019; Published: 22 July 2019

Abstract: The Power Domain-based Multiple Access (PDMA) scheme is considered as one kind of Non-Orthogonal Multiple Access (NOMA) in green communications and can support energy-limited devices by employing wireless power transfer. Such a technique is known as a lifetime-expanding solution for operations in future access policy, especially in the deployment of power-constrained relays for a three-node dual-hop system. In particular, PDMA and energy harvesting are considered as two communication concepts, which are jointly investigated in this paper. However, the dual-hop relaying network system is a popular model assuming an ideal linear energy harvesting circuit, as in recent works, while the practical system situation motivates us to concentrate on another protocol, namely non-linear energy harvesting. As important results, a closed-form formula of outage probability and ergodic capacity is studied under a practical non-linear energy harvesting model. To explore the optimal system performance in terms of outage probability and ergodic capacity, several main parameters including the energy harvesting coefficients, position allocation of each node, power allocation factors, and transmit signal-to-noise ratio (SNR) are jointly considered. To provide insights into the performance, the approximate expressions for the ergodic capacity are given. By matching analytical and Monte Carlo simulations, the correctness of this framework can be examined. With the observation of the simulation results, the figures also show that the performance of energy harvesting-aware PDMA systems under the proposed model can satisfy the requirements in real PDMA applications.

Keywords: ergodic capacity; non-linear energy harvesting; NOMA; outage probability

---

1. Introduction

As a key subject in the Fifth-Generation (5G) technology, the data rate is enhanced to adapt to the expected 1000-fold explosive growth of data traffic by 2020, and several great challenges are addressed in the development of 5G [1]. To meet the fast increasing traffic need caused by the explosion of wireless devices and improve the spectral effectiveness of the 5G cellular network, a promising technique is proposed and called NOMA [2]. By using the power domain, NOMA serves multiple users simultaneously, and such a technique is different from Orthogonal Multiple Access (OMA), which deploys the time and frequency domains. NOMA can exhibit a balance between network throughput and user fairness because the base station appropriately distributes the transmit power to multiple users under dissimilar channel conditions. To extend the coverage area and eliminate channel impairments, several relaying system models can be combined with NOMA, in which a relay is reflected as an efficient method to further signal processing in the intermediate node. As a result, the NOMA relaying architecture is introduced to address impairments, including shadowing, path loss, and fading. In dual-hop relaying, a relay operates to assist cooperative communications in which a source communicates with a destination via such a relay [3,4]. In the scenario of downlink or uplink in an NOMA, the user pairing problem was investigated in recent works [5]. On the other hand, other metrics are considered for system performance evaluation in cooperative relaying NOMA; for example, the authors in [6] examined two users, which were paired to establish NOMA based on an access scheme by using classic Decode-and-Forward (DF) relaying with a dual-hop and three-node model. In these models, the Maximal Ratio Combining (MRC) principle is employed to further obtain the space diversity and improve the achievable rate, as presented in [6], or secure performance in the context of NOMA, as achieved in [7]. In contrast with [6] and the full-duplex transmission presented in [8], these analytical expressions can be further extended to the NOMA scheme. As an important metric, a sum-throughput maximization problem was derived and explained with the optimal transmission policy, which is deployed in backscatter-assisted energy harvesting NOMA networks [9]. In [10], stochastic geometry was employed as a tool to evaluate the outage probability of the underlying NOMA-assisted cognitive radio in the scenario of a large-scale model. The cooperative NOMA in full-duplex was considered and surveyed in other findings [11,12]. The outage performance was characterized in full-duplex mode for Device-to-Device (D2D) transmission in cooperative NOMA [11]. Considering the influence of imperfect self-interference, the authors in [12] investigated the expressions of outage probability and the achievable sum rate for full-duplex NOMA relaying wherein DF mode was implemented to serve two NOMA users.

Unlike conventional OMA strategies, the implementation of NOMA may require more energy feeding than OMA, and the spectral efficiency in energy harvesting-based NOMA can always enhance system performance compared to OMA, as the proposed system given in [13]. The explicit evaluations of feasible rate calculation in NOMA-based systems and their prospective performance gains were studied in detail in four kinds of NOMA systems, including coding-assisted NOMA, scrambling-assisted NOMA, spreading-assisted NOMA, and interleaving-assisted NOMA. In addition, these models together with their performances were further compared with OMA [14]. In other trends of research, it was proven that NOMA achieves its prominent advantage with advanced features including the Successive Interference Cancellation (SIC) technique at the receivers, the multiplexing transmitting technique, and effective network resource distribution. The authors in [15] investigated random user deployment in the NOMA scenario to evaluate the network performance. The outage balancing in the downlink NOMA system can be jointly solved in terms of decoding order and the power allocation, as in the recent work [16]. The authors in [17] studied the multiple antenna-based mmwave model to be employed in the NOMA network. Nevertheless, to reduce the joint design of power allocation and decoding order, the fixed precoding beamformer is implemented in the design of NOMA. Unlike these works, the total achievable network rate is maximal to be applied in an uplink NOMA network, and it is expressed by the combined problem of power and sub-carrier allocation [18]. Inspired by [19], which examined a cooperative NOMA system, the work in [20] suggested a better trade-off between spectral efficiency and signal reception in an innovative mixture of downlink and uplink for the NOMA network. Interestingly, visible light communication systems could be able to employ the NOMA, as the potential system presented in [21]. In addition, Multiple-Input-Multiple-Output (MIMO) employing NOMA was introduced in the scenario of multiple antennas equipped at the transmitter or/and receiver, in which all the potential degrees of freedom can be obtained to maximize the performance improvement. To minimize the total expended power or maximize the sum spectral effectiveness, beamforming design in Multiple-Input-Single-Output (MISO) with the capability of the NOMA scheme was studied in [22,23]. The throughput maximization problem and two algorithms developed were determined as a particular case of a network deploying a two-antenna user [24].

Meanwhile, the ever-growing greenhouse gas emission challenge and explosive proliferation of lower-power devices are the main motivations for developing energy-efficient techniques for future wireless communication networks. In particular, energy-efficient techniques serving power-limited devices (i.e., sensors and mobile phones) can be separated into two main categories. One of the categories focuses on the techniques that can achieve outage probability in the system model where the relay is able to harvest energy from the source [25–29], while the other architecture aims to collect more energy, where the relay can be wirelessly charged via multiple antennas or co-channel interference [30,31], and these concerns are exhibited by the Simultaneous Wireless Information and Power Transfer (SWIPT)-assisted relaying network. In [25], the authors studied energy harvesting policies including a maximum harvested energy relay selection scheme, minimum self-interference relay selection, and an optimal relay selection scheme in a relaying network using full-duplex mode. In a specific condition related to imperfect Channel State Information (CSI), the degraded performance can be observed in a wireless-powered relaying network, as the interesting results presented in [26–28] show. A two-hop relaying network where the terminals and relay are affected by co-channel interference was investigated in [30]. With the existence of co-channel interference, the expressions of the outage probability are derived by considering time switching fractions, aiming at energy harvesting protocols related to making the analysis mathematically tractable [30]. The authors in [31] studied simple calculations in the relaying network, wherein the wireless-powered relay first harvests energy from both the received signal and co-channel interference, then the relay forwards the mixed signal to the destination.

Different from the traditional linear energy harvesting model presented in [30–32], this paper is motivated by a practical non-linear energy harvesting model introduced in [33–35]. Specifically, we focus on relays that are located close to the base station in NOMA systems, and such relays are used as energy harvesting-enabled equipment to help forward signals to the NOMA users at a far distance with poorer channel conditions. It is worth noting that Energy Harvesting (EH) circuits are known as a non-linear wireless power transfer scheme in practice. Therefore, the traditional Linear EH (LEH) models [30–32] are not suitable for making energy transfer enabling wireless relaying maintain its operations. Unfortunately, outage and ergodic analysis strategies following the LEH model may not illustrate the optimal performance in practical systems. In fact, the maximum harvested power value depends on the EH circuits. Thus, it is of great interest to address in this paper the Non-linear EH (NEH) model. For example, to maximize the harvesting power of each energy harvesting receiver, the multi-objective resource optimization problem is considered by using the weighted Tchebycheffmethod [33].

The implementations and contributions of this paper are summarized as follows:

- To the best of our knowledge, we are the first to consider a model where the impact of NEH is evaluated through two PDMA users' performance. Taking the advantages of wireless power transfer in the EH scheme, an NEH-PDMA scheme, which consists of two PDMA users that are served by a relay with the possible capability of wireless EH from the base station, is developed.
- In conventional NOMA, the channel gains are often ordered to perform SIC at the receiver. Interestingly, this paper proposes another approach in which we considered such a proposed NEH-PDMA model. Quality of Service (QoS)-based decoding order [36] is the criterion to eliminate interference and then to extract the main expected signal for each NOMA user.
- Outage probability and ergodic capacity are calculated under the impact of the target rates, SNR, and location of the node in such a network. To consider the role of non-linear wireless power transfer, the saturation threshold of the energy harvesting receiver is the priority factor to evaluate the influence on system performance.

The structure of the remaining parts of this study is arranged as follows. A unified NEH-PDMA framework in Section 2 is studied in wireless communication, where normal cellular users are ordered related to the decoding procedure based on their channel conditions. Next, Section 3 introduces the exact expressions in terms of outage probability to verify the performance of a pair of PDMA users, and these concerned expressions are carefully derived. To further evaluate the insights, Section 4 provides the impact of several parameters on system ergodic capacity. To determine the derived analytical results, Section 5 provides the numerical results. Further remarks and important results are given to conclude our paper in Section 6. Finally, the Appendix shows the proofs of the mathematically-raised problems in the main sections.

Notation: This paper needs some main notations to ease the understanding of the upcoming analyses. They are defined as follows: $E\{.\}$ shows the expectation computation. $f_X(.), F_X(.)$ denote the Probability Density Function (PDF) and Cumulative Distribution Function (CDF) of a random variable X, respectively. $Pr(.)$ represents the probability operation. $\mathbf{1}(C)$ denotes the identity function, $\mathbf{1}(C) = 1$ if C holds and $\mathbf{1}(C) = 0$ otherwise. $Ei(.)$ stands for the exponential integral function.

2. System Model

As illustrated in Figure 1, we considered a scenario where a PDMA-assisted Base Station (BS) wants to communicate with two cell-edge users simultaneously via an EH-assisted relay. The two users were assumed to have a similar channel condition. In addition, the relay utilized the DF relaying protocol to transmit the BS's superposed signal to both users. In addition, the relay was equipped with the EH mechanism using the Power-Splitting (PS) protocol to harvest energy from the received signal. The NEH architecture was deployed in this system. It was assumed that each node was equipped with a single antenna operating in half-duplex mode.

It was assumed that all channel coefficients in this model followed an independent and identical complex Gaussian distribution with zero mean and unit variance. Specifically, we have $g \sim \mathcal{CN}(0,1)$ and $g_i \sim \mathcal{CN}(0,1)$ for $i = 1,2$ as the channel coefficients from the BS to the relay and from the relay to user i, respectively.

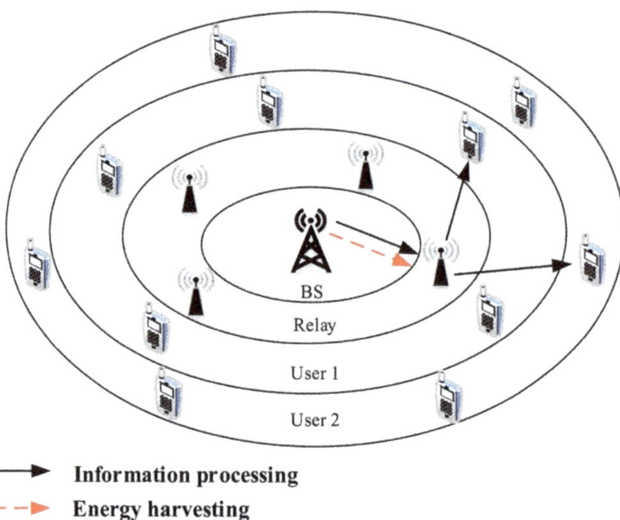

Figure 1. System model of downlink in the Non-linear Energy Harvesting (NEH)-NOMA network.

In some previous works, the decoding order of the SIC receivers was based on users' channel condition, in which the weaker user is served first [6,8,11]. However, in this paper, we assumed a QoS-based decoding order, in which the user with a lower required data rate is served first [36]. Let R_i (bits/s/Hz) be the target data rate for user i. Hence, without loss of generality, assume that $R_1 < R_2$.

Assume that the transmission from the BS to both users consumes a duration of T block time and is equally allocated for two phases. In the first phase, the BS transmits the superposed signal to the desired relay. Due to PS protocol, the received signal at the information receiver in the relay is given by:

$$y_R = \sqrt{(1 - E_R)P_S} \frac{g}{\sqrt{d^\alpha}} (\sqrt{a_1} x_1 + \sqrt{a_2} x_2) + n_R,$$ (1)

where $E_R \in (0, 1)$ is the power splitting ratio, P_S is the transmit power of the BS, d and α denote the distance from the BS to the relay and the path-loss exponent, respectively, x_i is the unit power information signal of user i and a_i for $i \in \{1, 2\}$ is the power allocation for user i, and $n_R \sim \mathcal{CN}(0, \sigma_R^2)$ denotes the Additive White Gaussian Noise (AWGN) at the relay.

The Signal-to-Interference-plus-Noise Ratios (SINRs) before and after SIC at the relay are given respectively by:

$$\gamma_{R,x_1} = \frac{(1 - E_R)a_1 p_S \ell(d)}{(1 - E_R)a_2 p_S \ell(d) + 1},$$ (2)

$$\gamma_{R,x_2} = (1 - E_R)a_2 p_S \ell(d),$$ (3)

where $\ell(d) \triangleq |g|^2/d^\alpha$ and $p_S \triangleq P_S/\sigma_R^2$.

Due to the nature of NEH, the transmit power at the relay in the linear and non-linear region is given by [35]:

$$P_R = E_E E_R \min\left(\frac{P_S |g|^2}{d^\alpha}, P_{th}\right),$$ (4)

where $E_E \in [0, 1]$ denotes the EH efficiency depending on the quality of the harvesting circuitry and P_{th} denotes the saturation threshold of EH receiver. Hence, the received signal at user i is given as

$$y_i = \frac{g_i}{\sqrt{d_i^\alpha}} (\sqrt{b_1} x_1 + \sqrt{b_2} x_2) \sqrt{P_R} + n_i,$$ (5)

where d_i, b_i, and $n_i \sim \mathcal{CN}(0, \sigma_i^2)$ denote the distance between the relay and the user i, the new power allocation for user i, and the AWGN at user i, respectively.

At both users, x_1 is decoded first due to the assumption $R_1 < R_2$; thus, the SINR to decode this signal is given by:

$$\gamma_{i,x_1} = \frac{b_1 p_R \ell(d_i)}{b_2 p_R \ell(d_i) + 1},$$ (6)

where $p_R \triangleq P_R/\sigma^2$, $\sigma_1^2 = \sigma_2^2 = \sigma^2$, $\ell(d_i) \triangleq |g_i|^2/d_i^\alpha$. At User 2, SIC is carried out to remove x_1 from the received superposed signal, and then, x_2 is decoded with the SINR and given by:

$$\gamma_{2,x_2} = b_2 p_R \ell(d_2).$$ (7)

3. Outage Probability Analysis

Since the capacity of the channel from the BS to the destination user is less than the required transmission rate, an outage event will occur. As a result, in this NEH-PDMA system model, the PDMA users cannot detect the information exactly. In this section, the outage probability is performed as a metric to examine the system performance of unified downlink NEH-PDMA networks. The outage event of the specific user in the typical cell is that the PDMA user is incapable of performing the signal detection operation. In particular, we characterize the performance in terms of outage probabilities for a pair of PDMA users who receive signal forwarding from an energy harvesting-aware relay as in the following. Before heading to the next section, Equations (6) and (7), i.e., the instantaneous SINR before and possibly after SIC (User 2) can be rewritten as

$$\gamma_{i,x_1} = \frac{a'_1 E_R \min\left(p_S \ell\left(d\right), p_{th}\right) \ell\left(d_i\right)}{a'_2 E_R \min\left(p_S \ell\left(d\right), p_{th}\right) \ell\left(d_i\right) + 1}, \tag{8}$$

$$\gamma_{2,x_2} = E_R a'_2 \min\left(p_S \ell\left(d\right), p_{th}\right) \ell\left(d_2\right), \tag{9}$$

where $a'_i \triangleq b_i E_E$.

3.1. Outage Probability at User 1

Let $v_i \triangleq 2^{2R_i} - 1$, the outage probability at User 1 is the probability an outage event occurs at either the relay or User 1, and it can be formulated as:

$$
\begin{aligned}
P_1 =& \mathbb{P}(\gamma_{R,x_1} < v_1 \text{ or } \gamma_{R,x_2} < v_2) \\
&+ \mathbb{P}(\gamma_{R,x_1} \geq v_1, \gamma_{R,x_2} \geq v_2)\mathbb{P}(\gamma_{1,x_1} < v_1 | \gamma_{R,x_1} \geq v_1, \gamma_{R,x_2} \geq v_2) \\
=& 1 - \mathbb{P}(\gamma_{R,x_1} \geq v_1, \gamma_{R,x_2} \geq v_2, \gamma_{1,x_1} \geq v_1),
\end{aligned}
\tag{10}
$$

where $(\gamma_{R,x_1} < v_1 \text{ or } \gamma_{R,x_2} < v_2)$ is the event that the relay cannot decode either x_1 or x_2 and $(\gamma_{1,x_1} < v_1)$ denotes the event that User 1 cannot decode its own signal. Subsequently, (10) can be evaluated as

$$
P_1 = 1 -
\begin{cases}
e^{-\frac{d_1^\alpha t'_1}{p_{th}} - \frac{d^\alpha p_{th}}{p_S}} + \Gamma\left(1, \frac{d^\alpha t_{\max}}{p_S}; \frac{d^\alpha d_1^\alpha t'_1}{p_S}\right) & \\
\quad - \Gamma\left(1, \frac{d^\alpha p_{th}}{p_S}; \frac{d^\alpha d_1^\alpha t'_1}{p_S}\right) & , p_{th} > t_{\max}, v_1 < v_{\min} \\
e^{-\frac{d_1^\alpha t'_1}{p_{th}} - \frac{d^\alpha t_{\max}}{p_S}} & , p_{th} \leq t_{\max}, v_1 < v_{\min} \\
0 & , \text{otherwise}
\end{cases}
\tag{11}
$$

where $\Gamma(\alpha, x; b) = \int_x^\infty t^{\alpha-1} e^{-t-bt^{-1}} dt$ denotes the generalized incomplete gamma (g.i.g.) function [37], $v_{\min} \triangleq \min\left(\frac{a_1}{a_2}, \frac{a'_1}{a'_2}\right)$, $t_{\max} \triangleq \max\left(t_1, t_2\right)$, $t_1 \triangleq \frac{v_1}{(a_1 - v_1 a_2)(1 - E_R)}$, $t_2 \triangleq \frac{v_2}{(1 - E_R)a_2}$, and $t'_1 \triangleq \frac{v_1}{(a'_1 - v_1 a'_2)E_R}$.

Proof. See Appendix A. $\square$

3.2. Outage Probability at User 2

The outage probability at User 2 is the probability in which an outage event occurs at either the relay or at this user, and it can be formulated as:

$$
\begin{aligned}
P_2 =& \mathbb{P}(\gamma_{R,x_1} < v_1 \text{ or } \gamma_{R,x_2} < v_2) \\
&+ \mathbb{P}(\gamma_{R,x_1} \geq v_1, \gamma_{R,x_2} \geq v_2)\mathbb{P}[(\gamma_{2,x_1} < v_1 \text{ or } \gamma_{2,x_2} < v_2) | \gamma_{R,x_1} \geq v_1, \gamma_{R,x_2} \geq v_2] \\
=& 1 - \mathbb{P}(\gamma_{R,x_1} \geq v_1, \gamma_{R,x_2} \geq v_2, \gamma_{2,x_1} \geq v_1, \gamma_{2,x_2} \geq v_2),
\end{aligned}
\tag{12}
$$

in which $(\gamma_{2,x_1} < v_1 \text{ or } \gamma_{2,x_2} < v_2)$ denotes the event that User 2 cannot decode x_1 nor decode its own signal after successful SIC. Subsequently, (14) can be expressed as

$$P_2 = 1 - \begin{cases} e^{-\frac{d_2^\alpha t'_{max}}{p_{th}} - \frac{d^\alpha p_{th}}{p_S}} + \Gamma\left(1, \frac{d^\alpha t_{max}}{p_S}; \frac{d^\alpha d_2^\alpha t'_{max}}{p_S}\right) \\ \qquad - \Gamma\left(1, \frac{d^\alpha p_{th}}{p_S}; \frac{d^\alpha d_2^\alpha t'_{max}}{p_S}\right) & , p_{th} > t_{max}, v_1 < v_{min} \\ e^{-\frac{d_2^\alpha t_{max}}{p_{th}} - \frac{d^\alpha t_{max}}{p_S}} & , p_{th} \leq t_{max}, v_1 < v_{min} \\ 0 & , \text{otherwise} \end{cases} \tag{13}$$

where $t_{max} \overset{\Delta}{=} \max(t_1, t_2)$, $t'_{max} \overset{\Delta}{=} \max(t'_1, t'_2)$, $t'_1 \overset{\Delta}{=} \frac{v_1}{(a'_1 - v_1 a'_2)E_R}$, and $t'_2 \overset{\Delta}{=} \frac{v_2}{E_R a'_2}$.

Proof. This can be rewritten (11) as

$$P_2 = 1 - \begin{cases} \mathbb{P}\left(p_S \ell(d) \geq \max(t_1, t_2), \ell(d_2) \geq \frac{\max(t'_1, t'_2)}{\min(p_S \ell(d), p_{th})}\right) & , v_1 < v_{min} \\ 0 & , \text{otherwise} \end{cases} \tag{14}$$

where $t_2 \overset{\Delta}{=} \frac{v_2}{(1 - E_R)a_2}$ and $t'_2 \overset{\Delta}{=} \frac{v_2}{E_R a'_2}$. When $v_1 < v_{min}$, it can be achieved:

$$P_2 = 1 - \mathbb{P}\left(p_S \ell(d) \geq \max(t_1, t_2), \ell(d_2) \geq \frac{\max(t'_1, t'_2)}{p_{th}}, p_S \ell(d) \geq p_{th}\right)$$

$$\qquad - \mathbb{P}\left(p_S \ell(d) \geq \max(t_1, t_2), \ell(d_2) \geq \frac{\max(t'_1, t'_2)}{p_S \ell(d)}, p_S \ell(d) < p_{th}\right). \tag{15}$$

Similar to (A2), we can obtain (13). $\square$

Proposition 1. *In order to minimize the outage at the relay, the optimal power allocation at the source is formulated by:*

$$a_2^* = \begin{cases} \frac{1}{2} & , \text{if } (v_1 < 1) \text{ and } (v_2 \geq \frac{v_1}{1 - v_1}), \\ \frac{v_2}{v_1 + v_2 + v_1 v_2} & , \text{if } \{(v_1 \geq 1)\} \text{ or } \{(v_1 < 1) \text{ and } (v_2 < \frac{v_1}{1 - v_1})\}. \end{cases} \tag{16}$$

Proof. Recall that the coverage probability at the source is equivalent to $(p_S \ell(d) \geq t_{max})$. Hence, by minimizing t_{max}, the optimal performance at the relay can be achieved. Thus, the optimization problem at the source is formulated as

$$a_2^* = \underset{a_2}{\arg\min} \max\{t_1(a_2), t_2(a_2)\} \overset{\Delta}{=} t_{max}(a_2),$$
$$\text{subject to} \quad a_2 \in \left[0, \min\left\{\frac{1}{2}, \frac{1}{1+v_1}\right\}\right] \tag{17}$$

in which $t_1(a_2) = \frac{v_1}{1 - E_R} \times \frac{1}{1 - (1 + v_1)a_2}$ and $t_2(a_2) = \frac{v_2}{1 - E_R} \times \frac{1}{a_2}$. Note that the points 0 and $\frac{1}{1+v}$ (in the case of $\frac{1}{1+v} \leq \frac{1}{2}$) are added for the ease of analysis and should be rejected later if a_2 takes these values. Otherwise outage will occur at the relay with a probability of 1. For convenience, the problem in (17) can be divided into two cases.

Case 1: $\frac{1}{1+v_1} \leq \frac{1}{2}$, which is equivalent to $v_1 \geq 1$ and $a_2 \in [0, \frac{1}{1+v_1}]$: $t_1(a_2)$ monotonically increases from $\frac{v_1}{1 - E_R}$ to $+\infty$, while $t_2(a_2)$ monotonically decreases from $+\infty$ to $\frac{v_2(1+v_1)}{1 - E_R} > \frac{v_1}{1 - E_R}$ since $v_2 > v_1$; thus, the curves $t_1(a_2)$ and $t_2(a_2)$ have a point of intersection that is also the global minimum of $t_{max} = \max\{t_1(a_2), t_2(a_2)\}$. Hence, by solving $\frac{v_1}{1 - (1 + v_1)a_2} = \frac{v_2}{a_2}$ to find a_2, the optimal a_2^* in this case is obtained as

$$a_2^* = \frac{v_2}{v_1 + v_2 + v_1 v_2}. \tag{18}$$

Case 2: $v_1 < 1$, which is equivalent to $a_2 \in [0, \frac{1}{2}]$: $t_1(a_2)$ increases from $\frac{v_1}{1-E_R}$ to $\frac{2v_1}{(1-v_1)(1-E_R)}$, while $t_2(a_2)$ decreases from $+\infty$ to $\frac{2v_2}{1-E_R}$. Further, if $v_2 \geq \frac{v_1}{1-v_1} \Rightarrow t_{\max}(a_2) = t_2(a_2)$, which accepts $a_2^* = \frac{1}{2}$ as its optimal point. For $v_2 < \frac{v_1}{1-v_1}$, using the same approach in Case 1, the optimal a_2 is obtained by solving $t_1(a_2) = t_2(a_2)$, resulting in (18). Combining all the above results, we achieve the optimal power allocation as in (16). $\square$

Remark 1. *In principle, if the signal decoding process for the considered user fails, the outage event is inevitable. It is expected that different outage probabilities can be addressed due to dissimilar power allocation factors for each PDMA user. In other words, the previous outage formulation makes the decoding procedure of the specific user highly dependent on the target rate. Such a concern will be checked in the simulation results.*

3.3. Asymptotic Analysis

In the high SNR regime, the outage probability at User 1 and User 2 becomes respectively:

$$P_1^\infty(p_S) \to 1 - \exp\left\{-\frac{d_1^\alpha t_1'}{p_{th}} - \frac{d^\alpha \max(p_{th}, t_{\max})}{p_S}\right\}, v_1 < v_{\min} \tag{19}$$

$$P_2^\infty(p_S) \to 1 - \exp\left\{-\frac{d_2^\alpha t_{\max}'}{p_{th}} - \frac{d^\alpha \max(p_{th}, t_{\max})}{p_S}\right\}, v_1 < v_{\min}. \tag{20}$$

Further, the diversity order in terms of outage probability at user i is defined as [38]:

$$D \triangleq - \lim_{p_S \to \infty} \frac{\log_{10}\{P_i^\infty(p_S)\}}{\log_{10}\{p_S\}}; \tag{21}$$

thus, both User 1 and User 2 have zero diversity order since at $p_S \to \infty$ and $v_1 < v_{\min}$, the term $\exp\{-\frac{d^\alpha}{p_S}\max(p_{th}, t_{\max})\}$ can be neglected, resulting in the outage floors at User 1 and User 2 being $\{1 - e^{-\frac{d_1^\alpha t_1'}{p_{th}}}\}$ and $\{1 - e^{-\frac{d_2^\alpha t_{\max}'}{p_{th}}}\}$, respectively.

4. Ergodic Capacity

The ergodic capacity for x_i is given as $C_i = \frac{1}{2}\mathbb{E}_{Z_i}[\log_2(1 + Z_i)]$, where we define $Z_1 \triangleq \min\left(\gamma_{R,x_1}, \gamma_{1,x_1}, \gamma_{2,x_1}\right)$ and $Z_2 \triangleq \min\left(\gamma_{R,x_2}, \gamma_{2,x_2}\right)$ [32]. In addition, C_i can be evaluated analytically as

$$C_i = \frac{1}{2\ln(2)} \int_0^\infty \frac{1 - F_{Z_i}(x)}{1 + x} dx, (i = 1, 2). \tag{22}$$

4.1. Ergodic Capacity for x_1

In order to derive (22) for $i = 1$, the CDF of Z_1 is calculated as

$$F_{Z_1}(\gamma) = 1 - \mathbb{P}(\min(\gamma_{R,x_1}, \gamma_{1,x_1}, \gamma_{2,x_1}) \geq \gamma)$$

$$= 1 - \begin{cases} \mathbb{P}\left(p_S\ell(d) \geq t_1(\gamma), \ell(d_1) \geq \frac{t_1'(\gamma)}{\min(p_S\ell(d), p_{th})}, \ell(d_2) \geq \frac{t_1'(\gamma)}{\min(p_S\ell(d), p_{th})}\right) & , \gamma < v_{\min} \\ 0 & , \text{otherwise} \end{cases} \tag{23}$$

where $t_1(\gamma) \triangleq \frac{\gamma}{(1-E_R)(a_1-a_2\gamma)}$ and $t'_1(\gamma) \triangleq \frac{\gamma}{(a'_1-a'_2\gamma)E_R}$. It can be evaluated analytically as

$$
F_{Z_1}(\gamma) = 1 - \begin{cases} \exp\left(-\frac{d^\alpha p_{th}}{p_S} - (d_1^\alpha + d_2^\alpha)\frac{t'_1(\gamma)}{p_{th}}\right) + P_{12}(\gamma) & , \gamma < \min(\frac{a'_1}{a'_2}, a_m) \\ \exp\left(-\frac{d^\alpha t_1(\gamma)}{p_S} - (d_1^\alpha + d_2^\alpha)\frac{t'_1(\gamma)}{p_{th}}\right) & , a_m \le \gamma < v_{\min} \\ 0 & , \text{otherwise} \end{cases} \tag{24}
$$

where:

$$
P_{12}(\gamma) = \Gamma\left(1, \frac{d^\alpha t_1(\gamma)}{p_S}; (d_1^\alpha + d_2^\alpha)\frac{d^\alpha t'_1(\gamma)}{p_S}\right) - \Gamma\left(1, \frac{d^\alpha p_{th}}{p_S}; (d_1^\alpha + d_2^\alpha)\frac{d^\alpha t'_1(\gamma)}{p_S}\right). \tag{25}
$$

Proof. See Appendix B. $\square$

Using the Gaussian–Chebyshev quadrature [39], the ergodic capacity of x_1 is approximated as

$$
\begin{aligned}
C_1 \approx &\sum_{n=0}^{N} k_{1,n}\left\{ \exp\left(-\frac{d^\alpha p_{th}}{p_S} - (d_1^\alpha + d_2^\alpha)\frac{t'_1(\gamma_{1,n})}{p_{th}}\right)\right. \\
& \left. - \Gamma\left(1, \frac{d^\alpha p_{th}}{p_S}; (d_1^\alpha + d_2^\alpha)\frac{d^\alpha t'_1(\gamma_{1,n})}{p_S}\right) + \Gamma\left(1, \frac{d^\alpha t_1(\gamma_{1,n})}{p_S}; (d_1^\alpha + d_2^\alpha)\frac{d^\alpha t'_1(\gamma_{1,n})}{p_S}\right) \right\} \\
& + 1\left(\frac{a'_1}{a'_2} \ge a_m\right) \sum_{n=0}^{N} k_{2,n} \exp\left(-\frac{d^\alpha t_1(\gamma_{2,n})}{p_S} - (d_1^\alpha + d_2^\alpha)\frac{t'_1(\gamma_{2,n})}{p_{th}}\right),
\end{aligned} \tag{26}
$$

in which:

$$
k_{i,n} \triangleq \frac{\pi}{2N}\Delta_i^- |\sin(\frac{2n-1}{N}\pi)|\frac{1}{1+\gamma_{i,n}}, \tag{27}
$$

$$
\gamma_{i,n} \triangleq \frac{1}{2}\Delta_i^+ + \frac{1}{2}\Delta_i^- \cos(\frac{2n-1}{N}\pi), \quad i \in \{1,2\}, \tag{28}
$$

where $a_m \triangleq \frac{a_1 p_{th}(1-E_R)}{1+a_2 p_{th}(1-E_R)}$, $\Delta_1^\pm \triangleq \min\left(\frac{a'_1}{a'_2}, a_m\right)$, $\Delta_2^\pm \triangleq v_{\min} \pm a_m$, and N is a coefficient reflecting the accuracy of the approximation.

4.2. Ergodic Capacity for x_2

Similar to (23), the CDF of Z_2 is calculated as

$$
\begin{aligned}
F_{Z_2}(\gamma) &= 1 - \mathbb{P}(\min(\gamma_{R,x_2}, \gamma_{2,x_2}) \ge \gamma) \\
&= 1 - \mathbb{P}\left(p_S \ell(d) \ge t_2(\gamma), \ell(d_2) \ge \frac{t'_2(\gamma)}{\min(p_S \ell(d), p_{th})}\right),
\end{aligned} \tag{29}
$$

where $t_2(\gamma) \triangleq \frac{\gamma}{(1-E_R)a_2}$ and $t'_2(\gamma) \triangleq \frac{\gamma}{E_R a'_2}$.

The above equation can be evaluated analytically as

$$
F_{Z_2}(\gamma) = 1 - \begin{cases} \exp\left(-\frac{d^\alpha p_{th}}{p_S} - \frac{d_2^\alpha t'_2(\gamma)}{p_{th}}\right) + P_{22}(\gamma) & , \gamma < p_{th}a_2(1-E_R) \\ \exp\left(-\frac{d^\alpha t_2(\gamma)}{p_S} - \frac{d_2^\alpha t'_2(\gamma)}{p_{th}}\right) & , \gamma \ge p_{th}a_2(1-E_R) \end{cases} \tag{30}
$$

where:

$$
P_{22}(\gamma) = \Gamma\left(1, \frac{d^\alpha t_2(\gamma)}{p_S}; \frac{d^\alpha d_2^\alpha t'_2(\gamma)}{p_S}\right) - \Gamma\left(1, \frac{d^\alpha p_{th}}{p_S}; \frac{d^\alpha d_2^\alpha t'_2(\gamma)}{p_S}\right). \tag{31}
$$

Proof. Equation (29) can be rewritten as

$$
\begin{aligned}
F_{Z_2}(\gamma) =&\, 1 - \mathbb{P}\left(p_S\ell(d) \geq t_2(\gamma), \ell(d_2) \geq \frac{t'_2(\gamma)}{p_{th}}, p_S\ell(d) \geq p_{th}\right) \\
&- \mathbb{P}\left(p_S\ell(d) \geq t_2(\gamma), \ell(d_2) \geq \frac{t'_2(\gamma)}{p_S\ell(d)}, p_S\ell(d) < p_{th}\right).
\end{aligned}
\tag{32}
$$

The first probability, denoted as $P_{21}(\gamma)$, can be evaluated analytically as

$$
\begin{aligned}
P_{21}(\gamma) =&\, \int_{\max(t_2(\gamma), p_{th})}^{\infty} f_{p_S\ell(d)}(x)dx \int_{t'_2(\gamma)/p_{th}}^{\infty} f_{\ell(d_2)}(y)dy \\
=&\, \exp\left(-\frac{d^\alpha}{p_S}\max(t_2(\gamma), p_{th}) - \frac{d_2^\alpha t'_2(\gamma)}{p_{th}}\right),
\end{aligned}
\tag{33}
$$

whereas the second probability, denoted as P_{22}, can be obtained as

$$
P_{22}(\gamma) = \int_{t_2(\gamma)}^{p_{th}} f_{p_S\ell(d)}(x)\int_{\frac{t'_2(\gamma)}{x}}^{\infty} f_{\ell(d_2)}(y)dx,
\tag{34}
$$

and can be calculated by using ([37], Equation (13)). Hence, the proof is the complete.

Further, by applying Gaussian–Chebyshev quadrature [39], the ergodic capacity for x_2 can be approximated as

$$
\begin{aligned}
C_2 \approx&\, \sum_{n=0}^{N} k_{3,n}\left\{\exp\left(-\frac{d^\alpha p_{th}}{p_S} - \frac{d_2^\alpha t'_2(\gamma_{3,n})}{p_{th}}\right) + \Gamma\left(1, \frac{d^\alpha t_2(\gamma_{3,n})}{p_S}; \frac{d^\alpha d_2^\alpha t'_2(\gamma_{3,n})}{p_S}\right)\right. \\
&\left. - \Gamma\left(1, \frac{d^\alpha p_{th}}{p_S}; \frac{d^\alpha d_2^\alpha t'_2(\gamma_{3,n})}{p_S}\right)\right\} - e^\mu \mathrm{Ei}(-\mu p_{th}a_2(1 - E_R) - \mu),
\end{aligned}
\tag{35}
$$

where $\mu \triangleq \frac{d^\alpha}{p_S a_2(1-E_R)} + \frac{d_2^\alpha}{p_{th} E_R a'_2}$ and:

$$
k_{3,n} \triangleq \frac{\pi}{2N}\Delta_3^-|\sin(\frac{2n-1}{N}\pi)|\frac{1}{1+\gamma_{3,n}},
\tag{36}
$$

$$
\gamma_{3,n} \triangleq \frac{1}{2}\Delta_3^+ + \frac{1}{2}\Delta_3^- \cos\left(\frac{2n-1}{N}\pi\right),
\tag{37}
$$

where $\Delta_3^\pm \triangleq p_{th}a_2(1 - E_R)$. It is noted that in (35), we utilized $\int_a^\infty \frac{e^{-\mu x}}{1+x}dx = -e^\mu \mathrm{Ei}(-\mu a - \mu)$ ([40], Equation (3.353.5)). $\square$

4.3. Asymptotic Analysis

In the high regime, $P_{12}(\gamma) \to 0$ and $P_{22}(\gamma) \to 0$. Hence, the analytical results of C_1 and C_2 with $\frac{a_1}{a_2} = \frac{a'_1}{a'_2} = \frac{b_1}{b_2}$ can be simplified as

$$C_1^\infty \to \frac{e^{-c_{th}c}}{2\ln 2} \left\{ \mathrm{Ei}\left(-\frac{c_1+c_2}{a_2}\right) - \mathrm{Ei}\left(-\frac{c_1+c_2}{a_2}(1+a_2 c_{th})\right) \right\} e^{\frac{c_1+c_2}{a_2}}$$
$$- \frac{e^{-c_{th}c}}{2\ln 2}\left\{ \mathrm{Ei}(-(c_1+c_2)) - \mathrm{Ei}(-(c_1+c_2)(1+c_{th})) \right\} e^{c_1+c_2}$$
$$+ \frac{1}{2\ln 2}\left\{ e^{\frac{c+c_1+c_2}{a_2}} \mathrm{Ei}\left(-\left(\frac{c+c_1+c_2}{a_2}\right)(1+a_2 c_{th})\right) \right.$$
$$\left. - e^{c+c_1+c_2} \mathrm{Ei}\left(-(c+c_1+c_2)(1+c_{th})\right) \right\}, \tag{38}$$

and:

$$C_2^\infty \to \frac{1}{2\ln 2} \exp\left(-c\times c_{th}+\frac{c_2}{a_2}\right)\left[\mathrm{Ei}\left(-\frac{c_2}{a_2}(1+a_2 c_{th})\right) - \mathrm{Ei}\left(-\frac{c_2}{a_2}\right)\right]$$
$$- \frac{1}{2\ln 2}\exp\left(\frac{c+c_2}{a_2}\right)\mathrm{Ei}\left(-\frac{c+c_2}{a_2}(1+a_2 c_{th})\right), \tag{39}$$

where $c = \frac{d^\alpha}{(1-E_R)\times p_S}$, $c_1 = \frac{d_1^\alpha}{E_E \times E_R \times p_{th}}$, $c_2 = \frac{d_2^\alpha}{E_E \times E_R \times p_{th}}$, and $c_{th} = p_{th} \times (1-E_R)$.

Proof. See Appendix C. $\square$

From the above results, it can be seen that at $p_S \to \infty$ or equivalently $c \to 0$, the ergodic capacities at User 1 and User 2 reach:

$$C_1^\infty \to -\frac{1}{2\ln 2}\left\{ e^{c_1+c_2}\mathrm{Ei}\left(-(c_1+c_2)\right) - e^{\frac{c_1+c_2}{a_2}}\mathrm{Ei}\left(-\frac{c_1+c_2}{a_2}\right) \right\}, \tag{40}$$

$$C_2^\infty \to -\frac{1}{2\ln 2}\exp\left(\frac{c_2}{a_2}\right)\mathrm{Ei}\left(-\frac{c_2}{a_2}\right), \tag{41}$$

respectively.

Remark 2. *In the LEH model, the harvested energy is unbounded, i.e., $P_R = E_E E_R \frac{P_S |g|^2}{d^\alpha}$. Subsequently, if the base station transmits with relatively large power so that the effect of path loss is insignificant, the transmit power at the relay is approximated to P_S. Hence, in high SNR regime, from (21) and (22), a diversity order of 1 is achieved at both users instead of zero. Further, from (42) and (43), no ceiling capacity is observed in such a model. In addition, in the lower transmit SNR regime, the probability for $\frac{P_S |g|^2}{d^\alpha}$ to exceed the threshold P_{th} becomes small, and the transmit power at the relay becomes approximated to that of the LEH model. As a result, the LEH model proves its capability to obtain tractable and accurate results in the low/middle transmit SNR regime. In conclusion, the non-linear energy harvesting model is proven to be more practical for the analysis in the high SNR regime with the drawbacks of mathematical complexity.*

5. Numerical Results

In this section, we evaluate the system performance by varying the main parameters such as power allocation factors on two PDMA users considering the design of the wireless-powered PDMA system based on the NEH model. For the NEH model parameters, the main parameters were chosen according to [34]. Given the following parameters: $d + d_i = 1$, $\alpha = 3.8$, the SNR is calculated by P/σ^2, while the variance of the additive noise is $\sigma^2 = 1$.

In Figure 2, we show the outage performance curves versus both R1 and R2. We set $a_1 = 0.8$, $a_2 = 1-a_1$, $d = 0.5$, $E_R = 0.6$, $E_E = 1$, $b_1 = 0.8$, $b_2 = 1-b_1$, SNR $= 20$ dB, and $p_{th} = 10$ dB. From Figure 2, we can observe that the performance of User 1 outperformed the performance of User 2 due to the different threshold rate constraints. The curves of outage probabilities versus SNR are illustrated in Figure 3. As can be seen, the simulated curves matched the analytical curves very tightly, highlighting the exactness of the proven expressions. The outage performance on User 1 was better than that on User 2 due to the different power allocation factors. However, a performance gap was only seen more clearly at high SNR, i.e., SNR > 15 dB. In fact, we know that this is because the outage probability depends on the target data rates, and varying target rates exhibit different performance and reach one in a high data rate region. The interesting point is that outage performance at User 2 remained at a stable level as the SNR level was higher than 30 dB. It can be observed that the different requirements at the receiver or different order in detecting the received signal made the outage performance for the two users become dissimilar, especially in the high SNR condition and fixed target rates. Figure 3 also shows that the outage probability would be worse as the target rates increased, especially as they were approximate to 1 bit/s; the worst cases that can be declared.

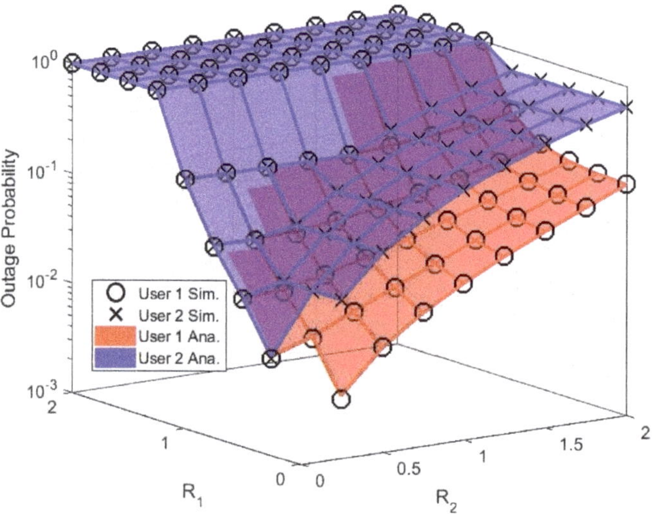

Figure 2. Outage probability of User 1 and User 2 versus R_1 and R_2.

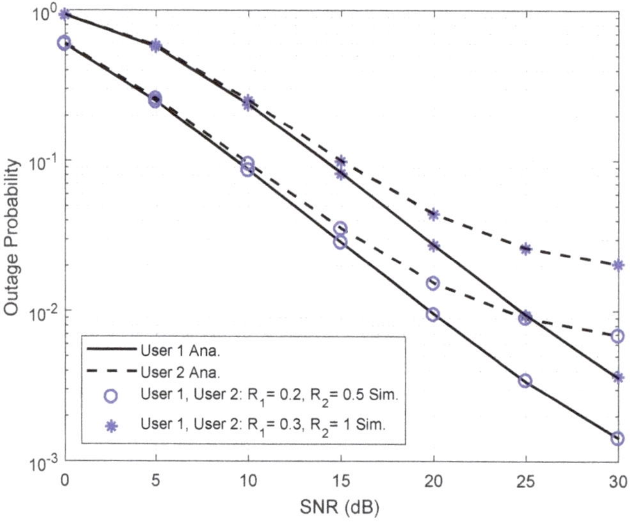

Figure 3. Outage probability of User 1 and User 2 versus SNR.

In Figure 4, it can be seen how the distances of the pair of nodes affected the outage performance. In addition, with the increase of d_{SR}, the difference of the outage probability among two users can be seen as d ranging from 0.1 to approximate 0.7. However, as d was greater than 0.7, these outage performances were definitely the same. The reason is that as the relay was placed very far from the BS, the different distances between the relay and two users were similar, and hence, outage performance was the same as well.

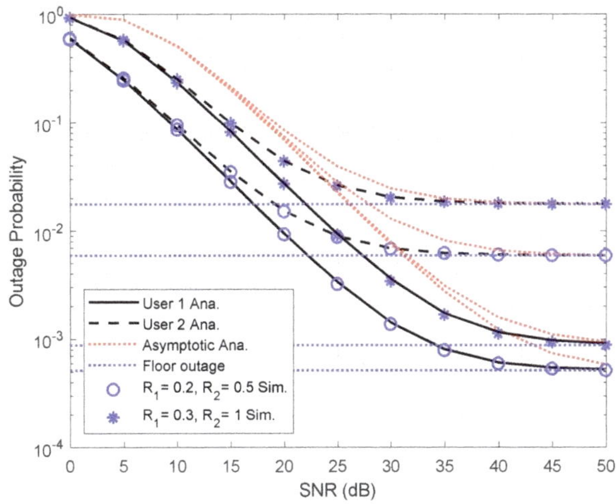

Figure 4. Outage probability of User 1 and User 2 versus SNR.

From Figure 5, we can see that with the increase of p_{th}, the outage probability of our scheme firstly decreased and then tended to be steady, which is because in our scheme, we considered the circuit sensitivity, so the outage probability would tend to be steady. The outage performances of these users were similar at high p_{th}. As a result, the outage probability was the smallest as enough energy was harvested at the relay. In this case, User 1 was sensitive with a small value of p_{th}, but also had optimal outage performance with a reasonable selection of p_{th}.

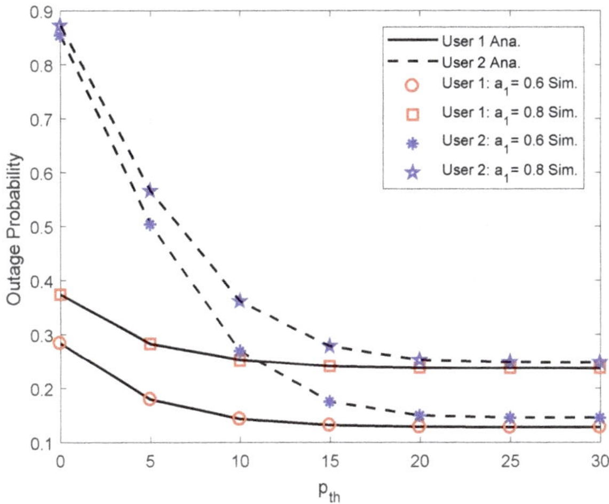

Figure 5. Outage probability of User 1 and User 2 versus distance.

In Figure 6, we plot the ergodic capacity versus SNR for different power allocation factors for each PDMA user. We set $d = 0.5$, $E_R = 0.6$, $E_E = 1$, $b_1 = 0.8$, $b_2 = 1 - b_1$, $R_1 = 0.5$, $R_2 = 1$, and $p_{th} = 20$ dB. Increasing the value of SNR in the concerned range, the ergodic capacity of User 2 grew faster than that of User 1. It is noted that two cases of power allocation factor affected the ergodic capacity performance significantly only at lower SNR, as an SNR lower than 30 dB, and then, ergodic capacity was not changed by the varying power allocation factor, as the SNR was greater than 30 dB. However, such an observation was different for User 1, where the performance gaps of two power allocation factors still existed in the whole SNR regime.

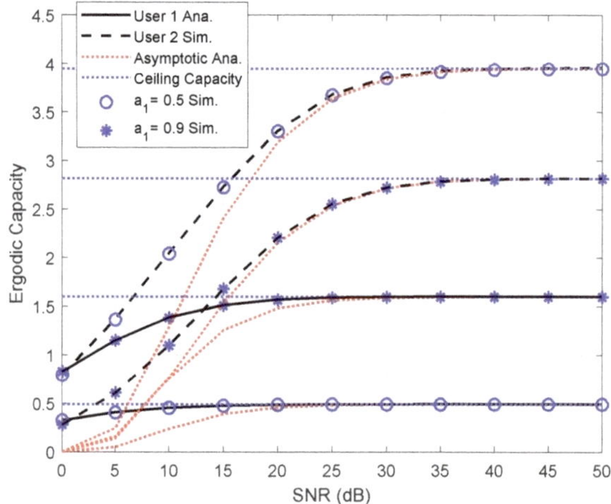

Figure 6. Ergodic capacity of User 1 and User 2 versus SNR.

From Figure 7, we can see that with the increase of the distance, the ergodic capacity of the two PDMA users in these schemes increased and tended to a maximal point at a specific location of the node. This comes from the fact that different SNR and then different ergodic capacity were obtained by exploiting channel condition disparity. Moreover, we can see that the optimal ergodic capacity could be achieved by User 2 in the numerical method.

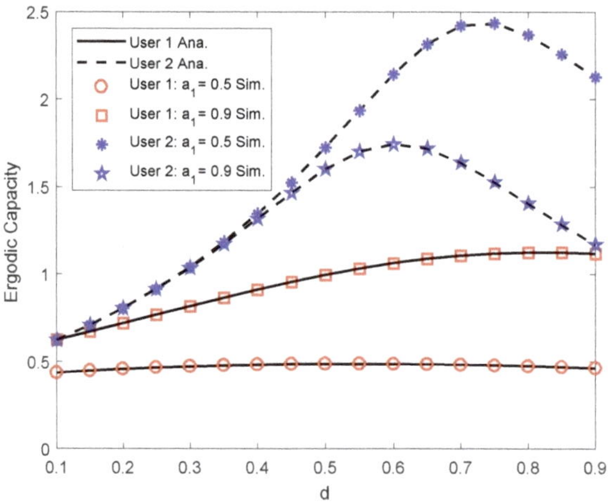

Figure 7. Ergodic capacity of User 1 and User 2 versus distance.

The impact of p_{th} on ergodic capacity is shown in Figure 8. In particular, we can see that the ergodic capacity of both PDMA users increased with increasing p_{th}. The impressive point is that the performance gap regarding the ergodic capacity of two PDMA users was very large at high p_{th}. Therefore, we can conclude that the energy harvested at the relay was an important factor for the ergodic capacity of User 2. Such characteristics of this figure can be seen considering the ergodic capacity of User 2 where p_{th} affected the related performance more remarkably. Moreover, the observation obtained in these figures verifies the correctness of the derived expressions.

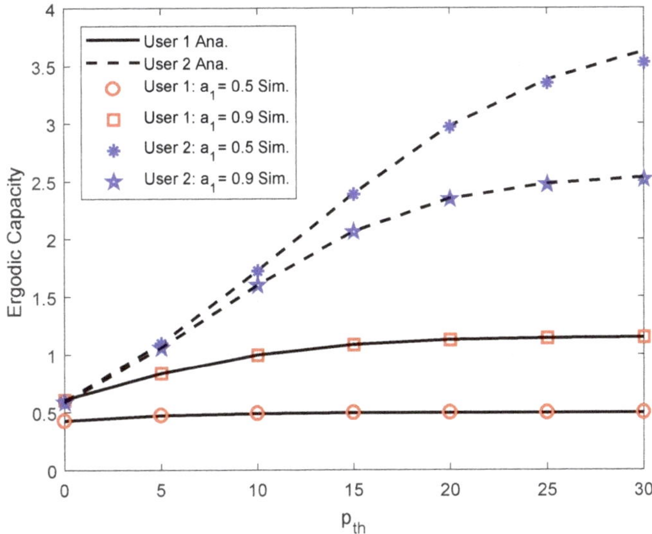

Figure 8. Ergodic capacity of User 1 and User 2 versus p_{th}.

6. Conclusions

We focused on the QoS-based ordered decoding scheme for downlink PDMA, assuming an energy constraint relay using the non-linear energy harvesting model. The outage probability and ergodic capacity performance problem were expressed in wirelessly-powered PDMA under the practical NEH model. The exact closed-form expressions of outage and ergodic capacity were proposed to evaluate system performance in the scenario of varying harvested power at a relay in the NEH model. By employing numerical simulation, the outage performance will be optimal by jointly optimizing the location of each node, the saturation threshold of the energy harvesting receiver, and power allocation coefficients. It was shown that there is a tradeoff between the outage and ergodic capacity performance versus SNR, and further study can consider which one is selected to achieve the optimal outage and ergodic capacity. Although the performance achieved under the NEH model may be dependent on the harvested power at the relay, power allocation factors in PDMA still lead to fluctuating performance in the considered conditions. These concerns were clearly illustrated in the simulation results, and it was confirmed that our derived expressions were correct.

Author Contributions: T.-L.N. designed the algorithm and performed the theoretical analysis, while D.-T.D. checked the results and wrote the manuscript. M.-S.V.N. implemented the simulation and contributed to the manuscript preparation. M.V. was responsible for formulating the research issues and revised the paper.

Funding: This research was funded by the Ministry of Education, Youth and Sport of the Czech Republic under SGSGrant No. SP2019/41 conducted at VSB, Technical University of Ostrava.

Conflicts of Interest: The authors declare no conflict of interest.

Appendix A

Substituting (2), (3), (8) with $i = 1$ and (9) into (10), we achieve:

$$P_1 = 1 - \begin{cases} \mathbb{P}\left(p_s\ell(d) \geq t_{\max}, \ell(d_1) \geq \frac{t_1'}{\min(p_s\ell(d),p_{th})}\right) & , v_1 < v_{\min} \\ 0 & , \text{otherwise} \end{cases}. \tag{A1}$$

Considering the case where $v_1 < v_{\min}$, the above equation can be further expressed as

$$\begin{aligned} P_1 =&1 - \mathbb{P}\left(p_s\ell(d) \geq t_{\max}, \ell(d_1) \geq \frac{t_1'}{p_{th}}, p_s\ell(d) \geq p_{th}\right) \\ &- \mathbb{P}\left(p_s\ell(d) \geq t_{\max}, \ell(d_1) \geq \frac{t_1'}{p_s\ell(d)}, p_s\ell(d) < p_{th}\right). \end{aligned} \tag{A2}$$

The above equation can be analyzed as

$$\begin{aligned} P_1 =&1 - \left(1 - F_{\ell(d_1)}\left(\frac{t_1'}{p_{th}}\right)\right) \int_{\max(t_{\max},p_{th})}^{\infty} f_{p_s\ell(d)}(x)dx \\ &- \mathbf{1}(p_{th} > t_{\max}) \int_{t_{\max}}^{p_{th}} \left(1 - F_{\ell(d_1)}\left(\frac{t_1'}{x}\right)\right) f_{p_s\ell(d)}(x)dx. \end{aligned} \tag{A3}$$

The CDF and PDF of $\ell(x)$ are $F_{\ell(x)}(\gamma) = 1 - \exp(-x^\alpha\gamma)$ and $f_{\ell(x)}(\gamma) = x^\alpha \exp(-x^\alpha\gamma)$, respectively. Note that $f_{k\ell(x)}(\gamma) = \frac{x^\alpha}{k}\exp(-\frac{x^\alpha}{k}\gamma)$. Hence, (A3) can be further derived as

$$\begin{aligned} P_1 =&1 - e^{-\frac{d_1^\alpha t_1'}{p_{th}} - \frac{d^\alpha}{p_S}\max(t_{\max},p_{th})} \\ &- \mathbf{1}(p_{th} > t_{\max}) \left[\Gamma\left(1, \frac{d^\alpha t_{\max}}{p_S}; \frac{d^\alpha d_1^\alpha t_1'}{p_S}\right) - \Gamma\left(1, \frac{d^\alpha p_{th}}{p_S}; \frac{d^\alpha d_1^\alpha t_1'}{p_S}\right)\right], \end{aligned} \tag{A4}$$

which is equivalent to (24).

Appendix B

Considering the case of $\gamma < \min(\frac{a_1}{a_2}, \frac{a_1'}{a_2})$, Equation (23) becomes:

$$\begin{aligned} F_{Z_1}(\gamma) =&1 - \mathbb{P}\left(p_s\ell(d) \geq t_1(\gamma), \ell(d_1) \geq \frac{t_1'(\gamma)}{p_{th}}, \ell(d_2) \geq \frac{t_1'(\gamma)}{p_{th}}, p_s\ell(d) \geq p_{th}\right) \\ &- \mathbb{P}\left(p_s\ell(d) \geq t_1(\gamma), \ell(d_1) \geq \frac{t_1'(\gamma)}{p_s\ell(d)}, \ell(d_2) \geq \frac{t_1'(\gamma)}{p_s\ell(d)}, p_s\ell(d) < p_{th}\right). \end{aligned} \tag{A5}$$

The first probability in (A5), denoted as $P_{11}(\gamma)$, can be evaluated analytically as

$$\begin{aligned} P_{11}(\gamma) =& \int_{\max(t_1(\gamma),p_{th})}^{\infty} f_{p_s\ell(d)}(x) \int_{t_1'(\gamma)/p_{th}}^{\infty} f_{\ell(d_1)}(y) \int_{t_1'(\gamma)/p_{th}}^{\infty} f_{\ell(d_2)}(z)dxdydz \\ =& e^{-\frac{d^\alpha}{p_S}\max(t_1(\gamma),p_{th})} e^{-\frac{d_1^\alpha t_1'(\gamma)}{p_{th}}} e^{-\frac{d_2^\alpha t_1'(\gamma)}{p_{th}}}, \end{aligned} \tag{A6}$$

whereas the second probability, denoted as $P_{12}(\gamma)$, is calculated as

$$P_{12}(\gamma) = \mathbf{1}(p_{th} > t_1(\gamma)) \int_{t_1(\gamma)}^{p_{th}} f_{ps\ell(d)}(x) \int_{t_1'(\gamma)/x}^{\infty} f_{\ell(d_1)}(y) \int_{t_1'(\gamma)/x}^{\infty} f_{\ell(d_2)}(z) dz dy dx$$

$$= \mathbf{1}(p_{th} > t_1(\gamma)) \int_{t_1(\gamma)}^{p_{th}} \frac{d^\alpha}{p_S} \exp\left(-\frac{d^\alpha}{p_S}x - t_1'(\gamma)(d_1^\alpha + d_2^\alpha)\frac{1}{x}\right) dx, \tag{A7}$$

by applying ([37], Equation (13)) and noticing that $\int_{t_1(\gamma)}^{p_{th}} f(x)dx = \int_{t_1(\gamma)}^{\infty} f(x)dx - \int_{p_{th}}^{\infty} f(x)dx$. Further, it is worth noticing that:

$$\gamma < v_{\min}, p_{th} > t_1(\gamma) \Leftrightarrow \gamma < \min\left(\frac{a_1'}{a_2'}, a_m\right) \tag{A8}$$

$$\gamma < v_{\min} \le t_1(\gamma) \Leftrightarrow (a_m \le \gamma < v_{\min}) \cap \left(\frac{a_1'}{a_2'} \ge a_m\right). \tag{A9}$$

Appendix C

Proof of Equation (38). When $a_1/a_2 = a_1'/a_2'$, (24) can be rewritten as

$$F_{Z_1}(x) = 1 - \begin{cases} e^{-c \times c_{th}} \exp\left\{-(c_1 + c_2)\frac{x}{a_1 - xa_2}\right\} & , x < a_m \\ \exp\left\{-(c + c_1 + c_2)\frac{x}{a_1 - xa_2}\right\} & , a_m \le x < a_1/a_2 \end{cases} . \tag{A10}$$

Substituting the above equation into (22) and using the change of variable $t \to x/a_1 - xa_2$, the ergodic capacity of User 1 in a high SNR regime is then given by:

$$C_1 \to \frac{e^{-c \times c_{th}}}{2\ln 2} \int_0^{c_{th}} \frac{a_1}{a_2} \frac{1}{t+1} \frac{1}{t + a_2^{-1}} e^{-(c_1 + c_2) \times t} dt$$

$$+ \frac{1}{2\ln 2} \int_{c_{th}}^{\infty} \frac{a_1}{a_2} \frac{1}{t+1} \frac{1}{t + a_2^{-1}} e^{-(c + c_1 + c_2) \times t} dt, \tag{A11}$$

and by applying the decomposition $\frac{a_1}{a_2}\frac{1}{t+1}\frac{1}{t+a_2^{-1}} = \frac{1}{t+1} - \frac{1}{t+a_2^{-1}}$, the first integral part becomes:

$$C_{11} \triangleq \int_0^{c_{th}} \left\{\frac{1}{t+1} - \frac{1}{t + a_2^{-1}}\right\} e^{-(c_1 + c_2) \times t} dt. \tag{A12}$$

Using the identity ([40], Equation (3.352.1)), C_{11} can be derived as

$$C_{11} = e^{\frac{c_1 + c_2}{a_2}} \left[\text{Ei}\left(-\frac{c_1 + c_2}{a_2}\right) - \text{Ei}\left(-\frac{c_1 + c_2}{a_2}(1 + a_2 c_{th})\right)\right]$$

$$+ e^{c_1 + c_2} \left[\text{Ei}(-c_1 - c_2) - \text{Ei}(-(c_1 + c_2)(1 + c_{th}))\right]. \tag{A13}$$

Further, using the same decomposition applied for C_{11} and with the help of ([40], Equation (3.352.2)), the second integral part is derived as

$$C_{12} \triangleq e^{\frac{c + c_1 + c_2}{a_2}} \text{Ei}\left(-\frac{c + c_1 + c_2}{a_2}(1 + a_2 c_{th})\right) - e^{c + c_1 + c_2} \text{Ei}(-(c + c_1 + c_2)(1 + c_{th})). \tag{A14}$$

Substituting (A14) and (A13) into (A11) completes the proof. $\square$

Proof of Equation (39). The Equation (30) can be rewritten as

$$
F_{Z_2} = 1 - \begin{cases} e^{-c \times c_{th}} \exp\left(-\frac{c_2}{a_2}x\right) & ,x < a_2 c_{th} \\ \exp\left(-\frac{c+c_2}{a_2}x\right) & ,x \geq a_2 c_{th} \end{cases}.
\tag{A15}
$$

By substituting the above equation into (22), it can be achieved that:

$$
\begin{aligned}
C_2 \rightarrow &\frac{e^{-c \times c_{th}}}{2\ln 2} \int_0^{a_2 c_{th}} \frac{1}{x+1} \exp\left(-\frac{c_2}{a_2}x\right) dx \\
&+ \frac{1}{2\ln 2} \int_{a_2 c_{th}}^{\infty} \frac{1}{x+1} \exp\left(-\frac{c+c_2}{a_2}x\right) dx.
\end{aligned}
\tag{A16}
$$

The derivation for the integrals in (A11) can be applied to solve the above integrals to complete the proof. $\square$

References

1. Chih-Lin, I.; Rowell, C.; Han, S.; Xu, Z.; Li, G.; Pan, Z. Toward green and soft: A 5G perspective. *IEEE Commun. Mag.* **2014**, *52*, 66–73.
2. Islam, S.R.; Avazov, N.; Dobre, O.A.; Kwak, K.S. Power-domain non-orthogonal multiple access (NOMA) in 5G systems: Potentials and challenges. *IEEE Commun. Surv. Tutor.* **2016**, *19*, 721–742. [CrossRef]
3. Do, D.T. Energy-aware two-way relaying networks under imperfect hardware: Optimal throughput design and analysis. *Telecommun. Syst.* **2016**, *62*, 449–459. [CrossRef]
4. Dinh-Thuan, D.O. Power switching protocol for two-way relaying network under hardware impairments. *Radioengineering* **2015**, *24*, 765–771.
5. Tabassum, H.; Ali, M.S.; Hossain, E.; Hossain, M.J.; Kim, D.-I. Non-orthogonal multiple access (NOMA) in cellular uplink and downlink: Challenges and enabling techniques. In Proceedings of the IEEE Vehicular Technology Conference (VTC'2017 Spring), Sydney, Australia, 4–7 June 2017.
6. Kim, J.; Lee, I. Capacity analysis of cooperative relaying systems using non-orthogonal multiple access. *IEEE Commun. Lett.* **2015**, *19*, 1949–1952. [CrossRef]
7. Do, D.-T.; Nguyen, Van M.-S.; Hoang, T.-A.; Voznak, M. NOMA-assisted multiple access scheme for IoT deployment: Relay selection model and secrecy performance improvement. *Sensors* **2019**, *19*, 736. [CrossRef] [PubMed]
8. Do, D.T. Optimal throughput under time power switching based relaying protocol in energy harvesting cooperative networks. *Wireless Personal Commun.* **2016**, *87*, 551–564. [CrossRef]
9. Lyu, B.; Yang, Z.; Gui, G. Backscatter assisted wireless powered communication networks with non-orthogonal multiple access. *IEICE Trans. Fundam. Electron. Commun. Comput. Sci.* **2017**, *100*, 1724–1728. [CrossRef]
10. Liu, Y.; Ding, Z.; Elkashlan, M.; Yuan, J. Non-orthogonal multiple access in large-scale underlay cognitive radio networks. *IEEE Trans. Veh. Technol.* **2016**, *65*, 10152–10157. [CrossRef]
11. Zhang, Z.; Ma, Z.; Xiao, M.; Ding, Z.; Fan, P. Full-duplex device-to-device aided cooperative non-orthogonal multiple access. *IEEE Trans. Veh. Technol.* **2017**, *66*, 4467–4471. [CrossRef]
12. Zhong, C.; Zhang, Z. Non-orthogonal multiple access with cooperative full-duplex relaying. *IEEE Commun. Lett.* **2016**, *20*, 2478–2481. [CrossRef]
13. Hedayati, M.; Kim, I.-M. On the Performance of OMA and NOMA in the Two-user SWIPT System. *IEEE Trans. Veh. Technol.* **2018**, *67*, 11258–11263. [CrossRef]
14. Wu, Z.; Lu, K.; Jiang, C.; Shao, X. Comprehensive Study and Comparison on 5G NOMA Schemes. *IEEE Access* **2018**, *6*, 18511–18519. [CrossRef]
15. Ding, Z.; Yang, Z.; Fan, P.; Poor, H.V. On the performance of non-orthogonal multiple access in 5G systems with randomly deployed users. *IEEE Signal Process. Lett.* **2014**, *21*, 1501–1505. [CrossRef]
16. Shi, S.; Yang, L.; Zhu, H. Outage balancing in downlink non-orthogonal multiple access with statistical channel state information. *IEEE Trans. Wirel. Commun.* **2016**, *15*, 4718–4731. [CrossRef]

17. Cui, J.; Liu, Y.; Ding, Z.; Fan, P.; Nallanathan, A. Optimal user scheduling and power allocation for millimeter wave noma systems. *IEEE Trans. Wirel. Commun.* **2018**, *17*, 1502–1517. [CrossRef]

18. Al-Imari, M.; Xiao, P.; Imran, M.A.; Tafazolli, R. Uplink non-orthogonal multiple access for 5G wireless networks. In Proceedings of the 2014 11th International Symposium on Wireless Communications Systems (ISWCS), Barcelona, Spain, 26–29 August 2014; pp. 781–785.

19. Ding, Z.; Peng, M.; Poor, H.V. Cooperative non-orthogonal multiple access in 5G systems. *IEEE Commun. Lett.* **2015**, *19*, 1462–1465. [CrossRef]

20. Wei, Z.; Dai, L.; Ng, D.W.K.; Yuan, J. Performance analysis of a hybrid downlink-uplink cooperative NOMA scheme. In Proceedings of the IEEE 86th Vehicular Technology Conference (VTC Fall), Toronto, ON, Canada, 24–27 September 2017; pp. 1–7.

21. Marshoud, H.; Kapinas, V.M.; Karagiannidis, G.K.; Muhaidat, S. Non-orthogonalmultiple access for visible light communications. *IEEE Photonics Technol. Lett.* **2016**, *28*, 51–54. [CrossRef]

22. Choi, J. Minimum power multicast beamforming with superposition coding for multiresolution broadcast and application to NOMA systems. *IEEE Trans. Commun.* **2015**, *63*, 791–800. [CrossRef]

23. Hanif, M.F.; Ding, Z.; Ratnarajah, T.; Karagiannidis, G.K. A minorization-maximization method for optimizing sum rate in the downlink of non-orthogonal multiple access systems. *IEEE Trans. Signal Process.* **2016**, *64*, 76–88. [CrossRef]

24. Sun, Q.; Han, S.; I, C.-L.; Pan, Z. On the ergodic capacity of MIMO NOMA systems. *IEEE Wirel. Commun. Lett.* **2015**, *4*, 405–408. [CrossRef]

25. Nguyen, X.-X.; Do, D.-T. Maximum Harvested Energy Policy in Full-Duplex Relaying Networks with SWIPT. *Int. J. Commun. Syst. (Wiley)* **2017**, *30*, e3359. [CrossRef]

26. Do, D.-T.; Nguyen, H.S.; Voznak, M.; Nguyen, T.S. Wireless powered relaying networks under imperfect channel state information: System performance and optimal policy for instantaneous rate. *Radioengineering* **2017**, *26*, 869–877. [CrossRef]

27. Nguyen, X.-X.; Do, D.-T. Optimal power allocation and throughput performance of full-duplex DF relaying networks with wireless power transfer-aware channel. *EURASIP J. Wirel. Commun. Netw.* **2017**, *2017*, 152. [CrossRef]

28. Nguyen, T.N.; Do, D.-T.; Tran, P.T.; Vozňák, M. Time Switching for Wireless Communications with Full-Duplex Relaying in Imperfect CSI Condition. *KSII Trans. Internet Inf. Syst.* **2016**, *10*, 4223–4239.

29. Nguyen, H.-S.; Do, D.-T.; Nguyen, T.-S.; Voznak, M. Exploiting hybrid time switching-based and power splitting-based relaying protocol in wireless powered communication networks with outdated channel state information. *Automatika* **2017**, *58*, 391–399. [CrossRef]

30. Nguyen, T.-L.; Do, D.T. Exploiting Impacts of Intercell Interference on SWIPT-assisted Non-orthogonal Multiple Access. *Wirel. Commun. Mob. Comput.* **2018**, *2018*, 2525492. [CrossRef]

31. Do, D.-T.; Nguyen, H.S. A Tractable Approach to Analyze the Energy-Aware Two-way Relaying Networks in Presence of Co-channel Interference. *EURASIP J. Wirel. Commun. Netw.* **2016**, *2016*, 271. [CrossRef]

32. Xiao, Y.; Hao, L.; Ma, Z.; Ding, Z.; Zhang, Z.; Fan, P. Forwarding strategy selection in dual-hop NOMA relaying systems. *IEEE Commun. Lett.* **2018**, *22*, 1644–1647. [CrossRef]

33. Wang, Y.; Wu, Y.; Zhou, F.; Wu, Y.; Chu, Z.; Wang, Y. Multi-Objective Resource Allocation in a NOMA Cognitive Radio Network With a Practical Non-Linear Energy Harvesting Model. In Proceedings of the 2017 9th International Conference on Wireless Communications and Signal Processing (WCSP), Nanjing, China, 11–13 October 2017; pp. 1–6.

34. Xie, X.; Chen, J.; Fu, Y. Outage Performance and QoS Optimization in Full-duplex System with Non-linear Energy Harvesting Model. *IEEE Access* **2018**, *6*, 44281–44290. [CrossRef]

35. Zhang, J.; Pan, G. Outage analysis of wireless-powered relaying MIMO systems with non-linear energy harvesters and imperfect CSI. *IEEE Access* **2016**, *4*, 7046–7053. [CrossRef]

36. Ding, Z.; Dai, H.; Poor, H.V. Relay selection for cooperative NOMA. *IEEE Commun. Lett.* **2016**, *5*, 416–419. [CrossRef]

37. Chaudhry, M.A.; Zubair, S.M. Generalized incomplete gamma functions with applications. *J. Comput. Appl. Math.* **1994**, *55*, 99–123. [CrossRef]

38. Zheng, L.; Tse, D.N.C. Diversity and multiplexing: A fundamental tradeoff in multiple-antenna channels. *IEEE Trans. Inf. Theory* **2003**, *49*, 1073–1096. [CrossRef]

39.   Hildebrand, E. *Introduction to Numerical Analysis*; Dover: New York, NY, USA, 1987.
40.   Jeffrey, A.; Zwillinger, D. *Table of Integrals, Series, and Products*; Academic Press: San Diego, CA, USA, 2007.

 *electronics*

Article

Low-Profile Frequency Reconfigurable Antenna for Heterogeneous Wireless Systems

Amjad Iqbal [1,], **Amor Smida** [2,3], **Lway Faisal Abdulrazak** [4], **Omar A. Saraereh** [5], **Nazih Khaddaj Mallat** [6], **Issa Elfergani** [7] **and Sunghwan Kim** [8,*]

[1] Centre For Wireless Technology, Faculty of Engineering, Multimedia University Cyberjaya, Selangor 63100, Malaysia
[2] Department of Medical Equipment Technology, College of Applied Medical Sciences, Majmaah University, AlMajmaah 11952, Saudi Arabia
[3] Unit of Research in High Frequency Electronic Circuits and Systems, Faculty of Mathematical, Physical and Natural Sciences of Tunis, Tunis El Manar University, Tunis 2092, Tunisia
[4] Computer Science Department, Cihan University Slemani, Sulaimaniya 46002, Iraq
[5] Department of Electrical Engineering, Hashemite University, Zarqa 13115, Jordan
[6] College of Engineering, Al Ain University (AAU), Al Ain 64141, UAE
[7] Instituto de Telecomunicações, Campus Universitário de Santiago, 3810-193 Aveiro, Portugal
[8] School of Electrical Engineering, University of Ulsan, Ulsan 44610, Korea
* Correspondence: sungkim@ulsan.ac.kr; Tel.: +82-52-259-1401

Received: 5 August 2019; Accepted: 30 August 2019; Published: 31 August 2019

Abstract: A low-profile ($0.21\lambda_g \times 0.35\lambda_g \times 0.02\lambda_g$) and a simply-structured frequency-switchable antenna with eight frequency choices is presented in this paper. The radiating structure (monopole) is printed on a 1.6-mm thicker, commercially-available substrate of FR-4 ($\epsilon_r = 4.4$, $\tan\delta = 0.020$). Specifically, it uses three PIN diodes in the designated places to shift the resonant bands of the antenna. The antenna operates at four different modes depending on the ON and OFF states of the PIN diodes. While in each mode, the antenna covers two unique frequencies (Mode 1 = 1.8 and 3.29 GHz, Mode 2 = 2.23 and 3.9 GHz, Mode 3 = 2.4 and 4.55 GHz, and Mode 4 = 2.78 and 5.54 GHz). The performance results show that the proposed antenna scheme explores significant gain (>1.5 dBi in all modes) and reasonable efficiency (>82% in all modes) for each mode. Using a high-frequency structure simulator (HFSS), the switchable antenna is designed and optimized. The fabricated model along with the PIN diode and biasing network is tested experimentally to validate the simulation results. The proposed antenna may also be combined in compact and heterogeneous radio frequency (RF) front-ends because of its small geometry and efficient utilization of the frequency spectrum.

Keywords: monopole antenna; S-parameters; frequency reconfigurable; 5G, 4/4.5G; LTE; ISM; WiFi; WiMAX; WLAN

1. Introduction

Recent advancements in wireless communication systems need transceivers to be efficiently utilized with heterogeneous systems such as wireless local area network (WLAN), Worldwide Interoperability for Microwave Access (WiMAX), Long-Term Evolution (LTE), Fourth-Generation (4G), LTE Advanced Pro (4.5G), Fifth-Generation (5G), and many more [1]. One of the important components of the transceiver is antenna. The antenna being an important component of the transceiver occupies a major portion of the system [2,3]. The size increases further by incorporating any tuning mechanism in it. Since, the antenna required for modern transceivers must be compact and frequency switchable to meet the standard of the transceivers. In this regards, many approaches have been used to reduce the size of the antenna and to make it tunable for many useful applications.

The antenna which is capable of switching its resonant band according to the user requirement, is known as a frequency-reconfigurable antenna. Various techniques have been adopted to change the surface current distribution of the antenna, which produces a highly-reconfigurable antennas. In [4,5], the lumped element between the monopole and the parasitic patch is used to switch the antenna from a single band to a dual band. A compact-sized frequency-reconfigurable antenna for mobile handsets is reported in [6]. The reported antenna in [6] covers Long-Term Evolution (LTE) LTE700, LTE2300, LTE2500, Global System for Mobile (GSM) GSM850, GSM900, GSM1800, GSM1900, and Universal Mobile Telecommunications System (UMTS) (1920–2170 MHz) bands for different switching conditions. A minimally-sized hexagonally-shaped reconfigurable slot antenna for WLAN is reported in [7]. The reported antenna covers dual bands (2.4 and 5.2 GHz) and a single band (2.4 GHz) for the PIN diode in the ON and OFF states, respectively. Two different types of frequency-reconfigurable antennas are reported in [8] for multiband applications. The reported antenna covers three bands: the Wireless Fidelity (WiFi) band, the Worldwide Interoperability for Microwave Access (WiMAX) band, and the WLAN band, for different biasing conditions of PIN diodes. Furthermore, a frequency-reconfigurable implantable antenna for the Medical Implant Communications Services (MICS: 402–405 MHz) and Industrial, Scientific, and Medical (ISM: 902–928 MHz) band is examined in [9]. Micro-electromechanical switches (MEMSs) are used in [10] to reconfigure the operating band of the antenna. However, the reported antenna cannot be efficiently used with any planner and compact components due to its large dimensions and 3D geometry. Following that, a polarization and frequency-reconfigurable antenna is reported in [11]. The optical switches are inserted between the main monopole and the parasitic patch in [12] to achieve different frequency responses with a wider bandwidth and high gain at the cost of a large antenna size. In [13], the PIN diode on the multi-layer antenna, control the patch and feeding lengths, generating two unique frequencies in the WLAN and 5G bands. In [14], two varactor diodes are connected with an F-shaped feed to reconfigure the operating band continuously in a wideband filtering patch antenna. Moreover, the hexa-band frequency-reconfigurable antenna for multi-standard wireless communication systems using PIN diodes is investigated in [15–17]. A fluidic channel-based frequency-reconfigurable monopole antenna is proposed in [18], which operates in three modes depending on the filled channels.

In the challenging world of communication, the size and performance of the reconfigurable antenna are important factors to be considered before using it in the RF front-end. So far, many antennas have been designed for reconfiguration of the frequency bands. However, their large dimensions and limited operational bands limit its applications in modern RF front-ends where limited space is available for the antennas. Therefore, the proposed antenna is designed with lower dimensions ($0.21\lambda_g \times 0.35\lambda_g \times 0.02\lambda_g$) and multiple frequency bands with high gain and efficiency for each operating band. The organization of the rest of the paper is as follows: Section 2 presents the antenna design equations and theoretical analysis and simulation results for each mode. Fabricated results and discussions, as well as a comparison with the state-of-the-art reconfigurable antennas are presented in Section 3.

2. Antenna Design Methodology

The geometry (top view, bottom view, and side view) of the proposed frequency-switchable monopole antenna is illustrated in Figure 1. The proposed frequency-switchable monopole antenna was printed on a 1.6-mm thicker, commercially-available substrate of FR-4 ($\epsilon_r = 4.4$, and $\tan\delta = 0.020$). The main radiating part of the antenna consisted of a hook-shaped monopole with three parasitic patches connected to it through three PIN diodes. A gap of 1 mm was kept between the parasitic patches to install the PIN diodes. The overall geometry of the proposed antenna was $0.21\lambda_g \times 0.35\lambda_g \times 0.02\lambda_g$. The effective length of the monopole was calculated using transmission model theory [19]. The effective length of the monopole was calculated using Equation (1).

$$L_{f_r} = \frac{c}{4f_r\sqrt{\epsilon_{eff}}} \tag{1}$$

and:

$$\epsilon_{eff} \approx \frac{\epsilon_r + 1}{2} + \frac{\epsilon_r - 1}{2}(1 + 12(\frac{w}{h}))^{-0.5} \tag{2}$$

where c is the speed of light in a vacuum, λ_g is the guided wavelength, ϵ_{eff} is the effective dielectric constant, w is the width of the substrate, and h is the thickness of the substrate.

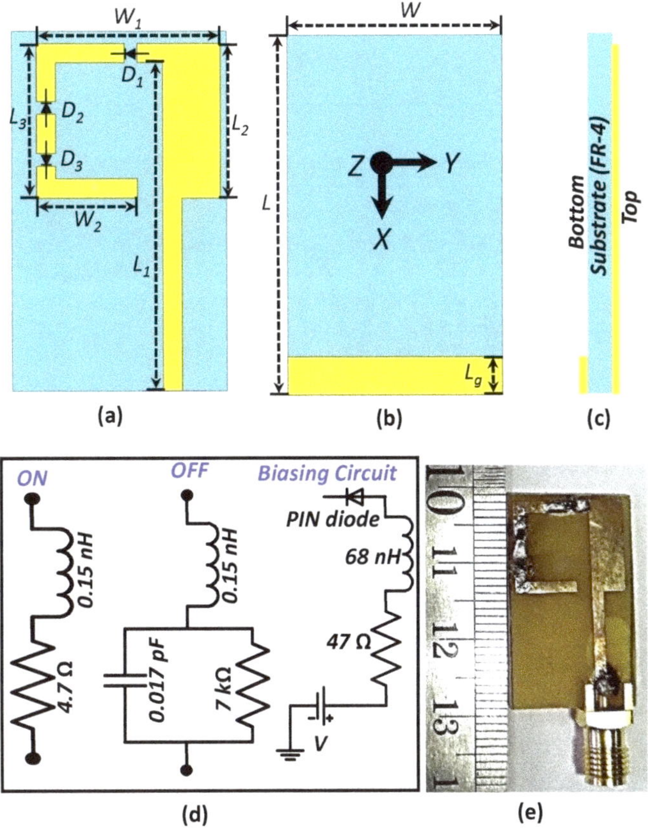

Figure 1. Geometry of the proposed antenna: (**a**) top view (W_1 = 14.5 mm, L_2 = 12 mm, W_2 = 8 mm, L_3 = 12 mm, and L_1 = 25.5 mm); (**b**) bottom view (W = 17 mm, L = 28 mm, and L_g = 2 mm); (**c**) side view, (**d**) equivalent circuit model of the PIN diode in the ON, OFF state and the biasing circuit, and (**e**) fabricated prototype.

2.1. Frequency Reconfigurability

The frequency reconfigurability in the antenna is obtained by adjusting the ON and OFF states of the PIN diodes used between the corresponding parasitic patches. The proposed antenna operates at four different modes. The states of the PIN diodes at each mode are summarized in Table 1. The proposed antenna operates at two unique bands for each operating mode. The antenna operates at Mode 1 when the PIN diode (D_1, D_2, and D_3) are in the ON state. The proposed antenna scheme operates at 1.8 GHz (GSM band) and 3.3 GHz (5G sub-6 GHz band) for operating mode 1, as illustrated in Figure 2. The same antenna operates at 2.25 GHz (3G advanced/LTE band) and 3.9 GHz (5G sub-6 GHz band), when it is operated in Mode 2 (D_1 = ON, D_2 = ON, and D_3 = OFF). The antenna covers 2.4 GHz (WiFi/WLAN/ISM/Bluetooth band) and 4.5 GHz (5G sub-6 GHz band) upon operating with Mode 3

(D_1 = ON, D_2 = OFF, and D_3 = OFF). The same antenna operates at 2.78 GHz (Airport surveillance radar band) and 5.54 GHz (WLAN band) when operated in Mode 4 (D_1 = OFF, D_2 = OFF, and D_3 = OFF).

Table 1. States of the PIN diodes for various operating modes.

	D_1	D_2	D_3	**Resonant Bands**
Mode 1	ON	ON	ON	1.8 and 3.3 GHz
Mode 2	ON	ON	OFF	2.25 and 3.9 GHz
Mode 3	ON	OFF	OFF	2.4 and 4.5 GHz
Mode 4	OFF	OFF	OFF	2.78 and 5.54 GHz

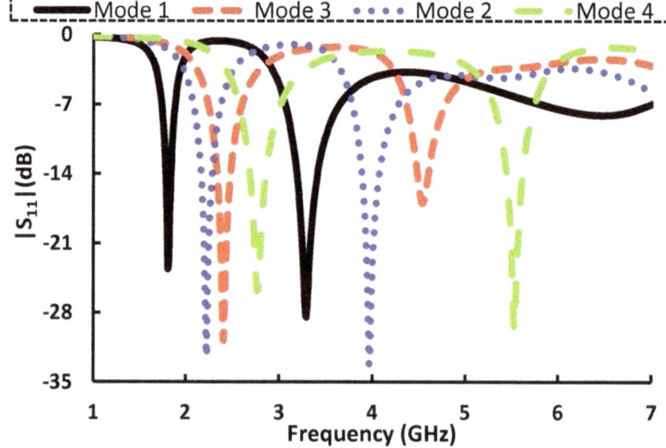

Figure 2. Reflection coefficient of the antenna for different operating modes.

2.2. Switching Technique

In this arrangement, the PIN diode acts as a variable resistor in the RF frequency range. The PIN diode Model HPND-4005 (640 μm $\times$ 220 μm) was used for frequency reconfiguration of the antenna. The equivalent circuit model of the PIN diode in the ON and OFF state is shown in Figure 1d [20]. The equivalent circuit model of the PIN diode is the series connection of the inductor (L = 0.15 nH) and low resistance resistor (R = 4.7 Ω) for the ON state of the PIN diode. The equivalent circuit model for OFF state of the diode was modeled as a parallel combination of the capacitor (C = 0.017 pF) and high-resistance resistor (R = 7 kΩ) in series connection with an inductor (L = 0.15 nH). For simplicity, the reconfigurability of the antenna was studied only in terms of the resistance, based on the concept that the PIN diode acts as an open circuit for a high resistor value and as a closed circuit for a lower value of the resistor.

3. Results and Discussions

The fabricated prototype of the antenna is shown in Figure 1e. Our simulation results are compared with the results obtained from the fabricated prototype. PIN diode Model HPND-4005 having dimensions of 640 μm $\times$ 220 μm was used for frequency reconfigurability in the antenna model. The proposed diode offered fewer insertion losses (0.4 dB) due to its small geometry and a small capacitance value (C = 0.017 pF). Conductive epoxy was used to mount the diodes on the antenna surface. Figure 1d shows the biasing circuitry of the antenna. It is illustrated from the biasing circuit that the PIN diode was excited through a current-limiting resistor (R = 47 Ω) and radio frequency (RF) chock (L = 68 nH).

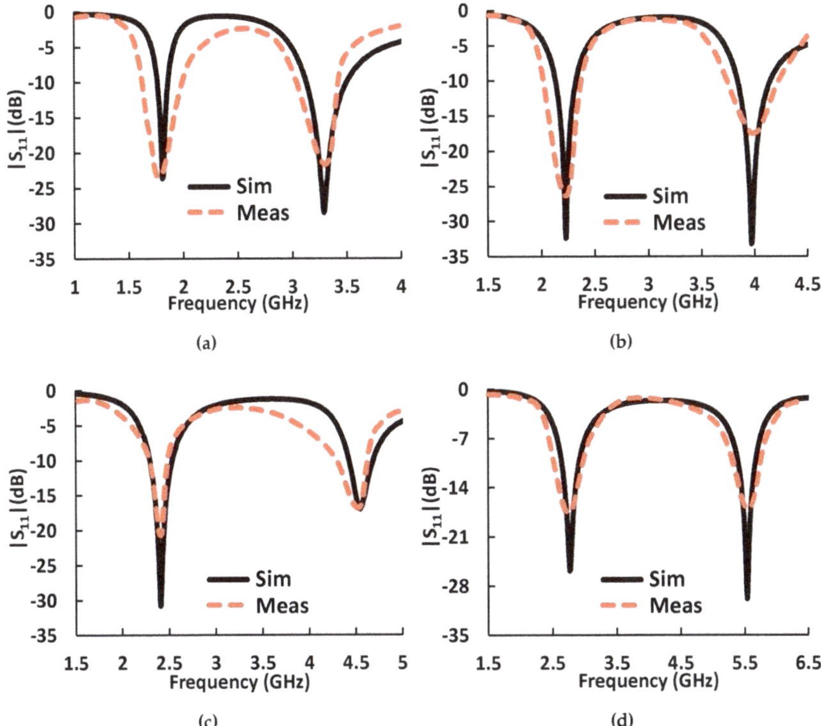

Figure 3. Simulated (Sim) and measured (Meas) reflection coefficient of the antenna: (**a**) Mode 1; (**b**) Mode 2; (**c**) Mode 3; and (**d**) Mode 4.

3.1. Reflection Coefficient

The reflection coefficient of the antenna was measured using a network analyzer for four operating modes of the antenna. The simulated and measured results for all operating modes of the antenna are compared in Figure 3. Figure 3a shows the reflection coefficient of the antenna when operated in Mode 1. The proposed antenna in Mode 1 operated at 1.8 and 3.3 GHz. The simulated 10-dB bandwidth of the antenna was 120 MHz (1.76–1.88 GHz) and 390 MHz (3.12–3.51 GHz) at the lower and higher resonant bands, respectively, for operating Mode 1. The measured 10-dB bandwidth for Mode 1 was recorded as 400 MHz (1.6–2 GHz) and 330 MHz (3.08–3.41 GHz) for the lower and upper operating bands, respectively. The same antenna operated at 2.25 and 3.9 GHz with different switching conditions (Mode 2). The simulated 10-dB bandwidth of 200 MHz (2.13–2.33 GHz) and 320 MHz (3.84–4.16 GHz) was quantified for lower and higher operating bands, respectively. The measured 10-dB bandwidth for Mode 2 was investigated as 300 MHz (2.05–2.35 GHz) and 420 MHz (3.78–4.2 GHz) for the lower and upper operating bands, respectively. In Mode 3, the antenna had resonant frequencies of 2.4 and 4.5 GHz with a simulated 10-dB bandwidth of 250 MHz (2.29–2.54 GHz) and 270 MHz (4.43–4.7 GHz). The measured 10-dB bandwidth for Mode 3 was recorded as 190 MHz (2.29–2.48 GHz) and 280 MHz (4.32–4.6 GHz) for the lower and upper operating bands, respectively. In Mode 4, the antenna operated at 2.78 and 5.54 GHz with a simulated 10-dB bandwidth of 310 MHz (2.63–2.94 GHz) and 320 MHz (5.37–5.69 GHz). The measured 10-dB bandwidth for Mode 4 was investigated as 350 MHz (2.6–2.95 GHz) and 360 MHz (5.35–5.71 GHz) for the lower and upper operating bands, respectively. The designed antenna can be used for different applications (GSM, 5G sub-6-GHz band, 3G Advanced/LTE, WiFi, WLAN, ISM, Bluetooth, and airport surveillance radars band) with different operating modes (Mode 1, Mode 2, Mode 3, and Mode 4).

3.2. Surface Current Distribution

The surface current distribution of each mode for their respective operating bands (Mode 1: 1.8 and 3.3 GHz, Mode 2: 2.25 and 3.9 GHz, Mode 3: 2.4 and 4.5 GHz, Mode 4: 2.78 and 5.54 GHz) is plotted in Figure 4. The surface current was distributed uniquely on the antenna surface for different operating bands. It is evident from the plots that the effective length was inversely proportional to the operating frequencies.

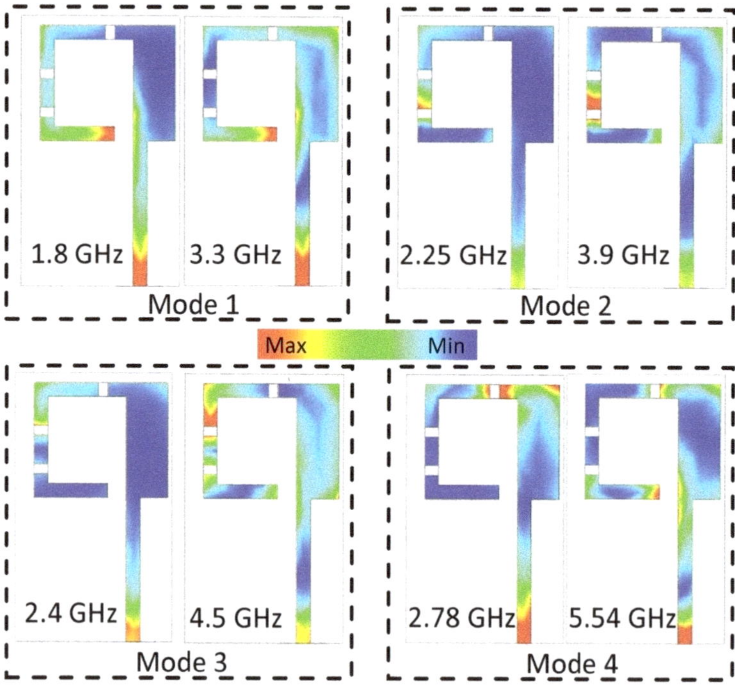

Figure 4. Surface current distribution of the antenna for different operating modes.

3.3. Gain and Efficiency

The simulated and measured antenna gains and efficiencies are presented in Figure 5. In Mode 1, the antenna's simulated (measured) gain was >1.5 dBi (>1.45 dBi) and >1.76 dBi (>1.72 dBi) for the lower and upper operating bands, respectively. The simulated (measured) antenna's efficiency in operating Mode 1 was >83% (>83.2%) and >82% (>82.1%) for the lower and upper bands, respectively. The simulated (measured) gain of the antenna in operating Mode 2 was >3.16 dBi (>3.14 dBi) and >3.22 dBi (>3.18 dBi) for the respective lower and upper bands. In Mode 2, the antenna's simulated (measured) efficiency was >88% (>87.5%) and >88% (>88.1%) for the lower and upper operating bands, respectively. The simulated (measured) antenna's efficiency in operating Mode 3 was >2.89 dBi (>2.85 dBi) and >2.75 dBi (>2.76 dBi) for the lower and upper bands, respectively. In Mode 3, the antenna's simulated (measured) efficiency was >88% (>87.5%) and >87% (>86.6%) for the lower and upper operating bands, respectively. The simulated (measured) antenna's efficiency in operating Mode 4 was >3.5 dBi (>3.48 dBi) and >3.79 dBi (>3.69 dBi) for the lower and upper bands, respectively. In Mode 4, the antenna's simulated (measured) efficiency was >88.3% (>88%) and >87% (>86.2%) for the lower and upper operating bands, respectively.

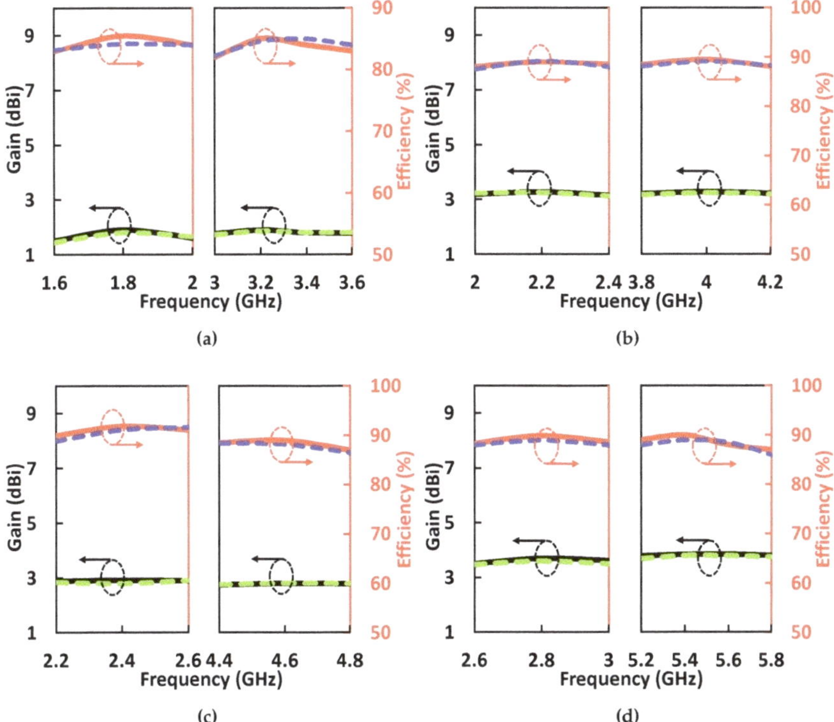

Figure 5. Gain and efficiency of the antenna (solid line = simulated, and dashed line = measured):
(**a**) Mode 1; (**b**) Mode 2; (**c**) Mode 3; and (**d**) Mode 4.

3.4. Radiation Pattern

The antenna's radiation characteristics were analyzed using 2D and 3D radiation patterns for each resonant frequency of the four operational modes. The 2D radiation pattern of the antenna on both principal planes ($\phi = 0°$ and $\phi = 90°$) was measured in an anechoic chamber. The simulated and measured 2D radiation pattern is illustrated in Figure 6. The proposed antenna had an omni-directional radiation pattern for a lower operating band at Phi = 0° in all four operating modes. At Phi = 90°, the antenna had the same radiation pattern as a figure of eight for all operating modes. For the upper operating band at Phi = 0°, the radiation pattern was omni-directional for Mode 1; however, the radiation pattern was distorted in Mode 2, Mode 3, and Mode 4. The 3D pattern of the antenna is illustrated in Figure 7.

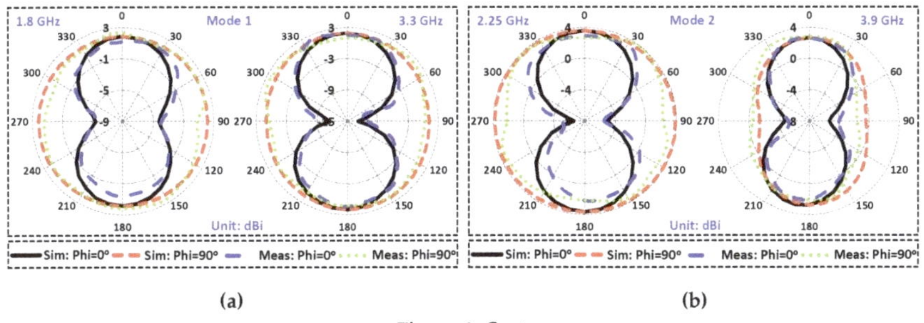

(**a**) (**b**)

Figure 6. *Cont.*

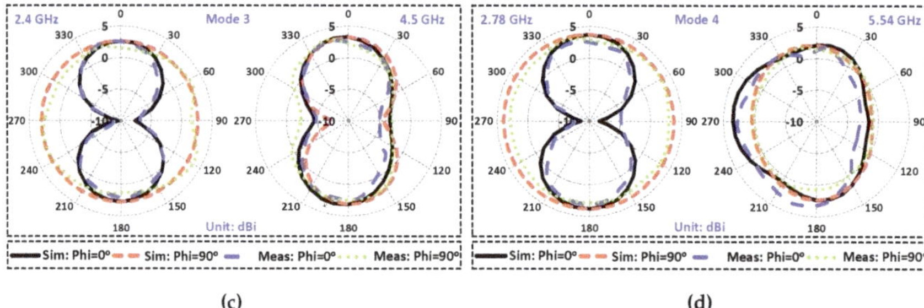

(c)   (d)

Figure 6. Simulated and measured radiation pattern at resonant frequencies (at Phi = 0° and at Phi = 90°) of the antenna: (**a**) Mode 1; (**b**) Mode 2; (**c**) Mode 3; and (**d**) Mode 4.

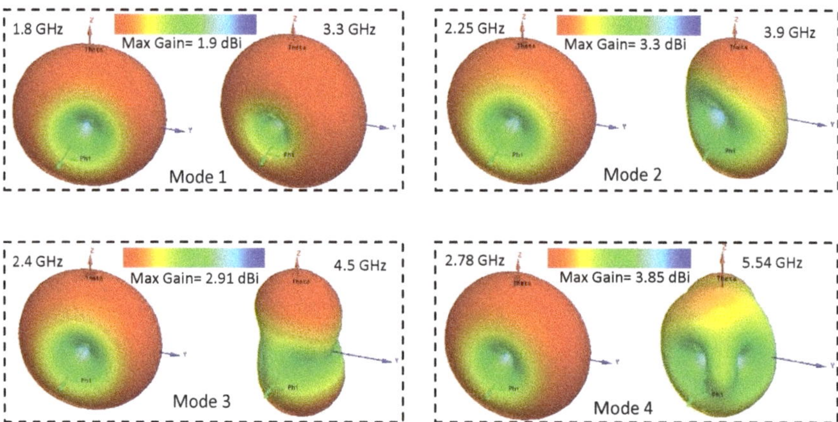

Figure 7. 3D pattern of the antenna for all operating bands.

3.5. Specific Absorption Rate Analysis

The proposed antenna covered different frequency bands such as GSM, 5G, 3G Advanced/LTE, WiFi, WLAN, ISM, and Bluetooth. The specific absorption rate (SAR) analysis was important for the proposed antenna, because most of the portable devices use these frequency bands. The SAR analysis was performed by placing the antenna on the three layers (skin, fat, and muscle) of a flat human model. The gap between the antenna and skin was kept as 4 mm. According to the International Commission of Non-Ionizing Radiation Protection (ICNIRP), the SAR value should not exceed 1.6 W/kg for one-gram and 2 W/kg for 10-gram standards [21,22]. Peak SAR values (10 grams) of 9.37, 8.99, 7.93, 5.98, 7.4, 7.11, 5.33, and 4.7 W/kg were noted for 1.8, 2.25, 2.4, 2.78, 3.3, 3.9, 4.5, and 5.54 GHz, respectively, for an input power of 1 W, as shown in Figure 8. The designed antenna exceeded the SAR limits (2 W/kg) for all operating bands using an input power of 1 W. However, many portable devices use power in the mW range [23]. Based on the above calculated SAR value, our antenna is safe if the input power is less than 213, 222, 252, 334, 270, 281, 375, and 425 mW for 1.8, 2.25, 2.4, 2.78, 3.3, 3.9, 4.5, and 5.54 GHz respectively. Hence, it is not recommended to install it directly into any portable device. An SAR reduction technique must be used before employing it in portable devices if the input power is more than the limit provided above.

An overall summary of the proposed antenna is presented in Table 2. Table 3 compares the performance of the proposed antenna with the state-of-the-art antennas. It can be noticed that the proposed antenna had smaller dimensions than the antennas presented in the literature. Furthermore, the proposed antenna used a fewer number of switches, which significantly reduce the system complexity and enhance the power efficiency. Additionally, the antenna had reasonably good gain and efficiency in all operating bands.

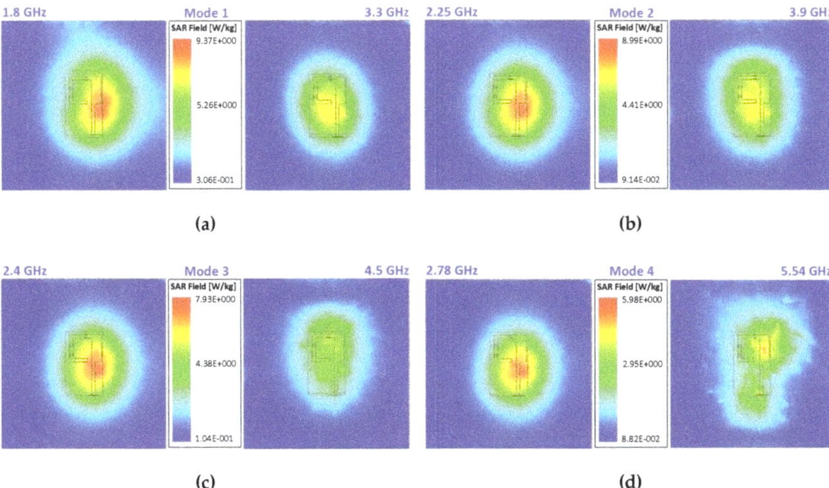

Figure 8. Simulated specific absorption rate (SAR) of the antenna: (**a**) Mode 1; (**b**) Mode 2; (**c**) Mode 3; and (**d**) Mode 4.

Table 2. Summary of the simulated antenna under different operating modes.

Parameters	Mode 1		Mode 2		Mode 3		Mode 4	
Frequencies (GHz)	1.8	3.3	2.25	3.9	2.4	4.5	2.78	5.54
Bandwidth (MHz)	120	390	200	320	250	270	310	320
VSWR	1.13	1.06	1.04	1.04	1.05	1.43	1.09	1.05
Gain (dBi)	>1.5	>1.76	>3.16	>3.22	>2.89	>2.75	>3.5	>3.79
Efficiency (%)	>83	>82	>88	>88	>88	>87	>88.3	>87
Peak SAR (W/kg)	9.37	7.4	8.99	7.11	7.93	5.33	5.98	4.7

Table 3. Comparison of the proposed antenna with state-of-the-art reconfigurable antennas.

Ref. []	Size $\lambda_g \times \lambda_g \times \lambda_g$	No. of Switches	No. of Bands	Operational Bands	Gain (dBi)	Efficiency (%)
[4]	0.62 × 0.65 × 0.02	1 PIN diode	3	(1.8–2.7 GHz), (2.49–3.84 GHz), (5.26–5.99 GHz)	NG	NG
[5]	0.62 × 0.65 × 0.02	1 PIN Diode	3	(2.06–3.14 GHz), (2.44–3.66 GHz), (5.11–5.66 GHz)	>1.2	>90
[6]	0.24 × 0.59 × 0.02	2 PIN diodes	8	(698–787 MHz), (2305–2400 MHz), (2500-2690 MHz), (824–894 MHz), (880–960 MHz), (1710–1880 MHz), (1850–1990 MHz), (1920–2170 MHz)	0.13–1.59	52.83–75.6
[7]	0.21 × 0.12 × 0.02	2 PIN diodes	4	(2.31–2.62 GHz), (5.13–5.32 GHz), (2.32–2.61 GHz), (5.12–5.33 GHz)	2.91–3.13	NG

Table 3. *Cont.*

Ref. []	Size $\lambda_g \times \lambda_g \times \lambda_g$	No. of Switches	No. of Bands	Operational Bands	Gain (dBi)	Efficiency (%)
[8]	0.58 × 0.88 × 0.02	1 PIN diode	3	3.5 %, 35.72 %, and 9.94 %	1.7–3.4	85–90
[12]	0.33 × 0.67 × 0.02	1 PIN diode	4	36 %, 15 %, 31 % and 22 %	1.76–2.91	76.43–84.2
[13]	1.68 × 1.68 × 0.09	2 PIN diodes	2	(2.37–2.67 GHz), (3.39–3.62 GHz)	6.51, 7.64	64.5, 69.5
[15]	0.22 × 0.46 × 0.02	3 PIN diodes	6	9.35, 13.2, 21.49, 11.71, 41.5 and 9.35%	1.85–3.46	80.41–96.75
[16]	0.58 × 0.67 × 0.02	2 PIN diodes	6	430, 1090, 1045, 2210, 1125, and 847 MHz	2.18–4.46	90–97
[17]	0.58 × 0.67 × 0.02	2 PIN diodes	6	1220, 335, 512, 1020, 486, and 463 MHz	2.20–4.01	92.5–97
This Work	0.21 × 0.35 × 0.02	3 PIN diodes	8	(1.76–1.88 GHz), (3.12–3.51 GHz), (2.13–2.33 GHz), (3.84–4.16 GHz), (2.29–2.54 GHz), (4.43–4.7 GHz), (2.63–2.94 GHz), (5.37–5.69 GHz)	1.5–3.85	82–89

λ_g = guided wavelength at the lower resonant frequency, NG = not given

4. Conclusions

A low-profile ($0.21\lambda_g \times 0.35\lambda_g \times 0.02\lambda_g$) and a simply-structured frequency-switchable antenna with eight frequency choices was presented in this paper. The antenna operated in four different modes depending on the ON and OFF states of the PIN diodes. In each mode, the antenna covered two unique frequencies (Mode 1 = 1.8 and 3.29 GHz, Mode 2 = 2.23 and 3.9 GHz, Mode 3 = 2.4 and 4.55 GHz, and Mode 4 = 2.78 and 5.54 GHz). The proposed antenna achieved high gain and reasonable efficiency in each mode. A fabricated model was tested experimentally to validate the simulation results. The proposed antenna can have a significant impact on compact and heterogeneous RF front-ends, because of its small geometry and efficient utilization of the frequency spectrum.

Author Contributions: Conceptualization, A.I.; methodology, A.I.; software, A.I.; validation, A.I., I.E. and A.S.; formal analysis, A.I.; investigation, A.I.; resources, L.F.A., O.A.S., S.K. and N.K.M.; data curation, A.I., N.K.M., I.E.; writing—original draft preparation, A.I.; writing—review and editing, A.I., A.S., L.F.A., I.E., O.A.S., S.K. and N.K.M.; visualization, A.I., S.K.; supervision, A.S., L.F.A., I.E., O.A.S., S.K. and N.K.M.; project administration, A.I., S.K.; funding acquisition, S.K.

Funding: Following are results of a study of the "Leaders in INdustry-university Cooperation +" Project, supported by the Ministry of Education and National Research Foundation of Korea. The authors extend their appreciation to the Deanship of Scientific Research at Majmaah University for funding this work under Project No. (RGP-2019-32).

Conflicts of Interest: The authors declare no conflicts of interest.

References

1. Elfergani, I.; Hussaini, A.S.; Rodriguez, J.; Abd-Alhameed, R. *Antenna Fundamentals for Legacy Mobile Applications and Beyond*; Springer: Bolingbrook, IL, USA, 2018.
2. Kunwar, A.; Gautam, A.K. Fork-shaped planar antenna for Bluetooth, WLAN, and WiMAX applications. *Int. J. Microw. Wirel. Technol.* **2017**, *9*, 859–864. [CrossRef]
3. Kunwar, A.; Gautam, A.K.; Kanaujia, B.K. Inverted L-slot triple-band antenna with defected ground structure for WLAN and WiMAX applications. *Int. J. Microw. Wirel. Technol.* **2017**, *9*, 191–196. [CrossRef]

4. Iqbal, A.; Saraereh, O.A. A compact frequency reconfigurable monopole antenna for Wi-Fi/WLAN applications. *Prog. Electromagnet. Res.* **2017**, *68*, 79–84.

5. Iqbal, A.; Ullah, S.; Naeem, U.; Basir, A.; Ali, U. Design, fabrication and measurement of a compact, frequency reconfigurable, modified T-shape planar antenna for portable applications. *J. Electr. Eng. Technol.* **2017**, *12*, 1611–1618.

6. Lee, S.; Sung, Y. Compact frequency reconfigurable antenna for LTE/WWAN mobile handset applications. *IEEE Trans. Antennas Propag.* **2015**, *63*, 4572–4577. [CrossRef]

7. Ali, T.; Biradar, R.C. A compact hexagonal slot dual band frequency reconfigurable antenna for WLAN applications. *Microw. Opt. Technol. Lett.* **2017**, *59*, 958–964. [CrossRef]

8. Ullah, S.; Hayat, S.; Umar, A.; Ali, U.; Tahir, F.A.; Flint, J.A. Design, fabrication and measurement of triple band frequency reconfigurable antennas for portable wireless communications. *AEU-Int. J. Electron. Commun.* **2017**, *81*, 236–242. [CrossRef]

9. Khan, O.M.; Raza, K.; Shubair, R.; Islam, Q.U. Frequency Reconfigurable Implant Antenna for MICS and ISM Band Applications. In Proceedings of the 18th International Symposium on Antenna Technology and Applied Electromagnetics (ANTEM). IEEE, Waterloo, ON, Canada,19–22 August 2018; pp. 1–2.

10. Ruvio, G.; Ammann, M.J.; Chen, Z.N. Wideband reconfigurable rolled planar monopole antenna. *IEEE Trans. Antennas Propag.* **2007**, *55*, 1760–1767. [CrossRef]

11. Bharathi, A.; Lakshminarayana, M.; Rao, P.S. A quad-polarization and frequency reconfigurable square ring slot loaded microstrip patch antenna for WLAN applications. *AEU-Int. J. Electron. Commun.* **2017**, *78*, 15–23. [CrossRef]

12. Shah, S.A.A.; Khan, M.F.; Ullah, S.; Basir, A.; Ali, U.; Naeem, U. Design and Measurement of Planar Monopole Antennas for Multi-Band Wireless Applications. *IETE J. Res.* **2017**, *63*, 194–204. [CrossRef]

13. Jin, G.P.; Deng, C.H.; Yang, J.; Xu, Y.C.; Liao, S.W. A New Differentially-Fed Frequency Reconfigurable Antenna for WLAN and Sub-6GHz 5G Applications. *IEEE Access* **2019**. [CrossRef]

14. Hu, P.F.; Pan, Y.M.; Zhang, X.Y.; Hu, B.J. A Filtering Patch Antenna With Reconfigurable Frequency and Bandwidth Using F-Shaped Probe. *IEEE Trans. Antennas Propag.* **2019**, *67*, 121–130. [CrossRef]

15. Shah, I.; Hayat, S.; Basir, A.; Zada, M.; Shah, S.; Ullah, S. Design and analysis of a hexa-band frequency reconfigurable antenna for wireless communication. *AEU-Int. J. Electron. Commun.* **2019**, *98*, 80–88. [CrossRef]

16. Ullah, S.; Ahmad, S.; Khan, B.; Ali, U.; Tahir, F.; Bashir, S. Design and Analysis of a Hexa-Band Frequency Reconfigurable Monopole Antenna. *IETE J. Res.* **2018**, *64*, 59–66. [CrossRef]

17. Ullah, S.; Ahmad, S.; Khan, B.A.; Flint, J.A. A multi-band switchable antenna for Wi-Fi, 3G Advanced, WiMAX, and WLAN wireless applications. *Int. J. Microw. Wirel. Technol.* **2018**, *10*, 991–997. [CrossRef]

18. Singh, A.; Goode, I.; Saavedra, C.E. A Multi-State Frequency Reconfigurable Monopole Antenna using Fluidic Channels. *IEEE Antennas Wirel. Propag. Lett.* **2019**. [CrossRef]

19. Balanis, C.A. *Antenna Theory: Analysis and Design*; John wiley & sons, 2016.

20. Iqbal, A.; Smida, A.; Mallat, N.K.; Ghayoula, R.; Elfergani, I.; Rodriguez, J.; Kim, S. Frequency and pattern reconfigurable antenna for emerging wireless communication systems. *Electronics* **2019**, *8*, 407. [CrossRef]

21. Bouazizi, A.; Zaibi, G.; Iqbal, A.; Basir, A.; Samet, M.; Kachouri, A. A dual-band case-printed planar inverted-F antenna design with independent resonance control for wearable short range telemetric systems. *Int. J. RF Microw. Comput.-Aided Eng.* **2019**, *29*, e21781. [CrossRef]

22. Basir, A.; Bouazizi, A.; Zada, M.; Iqbal, A.; Ullah, S.; Naeem, U. A dual-band implantable antenna with wide-band characteristics at MICS and ISM bands. *Microw. Opt. Technol. Lett* **2018**, *60*, 2944–2949. [CrossRef]

23. Iqbal, A.; Basir, A.; Smida, A.; Mallat, N.K.; Elfergani, I.; Rodriguez, J.; Kim, S. Electromagnetic Bandgap Backed Millimeter-Wave MIMO Antenna for Wearable Applications. *IEEE Access* **2019**, *7*, 111135–111140. [CrossRef]

Article

High-Performance Multiple-Input Multiple-Output Antenna System For 5G Mobile Terminals

Mujeeb Abdullah [1], Saad Hassan Kiani [2], Lway Faisal Abdulrazak [3], Amjad Iqbal [4,*], M. A. Bashir [5], Shafiullah Khan [6] and Sunghwan Kim [7,*]

[1] Department of Computer Science, Bacha Khan University, Charsadda 24420, Pakistan; mujeeb.abdullah@gmail.com

[2] Electrical Engineering Department, Iqra National University, Peshawar 25000, Pakistan; iam.kiani91@gmail.com

[3] Computer Science Department, Cihan University Slemani, Sulaimaniya 46002, Iraq; lway.faisal@sulicihan.edu.krd

[4] Centre For Wireless Technology, Faculty of Engineering, Multimedia University, Cyberjaya 63100, Malaysia

[5] School of Electronic Sicence and Engineering, University of Electronic Science Technology China, Chengdu 611731, China; adil.bashir@yahoo.com

[6] Department of Electronics, Islamia College University, Peshawar 25000, Pakistan; shafielectron@yahoo.com

[7] School of Electrical Engineering, University of Ulsan, Ulsan 44610, Korea

* Correspondence: aiqbal@ieee.org (A.I.); sungkim@ulsan.ac.kr (S.K.); Tel.: +82-52-259-1401 (S.K.)

Received: 25 August 2019; Accepted: 22 September 2019; Published: 25 September 2019

Abstract: In this paper, the systematic design of a multiple antenna system for 5G smartphone operating at 3.5 GHz for multiple-input multiple-output (MIMO) operation in smartphones is proposed. The smartphone is preferred to be lightweight, thin, and attractive, and as a result metal casings have become popular. Using conventional antennas, such as a patch antenna, Inverted-F antennas, or monopole, in proximity to metal casing leads to decreasing its total efficiency and bandwidth. Therefore, a slot antenna embedded in the metal casing can be helpful, with good performance regarding bandwidth and total efficiency. The proposed multiple antenna system adopted the unit open-end slot antenna fed by Inverted-L microstrip with tuning stub. The measured S-parameters results agree fairly with the numerical results. It attains 200 MHz bandwidth at 3.5 GHz with ports isolation of (≤ -13 dB) for any two antennas of the system. The influence of the customer's hand for the proposed multiple antenna system is also considered, and the MIMO channel capacity is computed. The maximum achievable MIMO channel capacity based on the measured result is 31.25 bps/Hz and is about 2.7 times of 2×2 MIMO operation.

Keywords: 5G antenna; slot antenna; mobile terminal antenna; MIMO antenna

---

1. Introduction

With evolving semiconductor technology, the electronic communication components can be easily packed closely to design compact structures with high processing capabilities, such as a smartphones and wireless routers [1]. Hence, multiple antenna structures use high processing capability efficiently and boost data throughput. As such, 2×2 multiple-input multiple-output (MIMO) devices are extensively studied and commercialized for modern cellular technology. Moreover, most of the focus of the research was dedicated to the reliability of the link. However, the availability and advancement of computing combined with the upgraded radio propagation system or antenna system will open up the mobile internet of things (IoT) and enhance the data rate. Most recently, Ericsson, Telstra, and Qualcomm Technologies have demonstrated 4×4 MIMO operation, an opening step towards download speeds of 1 Gbps in the commercial network [2]. This processing capacity can be further enhanced to 8 × 8 MIMO and 16 × 16 MIMO configuration for a future 5G smartphone. Such configurations will require multiple antenna systems accommodating more than four antenna elements. As a result, most of the reported work in [3–6] on multiple antenna structure work is limited to the frequency band for 2G/3G/4G. In [3], the two antenna elements are oriented diagonally at the printed circuited board of volume of 110 × 65 × 0.8 mm^3, and each antenna has dimensions of 24 × 14.5 mm^2. The two antennas are decoupled by employing ground slots and inverted-L ground branches etched at the bottom layer of the substrate. As the antenna decoupling mechanism required a large space and extended microstrip feeding mechanism, this resulted in limited prospects to be employed for massive MIMO.

In [4], the four-antenna structure is implemented with the main antenna operating for LTE 700/2300/2700 MHz UMTS/GSM 800 MHz /WLAN bands with dimensions of 65 × 9.5 mm^2 and three auxiliary antennas operating at 1800 to 2700 MHz with a dimension of 40 × 6 mm^2. Also in [5], a multiple antenna system is realized by employing four printed monopoles at the four corners of the printed circuit board for the UMTS frequency band. The printed circuit board dimensions are 95 × 60 × 0.8 mm^3. The occupied area for each antenna element is comparable to the operating wavelength, and the minimum isolation between antenna elements is less than 11 dB. As a result, the above-reported structure [3–5] has an inherent limitation of its size and a complicated antenna isolation structure. An attempt is made to realize a multiple antenna system as discussed in [6–8]. In [6], a multiple antenna system is realized on a circuit board of size 115 × 65 × 0.8 mm^3, with each unit antenna having dimensions of 16 × 7.42 mm^2 and a minimum isolation of greater than 10 dB with the operating frequency band of 2 GHz. In [7], the multiple antenna system is investigated by using an unit antenna element with dimensions of dielectric cube 10 × 10 × 5 mm^3 with the printed circuit board of FR-4 substrate of the dimensions 136 × 68.8 × 1 mm^3. The antenna element is a three-dimensional inverted-F antenna wrapped around the said cube. The unit antenna operates at frequency bands for GSM1900, LTE2300, 2.4-GHz WLAN, and LTE2500. Small antenna cubic elements were proposed for MIMO operation at LTE bands [9]. As the maximum isolation between the middle and the top antenna element is 10 dB, this results in limited practical application. In summary, the antenna system discussed mainly focused on the GSM/UMTS frequency band of operation in low radio frequency bands [3–8,10–14]. Moreover, accommodating multiple antennas at the low-frequency band is a daunting task on a limited footprint area of a smartphone due to the transfer of power among ports.

To counter this difficult task, the World radio conference 2015 (WRC-2015) allocated new frequency bands for 5G cellular technology consisting of sub-6 GHz or below 6 GHz (2.6/3.5 GHz bands [15]. Slot antennas have been proposed for multiband operation below 3GHz [16–19]. Therefore, it motivates this work to propose new multiple antenna systems for smartphone operating in the radio spectrum Sub-6 GHz band (3.5 GHz). The proposed multiple antenna system is based on a unit slot antenna structure with ease of integration in the modern trend of smartphone and adequate performance measure compared to the inverted-F antenna [20]. The MIMO performance measures, such as envelope correlation coefficient (ECC), mean effective gain (MEG), and MIMO channel capacity, and customer's hand effect are studied extensively for the proposed multiple antenna system.

2. Antenna Geometry

The detailed design procedure of the six open-end slot antenna system is discussed in this section. The main circuit board or Printed circuit board (PCB) supporting the proposed multiple antenna system consists of FR-4 substrate with relative permittivity of 4.4 and loss tangent of 0.02. The dimensions of PCB are 136 × 68 × 1.6 mm^3, as shown in Figure 1. To swiftly accommodate (2G/3G/4G) antennas, space reservation is made. The dimensions of each open-end slot antenna are 8.5 × 3 mm^2 fed by microstrip with tuning stub, also referred to as inverted-L microstrip feed. The tuning stub is helpful to effectively couple electromagnetic energy to the antenna.

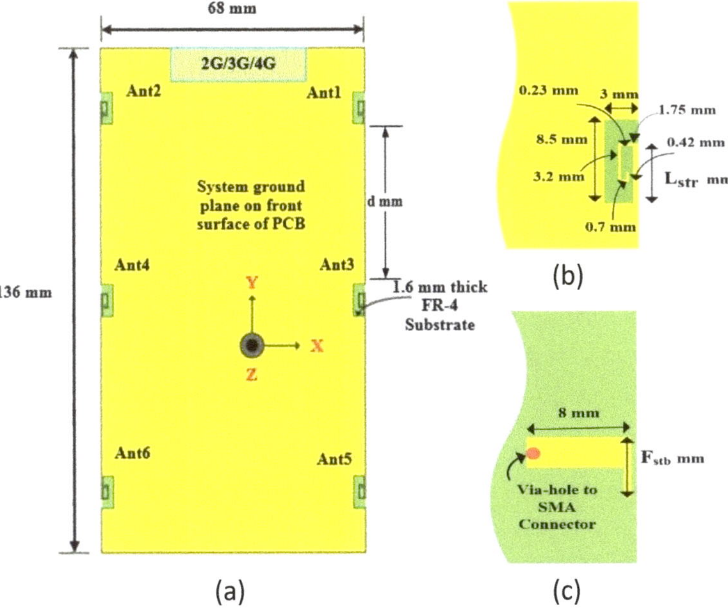

Figure 1. (**a**) The multiple antenna system for multiple-input multiple-output (MIMO) application in a smartphone. (**b**) Geometric configuration of unit slot antenna. (**c**) Inverted-L feeding strip.

3. Results and Discussions

The detailed analysis of the proposed multiple antenna system or multiple antenna system is carried out with commercial electromagnetic software, namely Computer Simulation Technology (CST) Microwave Studio. The design of the proposed multiple antenna system evolved from an open-end rectangular shape slot etched at the top edge, called the reference antenna or Ref_Ant, and fed by the inverted-L microstrip printed on the opposite side of the substrate. The fabricated MIMO antenna system is shown in Figure 2 with corresponding simulated scattering parameters are depicted in Figure 3. As evident with the length of 8 mm, the unit slot or the Ref_Ant1 failed to operate for the desired band of 3.5 GHz for Figure 3a.The slot antenna is usually treated as a magnetic dipole, with its first resonating frequency depending on its electric length. The resonating behavior of the unit slot antenna can further be studied by considering the input impedance depicted in Figure 3b. The dominating inductive impedance is effectively subdued by adding a parasitic strip with more capacitive behavior and referred to as Prop_Ant.

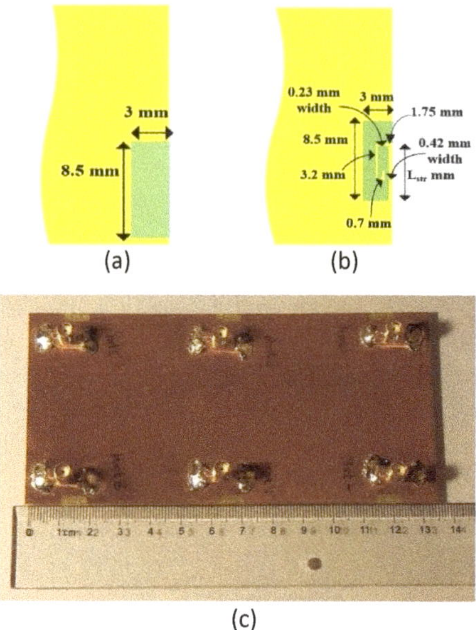

Figure 2. (**a**) The unit slot antenna Reference Antenna or Ref. Ant. (**b**) Proposed Antenna or Prop_Ant (Ant1). (**c**) Fabricated multiple antenna system.

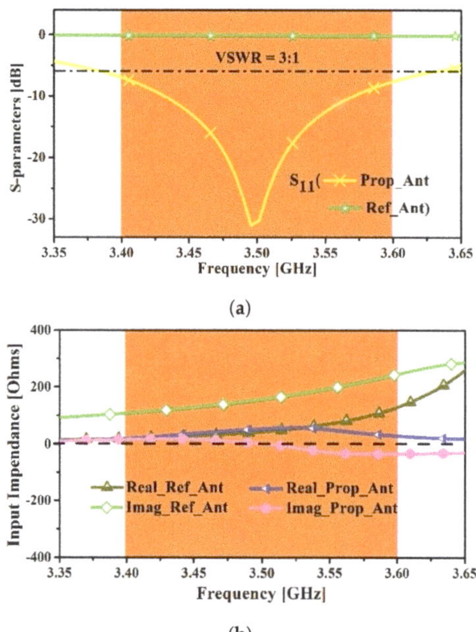

Figure 3. (**a**) Simulated Scattering S-Parameters for the unit slot antenna. (**b**) Input impedance curve.

Initially, for a unit slot or Ref_Ant, pivotal geometric parameters such as L_{str} mm (parasitic strip length) and F_{stb} (Feeding strip length) are studied extensively. The design process of the multiple antenna system is further studied extensively by considering parameters F_{stb}, L_{str}, and d mm. The parameter F_{stb} of the feeding strip plays an important role together with the parasitic strip.

Varying the feeding strip length of F_{stb} = 4.5 mm to 5.5 mm improves the electromagnetic coupling, or more specifically Prop_Ant, which will resonate at 3.5 GHz, as shown in Figure 4a. Also, when the value of L_{str} is varied from 5 mm to 5.8 mm the resonating frequency of the unit slot, open-end antenna, decreases as depicted in Figure 4b. The mutual interaction between antenna elements in the system is quantified by studying the parameter d mm for Ant1, Ant3, and Ant5. The scattering parameters of interest, S_{31} and S_{51}, are given in Figure 4c. The other S-parameters are well below -25 dB. Hence, they are not plotted for clarity. With a minimum value of d = 15.25 mm, the S_{31} is more than -9.8 dB. Increasing the distance between two antennas, with d = 43.5 mm or the middle position of the ground plane, can help to increase the dominating S_{31} to -13 dB, as illustrated in Figure 4c. Furthermore, an inter-spacing between antenna elements greater than $\lambda_g/2$ (λ_g is guided wavelength at 3.5 GHz) results in better isolation, which is enough to ensure independent behavior of each antenna with less mutual coupling.

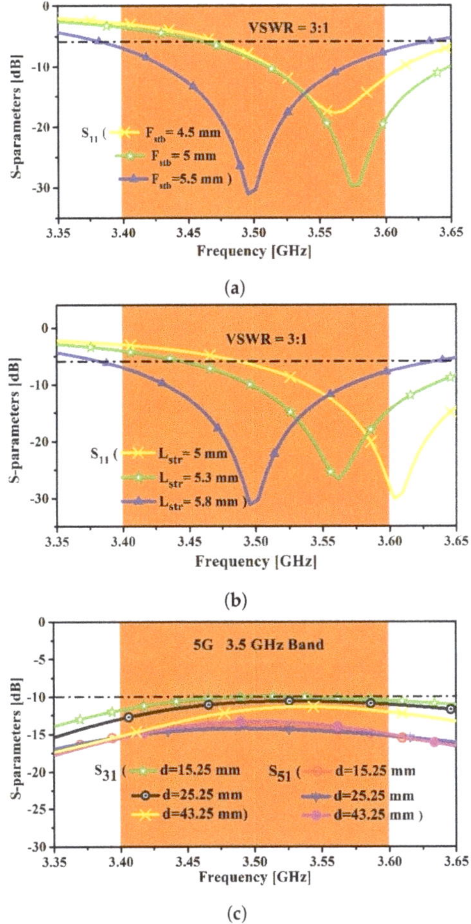

Figure 4. The simulated S-parameter as a function of (a) F_{stb}, (b) L_{str}, and (c) d.

4. Measured Results

4.1. S-Parameters

The proposed multiple antenna system for MIMO operation is fabricated as depicted in Figure 2c. The measurement of scattering parameters is acquired with Agilent Network analyzer N5247A, in such a manner that the two antennas under study are connected to a vector network analyzer while the corresponding antennas are connected to an impedance load of 50 Ω. Due to the symmetric configuration of the structure and for clarity, the scattering parameters or (S-parameters) are only discussed for Ant1, Ant3, Ant5, which are shown in Figure 5. Fair agreement between the simulated and measured results can be observed.

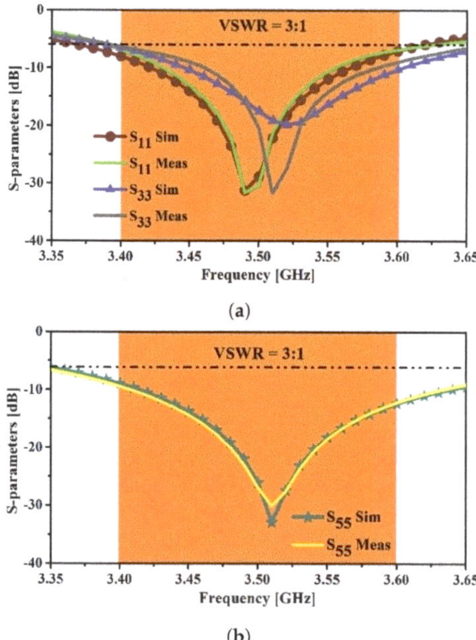

Figure 5. The S-Parameters; Simulated is denoted by (Sim) and Measured by (Meas). (**a**) Ant1 and Ant3. (**b**) Ant5.

The slight difference between measured and simulated results for Ant1 and Ant3 may be attributed to the tolerance of the SMA connector, the termination resistance, and the contribution of hand soldering. The measured impedance bandwidth obtained based on V.S.W.R 3:1 is 200 MHz from 3400 MHz to 3600 MHz. The mutual coupling between unit open-end slot antennas with other antennas is better than 12.5 dB, as shown in Figure 6a.

The measured total efficiencies at Ant1, Ant3, and Ant5 are 58% to 50%, 55% to 48%, and 55% to 49%. The corresponding gains at the Ant1, Ant3 are 3.2 and 4.8 dBi respectively, and for Ant5 it is 3 dBi, as shown in Figure 6b.

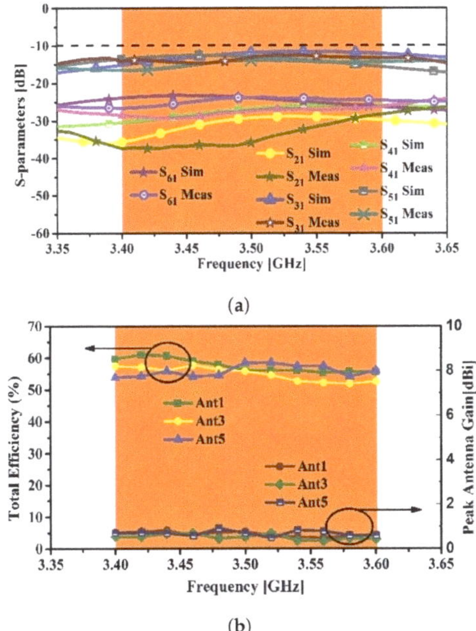

(a)

(b)

Figure 6. (**a**) The measured and simulated S-Parameters for port isolation. (**b**) Measured gain and total efficiency for Ant1, Ant3, and Ant5.

4.2. Radiation Pattern

The far-field measurements such as total efficiencies (including the mismatching losses) of the proposed multiple antenna system are carried out in the anechoic Star lab SATIMO chamber. For each total efficiency measurement, the other corresponding antenna elements were connected to a 50 Ω load. As discussed earlier, due to the symmetric configuration the measured radiation patterns for Ant1, Ant3, and Ant5 are discussed.

The radiation pattern with directive gains is depicted in the x-y plane in Figure 7 for Ant1, Ant3, and Ant5. In the x-y plane for the Ant1 and for Ant5, the radiation pattern is nearly quasi-omnidirectional, with a dip null at $\phi = 210$ and $\phi = 150$ respectively. The Ant3 radiation pattern is nearly omnidirectional. The Ant1, Ant3, and Ant5 are directives in the +x-axis. The slot and antennas are planar and by necessity have a pure linear polarization in the plane of the antenna, in this case, the x-y plane. The measured level of the cross polarization is less than -28 dB in the main radiation direction.

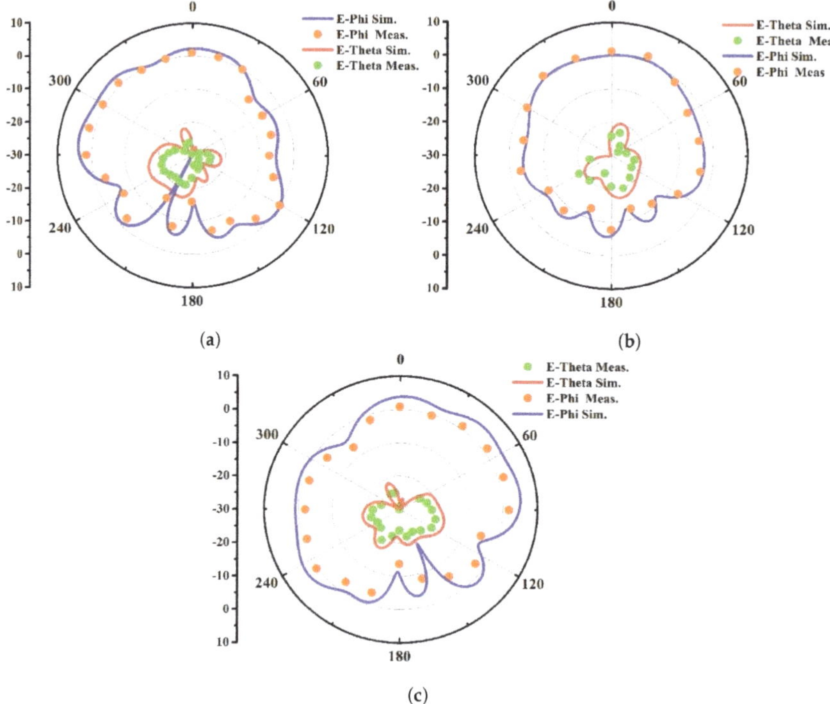

(a)

(b)

(c)

Figure 7. Measured and simulated radiation pattern x-y plane for (**a**) Ant1, (**b**) Ant3, and (**c**) Ant5.

4.3. Surface Current and Electric Field Distribution

The study of surface current distribution is helpful to grasp the radiating behavior of the unit antenna and is discussed in this section. The unit slot antenna is excited by the inverted-L shape microstrip feed printed on the back of PCB. As strong current intensity is observed at the parasitic strip, the close and inner end of the slot is shown in Figure 8a.

Also, depicted in Figure 8b, for Port 1 excitation the electric field anti-nodes will be located in the middle of the slot and the field node position is at the end of the slot. Hence, the antenna evolves into the open-end slot monopole, and the antenna size is reduced.

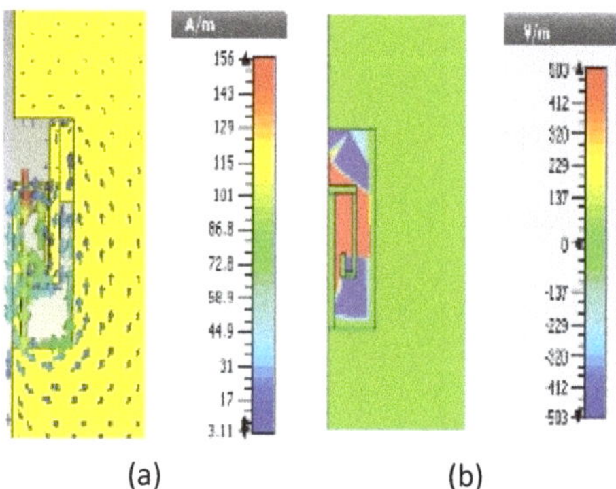

(a)

(b)

Figure 8. The unit antenna element or Ant1 simulated at 3.5 GHz. (**a**) Current distribution. (**b**) Electric field distribution.

4.4. Customer's Hand Effect

In this section, the customer's hand effect on the performance of the multiple antenna system is studied, with a focus on data mode operation. Furthermore, this section will also consider the Envelope Correlation Coefficient (ECC) for the customer hand effect. The study of the customer hand effect is an essential requirement to assess its influence on the performance antenna elements. The main factors affecting the hand are as follows: antenna (design, size, location) and handgrip (position of the fingers concerning antenna, obstructed antenna area, palm-hand distance). The positions or grip style used in this study are by the Cellular Telecommunications Industry Association (CTIA) standards version 3.4 [21].

Accordingly, the electric properties of hand phantom or customer's hand are modeled as reported in [22] across the desired 3.5 GHz Band. The target value is a real part of permittivity 28 to 32 and the effective conductivity is 0.7 to 0.9 S/m for hand phantom, depicted in Figure 9. However, for this study, we model the customer's or user's hand with a constant value interpolated at the center frequency of 3.5 GHz with effective permittivity of 29 and effective conductivity of 0.8 S/m. The two postures considered for this study are Single-Hand Operation (SHO) and Two-Hand Operation (THO), depicted in Figure 9. Two- or dual-hand operation is not yet standardized, hence the posture adopted in this study is inspired by that which is mostly used by customers.

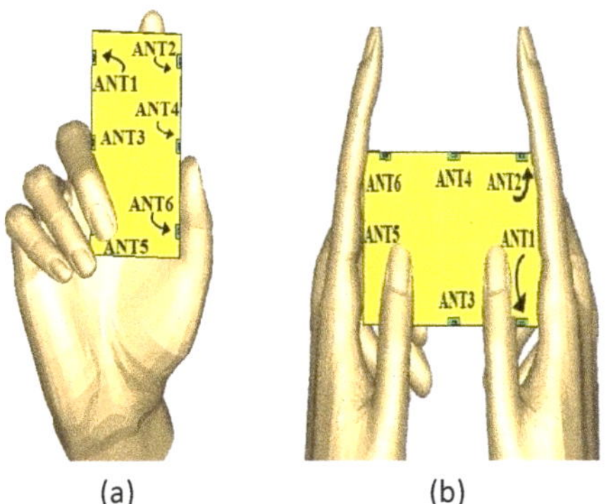

Figure 9. The customer's hand usage scenarios: (**a**) Single-Hand Operation (SHO) and (**b**) Two-hand Operation (THO).

The total efficiency of the proposed multiple antenna system in free space is illustrated in Figure 10 and a comparison of the total efficiency of the single antenna element with reference to a multiple antenna system is presented. The single antenna element total efficiency varies from 60% to 70%. On the other hand, for the antenna elements in the other system the average total efficiency varies on average from 45% to 57%. The corresponding total efficiencies and ECC for the SHO and THO are presented in Figures 11 and 12 respectively. As for the SHO mode, the customer's palm being near a smartphone resulted in a decrease in total efficiency from Ant3 to Ant6 to below 50%. The customer's hand will lead to dielectric loading of the antenna system. Thus, it reduces the total efficiency depending on the proximity of the hand to the antenna, with the details reported in [23]. Recently, different techniques have been proposed to mitigate the user hand effect. In [24,25], a thick buffer material with high permittivity or dielectric constant with low loss or low conductivity was used to counter the near-field coupling of the radiating antenna with the hand. This technique needs extra volume.

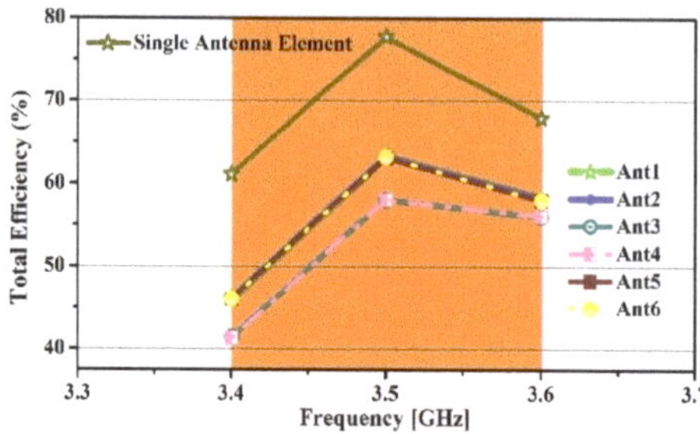

Figure 10. The total efficiency of the multiple antenna system free spaces (without the customer's hand).

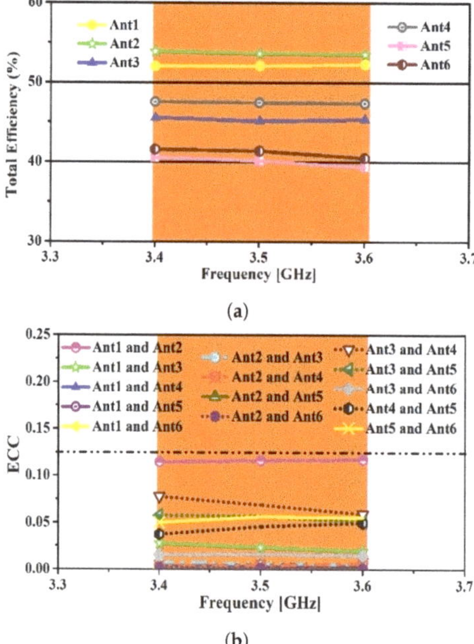

(a)

(b)

Figure 11. Single-Hand Operation (SHO) for the proposed multiple antenna system. (**a**) total efficiency. (**b**) Envelope correlation coefficient.

Furthermore, in [26] the authors dynamically selected the antenna with the best performance metrics in the presence of the user's hand for a mobile terminal. The simulated Envelope correlation coefficient for the desired scenarios of SHO and THO is depicted in Figure 12b, and is lower than 0.5 as required for MIMO operation. Aside from the above-mentioned techniques, the user hand effect mitigated by using passive circuit components or tuning circuit is classified as adaptive impedance matching (AIM) [27].

AIM with variable impedance is a separate circuit module attached to the antenna to detect the time-varying mismatch. This work studied the direct interaction of the antenna system with the user's hand and the method discussed in [23,24,26] was applied to reduce the effect of user's hand dielectric loading of the proposed antenna system.

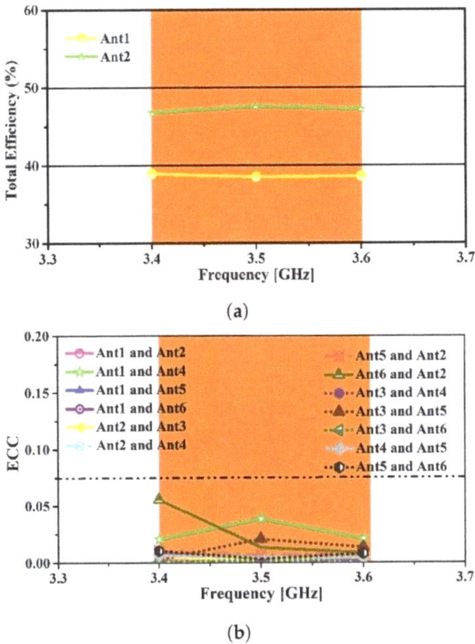

(a)

(b)

Figure 12. Two-Hand Operation (THO) for the proposed multiple antenna system. (**a**) total efficiency. (**b**) Envelope correlation coefficient.

4.5. Key Performance Metrics Evaluation for MIMO Systems

The performance of a smartphone is significantly affected by different hand postures in rich scattering propagation. The critical performance metrics required in such prorogation scenarios are Envelope correlation coefficient (ECC) and Mean effective gain (MEG). The propagation environment effects are expressed as a statistical distribution function as explained in [28]. As for optimal MIMO operation, the ECC should be less than 0.5 to quantify the independent behavior of each antenna in the multiple antenna system. It is calculated from either S-parameters [29] or 3-D radiation pattern [30,31] for the far-field zone of multiple antenna structure. In this work, the ECC, MEG, and MIMO channel capacities are calculated based on the measured result conducted from the far-field measurements in the starlab anechoic chamber. ECC and MEG are calculated from the measured results according to [28]. Most studies reported in [32] considered $\Gamma = 0$ dB for indoor and $\Gamma = 5$ dB for outer door fading environment. The ECC as given in (1) and MEG in (2) are calculated from the measured results according to [32], given as:

$$ECC = \frac{|\iint_{4\pi}(\vec{F_i}(\theta,\phi)) \times (\vec{F_j}(\theta,\phi))\, d\Omega|^2}{\iint_{4\pi}|(\vec{F_i}(\theta,\phi))|^2\, d\Omega \iint_{4\pi}|(\vec{F_j}(\theta,\phi))|^2\, d\Omega} \tag{1}$$

where $\vec{F_i}(\theta,\Phi)$ describe the 3D radiation pattern when antenna i is excited and $\vec{F_j}(\theta,\Phi)$ describe the 3D radiation pattern when antenna j is excited. Solid angle in above Equation (1) is represented as Ω.

$$MEG = \int_{-\pi}^{\pi}\int_{0}^{\pi}[\frac{r}{r+1}G_\theta(\theta,\phi)P_\theta(\theta,\phi) + \frac{1}{1+r}G_\phi(\theta,\phi)P_\phi(\theta,\phi)]sin\theta d\theta d\phi \tag{2}$$

where $G_\phi(\theta,\phi)$ and $P_\theta(\theta,\phi)$ are angle of arrival and r is the cross polar ratio which can be expressed as Equation (3).

$$r = 10log_{10}\left(\frac{P_{vpa}}{P_{hpa}}\right) \tag{3}$$

where the power received by vertically polarized antenna and horizontally polarized antenna are represented as P_{vpa} and P_{hpa}, respectively.

The ECC calculated based on the measured far-field measurement for the proposed multiple antenna system is well below 0.27, as depicted in Figure 13. The MEG calculation accounts for total efficiency, gain, and the wireless propagation environment to measure the antenna–channel mismatch. For good power balance, the quantity of (MEGi/MEGj) or the difference between MEGs should be less than 3 dB for diversity performance. Terms i and j represent antennas. The MEG calculated based on the measured result is well below 3 dB, as shown in Table 1.

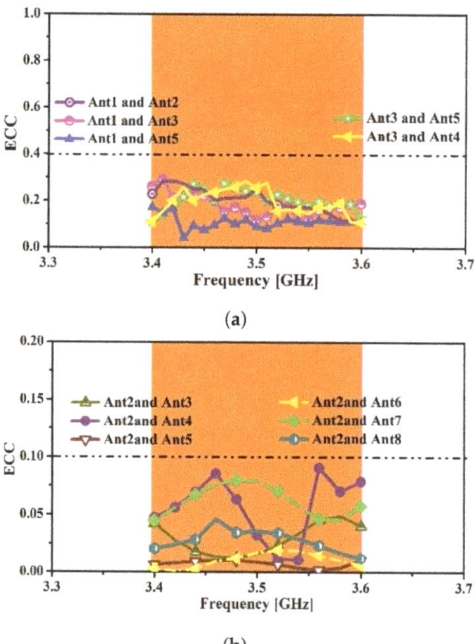

Figure 13. The Envelope Correlation Coefficient (ECC) based on measured far-field results (**a**) ECC for Ant1 with Ant2,Ant3,Ant4 and Ant5, and (**b**) ECC for Ant2 with Ant3,Ant4,Ant5,Ant6,Ant7 and Ant8.

The channel capacity is an important parameter to estimate MIMO system performance. The channel capacities for the proposed multiple antenna system to support 6 × 6 MIMO based on measured total efficiency and ECC is 31.25 bps/Hz. The MIMO channel capacity was calculated based on [32–34]. The MIMO channel capacity was calculated by averaging the 10,000 Rayleigh fading realization with a reference signal to noise ratio (SNR) of 20 dB. The customer hand effect on the operation of the proposed multiple antenna system with respect to ECC and MIMO channel capacity was tabulated in Table 2, based on the simulated data using CST Microwave Studio.

Table 1. Mean Effective Gain (MEG) for Ant1 to Ant6.

Frequency (GHz)	MEG Ant1 (dB)	MEG Ant2 (dB)	MEG Ant3 (dB)	MEG Ant4 (dB)	MEG Ant5 (dB)	MEG Ant6 (dB)
Indoor XPR = 1	−3.20	−3.15	3.20	−3.18	−3.22	−3.13
Indoor XPR = 5	−3.40	−3.50	−3.64	−3.48	−3.50	−3.70

Table 2. MIMO channel capacities and ECC for free space (without hand) and customer's hand scenarios based on simulated results.

MIMO Configuration	Scenario	ECC	Channel Capacity (bps/Hz)
	Free Space	0.012	34.25
6 × 6	Single-Hand Mode	0.10	27.43
	Two-Hand Mode	0.07	23.24

4.6. Impact of Display Module (Metal Casing and Liquid Crystal Display)

The impact of the display module consists of a metal frame and the Liquid Crystal Display (LCD) is studied in this subsection. The metal frame adheres to the ground and the LCD display is made of glass with reflective permittivity of 7 and loss tangent of 0.02 with overall dimensions 136 mm × 68 mm × 1 mm, as shown in Figure 14. The metal frame section in the display module is extended to the slot radiator due to the fact that it is connected to the ground plane of Printed Circuit board (PCB) can be considered a ground for real phone. As evident from Figure 15, the display module will affect the resonating frequency of the unit antenna or Ant1 and it still meets the required standard of VSWR of 3:1. The ports isolation for the proposed multiple antenna system is greater the 13.5 dB. The tuning parameter of F_{stb} can be used for adjusting the resonating frequency, as given in Figure 16, for a unit slot antenna. Similarly, the other antenna elements can also be tuned to the center frequency of 3.5 GHz.

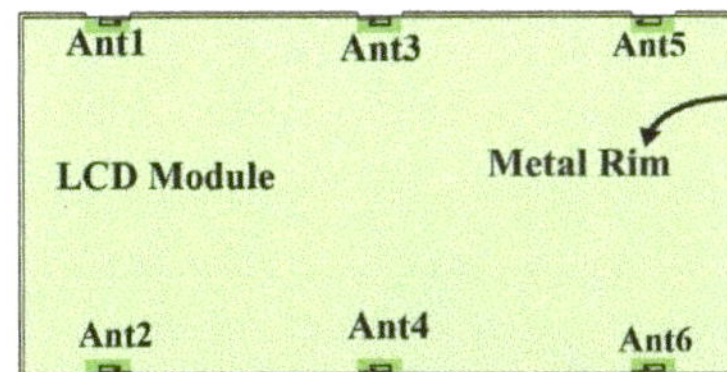

Figure 14. Simulation model for the proposed multiple antenna system with the display module.

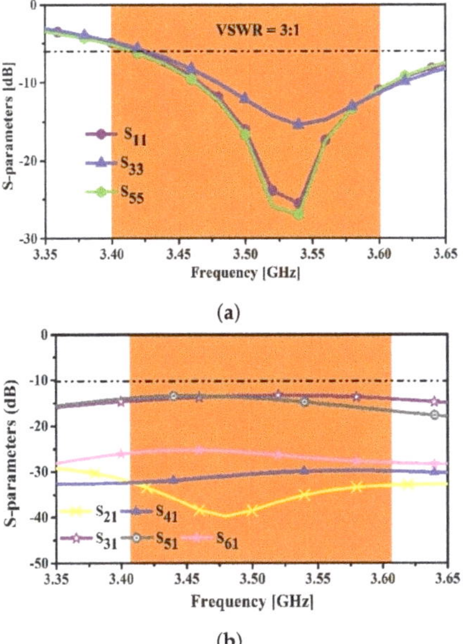

Figure 15. Simulated S-Parameters (**a**) Impedance Bandwidth, (**b**) Ports Isolation.

Table 3 shows the performance comparison between the proposed works and the previous report. As discussed earlier that the most of reported focused on Long term Evolution (LTE) and Global System of Mobile communication frequency band of 700 MHZ to 1800 MHz and also in 2 GHz bands. Furthermore, the reported structure is large in dimensions has limited applications to be integrated for 5G Mobile terminal. Hence, the proposed multiple antenna system using the ground plane structure of mobile or slot antenna is easily designed and implemented on the longer edges or sided edges. With the modern trend, od smartphone requires a metal frame structure. Thus the proposed multiple antenna system can be easily embedded in it. As for 2G/ LTE, smooth operation space reservation is made at the top and lower short edges of PCB of the smartphone.

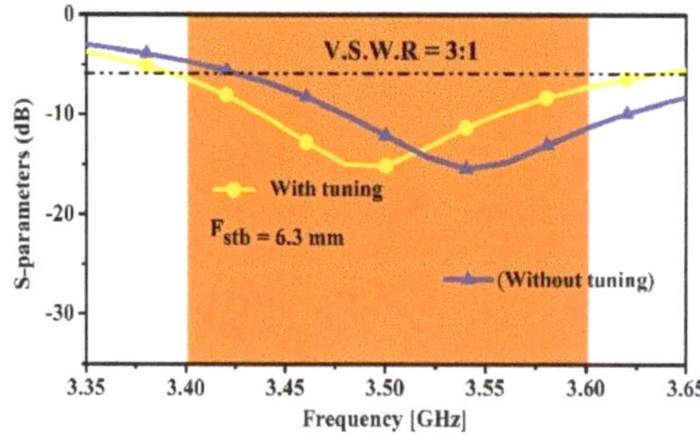

Figure 16. Comparison of simulated S_{11} of Ant1 by tuning the F_{stb} parameter.

Table 3. Performance comparison with smartphone antenna. Abbreviation: Number of Antenna elements = # Ant, Total Efficiency = T.E%, Peak Channel Capacity (bps/Hz) = PCC, Envelope correlation Coefficient = ECC.

Ref.	Bandwidth (GHz)	#Ant	T.E% (meas)	PCC	ECC (required <0.5)
Proposed	3.4–3.6	6	50 to 60	32	0.15
[3]	1.67–2.76	2	70 to 80	Not Given	0.05
[4]	1.7–2.2	4	83 to 89	Not Given	0.02
[5]	0.6-0.9/1.6–2.6	4	40 to 70	Not Given	0.25
[7]	1.8–1.9/2.3–2.6	6	37 to 79	Not Given	0.16
[8]	0.73–0.79/2.3–2.4	4	~70	Not Given	0.4
[13]	0.6–0.9	4	40 to 50	17	0.3

5. Conclusions

This work presented a multiple-input multiple-output, six element, open-ended slot antenna system fed by Inverted-L shaped microstrip with tuning stub for 5G mobile terminals operating at a single band of 3.5 GHz. Three identical antenna elements (open-ended slot) were etched across each length of the chassis and investigated. The impedance bandwidth of 200 MHz achieved, base on −6 dB criteria, enough for a future 5G cellular wireless communication system, with good isolation of <−13 dB using a spacing of half a wavelength among any two radiating elements and having an ECC less than 0.15 and a peak gain of 4.8 dBi. Meanwhile, the effects of the user's hand in both single-hand and two-hand scenarios was studied and showed good results, with ECCs of 0.10 and 0.007 and channel capacities of 27.43 and 23.24 bps/Hz, respectively. The measured results agreed fairly well with the simulated results. The free space ergodic MIMO channel capacity calculated for the proposed antenna system using Matlab was about 31.25 bps/Hz, approximately 2.7 times higher than 2×2 MIMO operations with a 20 dB SNR reference level in a Rayleigh fading environment. Due to better performances in bandwidth, isolation, total efficiency, channel capacity and radiation patterns, the proposed six-element MIMO array is a potential applicant for 5G smartphone systems.

Author Contributions: Conceptualization, M.A., S.H.K. and A.I.; methodology, M.A.; software, M.A, S.H.K. and A.I.; validation, M.A., L.F.A. and S.K. (Sunghwan Kim); formal analysis, A.I., M.A.B., S.K. (Shafiullah Khan) and S.K. (Sunghwan Kim); investigation, M.A., S.H.K., A.I. and L.F.A.; resources, L.F.A., A.I. and S.K. (Sunghwan Kim); data curation, A.I.; writing—original draft preparation, M.A. and S.H.K.; writing—review and editing, M.A., S.H.K., L.F.A., A.I., M.A.B., S.K. (Shafiullah Khan) and S.K. (Sunghwan Kim); visualization, A.I.; supervision, L.F.A. and S.K. (Sunghwan Kim); project administration, L.F.A., A.I. and S.K. (Sunghwan Kim); funding acquisition, S.K. (Sunghwan Kim).

Funding: This work (S2666095) was supported by project for Cooperative R&D between Industry, Academy, and Research Institute funded Korea Ministry of SMEs and Startups in 20.

Conflicts of Interest: The authors declare no conflict of interest.

References

1. Iqbal, A.; Smida, A.; Mallat, N.K.; Ghayoula, R.; Elfergani, I.; Rodriguez, J.; Kim, S. Frequency and pattern reconfigurable antenna for emerging wireless communication systems. *Electronics* **2019**, *8*, 407. [CrossRef]
2. Werner, K.; Furuskog, J.; Riback, M.; Hagerman, B. Antenna configurations for 4x4 MIMO in LTE-field measurements. In Proceedings of the 2010 IEEE 71st Vehicular Technology Conference, Taipei, Taiwan, 16–19 May 2010; pp. 1–5.
3. Wang, Y.; Du, Z. A wideband printed dual-antenna system with a novel neutralization line for mobile terminals. *IEEE Antennas Wirel. Propag. Lett.* **2013**, *12*, 1428–1431. [CrossRef]
4. Shoaib, S.; Shoaib, I.; Shoaib, N.; Chen, X.; Parini, C.G. MIMO antennas for mobile handsets. *IEEE Antennas Wirel. Propag. Lett.* **2014**, *14*, 799–802. [CrossRef]
5. Guo, J.; Fan, J.; Sun, L.; Sun, B. A four-antenna system with high isolation for mobile phones. *IEEE Antennas Wirel. Propag. Lett.* **2013**, *12*, 979–982. [CrossRef]
6. Ikram, M.; Hussain, R.; Sharawi, M.S. Low profile 6-element modified-monopole MIMO antenna system for mobile applications. In Proceedings of the 2016 IEEE International Symposium on Antennas and Propagation (APSURSI), Fajardo, Puerto Rico, 26 June–1 July 2016; pp. 73–74.
7. Chen, Z.; Geyi, W.; Zhang, M.; Wang, J. A study of antenna system for high order MIMO device. *Int. J. Antennas Propag.* **2016**, *2016*. [CrossRef]
8. Sharawi, M.; Jan, M.; Aloi, D. Four-shaped 2× 2 multi-standard compact multiple-input–multiple-output antenna system for long-term evolution mobile handsets. *IET Microw. Antennas Propag.* **2012**, *6*, 685–696. [CrossRef]
9. Anguera, J.; Andújar, A.; Mateos, R.M.; Kahng, S. A 4× 4 MIMO multiband antenna system with non-resonant elements for smartphone platforms. In Proceedings of the 2017 11th European Conference on Antennas and Propagation (EUCAP), Paris, France, 19–24 March 2017; pp. 2705–2708.
10. Moradikordalivand, A.; Leow, C.Y.; Rahman, T.A.; Ebrahimi, S.; Chua, T.H. Wideband MIMO antenna system with dual polarization for WiFi and LTE applications. *Int. J. Microw. Wirel. Technol.* **2016**, *8*, 643–650. [CrossRef]
11. Malathi, C.; Thiripurasundari, D. CSRR Loaded 2x1 Triangular MIMO Antenna for LTE Band Operation. *Adv. Electromagn.* **2017**, *6*, 78–83. [CrossRef]
12. Yang, B.; Chen, M.; Li, L. Design of a four-element WLAN/LTE/UWB MIMO antenna using half-slot structure. *AEU-Int. J. Electron. Commun.* **2018**, *93*, 354–359. [CrossRef]
13. Wong, K.L.; Chen, Y.C.; Li, W.Y. Four LTE low-band smartphone antennas and their MIMO performance with user's hand presence. *Microw. Opt. Technol. Lett.* **2016**, *58*, 2046–2052. [CrossRef]
14. Wong, K.L.; Kang, T.W.; Tu, M.F. Internal mobile phone antenna array for LTE/WWAN and LTE MIMO operations. *Microw. Opt. Technol. Lett.* **2011**, *53*, 1569–1573. [CrossRef]
15. Wang, T.; Li, G.; Huang, B.; Miao, Q.; Fang, J.; Li, P.; Tan, H.; Li, W.; Ding, J.; Li, J.; et al. Spectrum analysis and regulations for 5G. In *5G Mobile Communications*; Springer: Berlin/Heidelberg, Germany, 2017; pp. 27–50.
16. Anguera, J.; Sanz, I.; Mumbrú, J.; Puente, C. Multiband handset antenna with a parallel excitation of PIFA and slot radiators. *IEEE Trans. Antennas Propag.* **2009**, *58*, 348–356. [CrossRef]

17. Abedin, M.; Ali, M. Modifying the ground plane and its effect on planar inverted-F antennas (PIFAs) for mobile phone handsets. *IEEE Antennas Wirel. Propag. Lett.* **2003**, *2*, 226–229. [CrossRef]

18. Anguera, J.; Cabedo, A.; Picher, C.; Sanz, I.; Ribó, M.; Puente, C. Multiband handset antennas by means of groundplane modification. In Proceedings of the 2007 IEEE Antennas and Propagation Society International Symposium, Honolulu, HI, USA, 9–15 June 2007; pp. 1253–1256.

19. Hossa, R.; Byndas, A.; Bialkowski, M. Improvement of compact terminal antenna performance by incorporating open-end slots in ground plane. *IEEE Microw. Wirel. Components Lett.* **2004**, *14*, 283–285. [CrossRef]

20. Chang, T.H.; Kiang, J.F. Compact multi-band H-shaped slot antenna. *IEEE Trans. Antennas Propag.* **2013**, *61*, 4345–4349. [CrossRef]

21. Gabriel, C. Tissue equivalent material for hand phantoms. *Phys. Med. Biol.* **2007**, *52*, 4205. [CrossRef] [PubMed]

22. Certification, C. Test Plan for mobile station over the air performance. *Method Meas. Radiat. Power Receiv. Perform. Revis.* **2005**, *2*, 273–275.

23. Pelosi, M.; Franek, O.; Knudsen, M.; Pedersen, G. Influence of dielectric loading on PIFA antennas in close proximity to user's body. *Electron. Lett.* **2009**, *45*, 246–248. [CrossRef]

24. Plicanic, V.; Lau, B.K.; Ying, Z. Performance of a multiband diversity antenna with hand effects. In Proceedings of the 2008 International Workshop on Antenna Technology: Small Antennas and Novel Metamaterials, Chiba, Japan, 4–6 March 2008; pp. 534–537.

25. Bouazizi, A.; Zaibi, G.; Iqbal, A.; Basir, A.; Samet, M.; Kachouri, A. A dual-band case-printed planar inverted-F antenna design with independent resonance control for wearable short range telemetric systems. *Int. J. Microw. Comput. Eng.* **2019**, *29*, e21781. [CrossRef]

26. Ilvonen, J.; Valkonen, R.; Holopainen, J.; Kivekäs, O.; Vainikainen, P. Reducing the interaction between user and mobile terminal antenna based on antenna shielding. In Proceedings of the 2012 6th European Conference on Antennas and Propagation (EUCAP), Prague, Czech Republic, 26–30 March 2012; pp. 1889–1893.

27. Zhang, S.; Zhao, K.; Ying, Z.; He, S. Adaptive quad-element multi-wideband antenna array for user-effective LTE MIMO mobile terminals. *IEEE Trans. Antennas Propag.* **2013**, *61*, 4275–4283. [CrossRef]

28. Ding, Y.; Du, Z.; Gong, K.; Feng, Z. A novel dual-band printed diversity antenna for mobile terminals. *IEEE Trans. Antennas Propag.* **2007**, *55*, 2088–2096. [CrossRef]

29. Iqbal, A.; Saraereh, O.A.; Ahmad, A.W.; Bashir, S. Mutual coupling reduction using F-shaped stubs in UWB-MIMO antenna. *IEEE Access* **2017**, *6*, 2755–2759. [CrossRef]

30. Iqbal, A.; A Saraereh, O.; Bouazizi, A.; Basir, A. Metamaterial-based highly isolated MIMO antenna for portable wireless applications. *Electronics* **2018**, *7*, 267. [CrossRef]

31. Iqbal, A.; Basir, A.; Smida, A.; Mallat, N.K.; Elfergani, I.; Rodriguez, J.; Kim, S. Electromagnetic Bandgap Backed Millimeter-Wave MIMO Antenna for Wearable Applications. *IEEE Access* **2019**, *7*, 111135–111144. [CrossRef]

32. Ying, Z.; Bolin, T.; Plicanic, V.; Derneryd, A.; Kristensson, G. Diversity antenna terminal evaluation. In Proceedings of the 2005 IEEE Antennas and Propagation Society International Symposium, Washington, DC, USA, 3–8 July 2005; Volume 2, pp. 375–378.

33. Tian, R.; Lau, B.K.; Ying, Z. Multiplexing efficiency of MIMO antennas. *IEEE Antennas Wirel. Propag. Lett.* **2011**, *10*, 183–186. [CrossRef]

34. Kildal, P.S.; Rosengren, K. Correlation and capacity of MIMO systems and mutual coupling, radiation efficiency, and diversity gain of their antennas: simulations and measurements in a reverberation chamber. *IEEE Commun. Mag.* **2004**, *42*, 104–112. [CrossRef]

Article

A Compact Semi-Circular and Arc-Shaped Slot Antenna for Heterogeneous RF Front-Ends

Chemseddine Zebiri [1,2], Djamel Sayad [3], Issa Elfergani [4,*], Amjad Iqbal [5], Widad F.A. Mshwat [2], Jamal Kosha [2], Jonathan Rodriguez [4] and Raed Abd-Alhameed [2,6]

[1] Department of Electronics, University of Ferhat Abbas, Sétif 1, Sétif 19000, Algeria; czebiri@univ-setif.dz
[2] School of Electrical Engineering and Computer Science, University of Bradford, Bradford BD71DP, UK; W.F.A.AMshwat@bradford.ac.uk (W.F.A.M.); J.S.M.Kosha@bradford.ac.uk (J.K.); R.A.A.Abd@bradford.ac.uk (R.A.-A.)
[3] Department of Electrical Engineering, University of 20 Aout 1955—Skikda, Skikda 21000, Algeria; dsayad2002@yahoo.fr
[4] Instituto de Telecomunicações, Campus Universitário de Santiago, 3810-193 Aveiro, Portugal; Jonathan@av.it.pt
[5] Centre for Wireless Technology, Faculty of Engineering, Multimedia University, Cyberjaya 63100, Malaysia; amjad730@gmail.com
[6] Information and Communication Engineering Department, Basrah University College of Science and Technology, Basrah 24001, Iraq
* Correspondence: i.t.e.elfergani@av.it.pt; Tel.: +351-234-377-900

Received: 22 August 2019; Accepted: 1 October 2019; Published: 6 October 2019

Abstract: In this paper, a new miniaturized compact dual-band microstrip slot antenna is presented. To achieve the dual-band characteristics, two adjunct partial arc-shaped small slots are joined to two main circular slots embedded in the ground of the antenna structure. With a reduced size of $30 \times 28.5 \times 0.8$ mm^3, the proposed antenna presents a dual-band characteristic. The design is optimized using a High Frequency Structure Simulator (HFSS) followed by experimental verifications. An impedance bandwidth, for $S_{11} \leq 10$ dB, that covers the 1.8 GHz and 2.4 GHz bands is accomplished, which makes the proposed antenna basically suitable for hand-held devices and medical applications. More applications such as digital communication system (DCS) 1.71–1.88 GHz, personal communication services (PCS) 1.85–1.99 GHz, Universal and mobile telecommunications system UMTS 1.92–2.17 GHz, Bluetooth 2.4–2.5 GHz, and Wi-Fi 2.4–2.454 GHz, Industrial Scientific and Medical radio frequency (RF) band ISM-2.4 GHz, Wireless Local Area Network (WLAN-2.4) are possible by simply changing one of the geometrical antenna dimensions. The antenna is characterized by stable radiation patterns as well.

Keywords: slot antenna; medical applications; miniaturized antenna; arc-shaped; dual-band

---

1. Introduction

Conventional microstrip antennas have the attractive features of a low profile, light weight, ease of analysis and fabrication, and ease of integration into microwave devices. However, they exhibit some limitations such as single resonance frequency, low impedance bandwidth, low gain, larger size, and polarization impurity. Among the planar antenna structures, the slot antenna is one of the most promising candidates for multi-band antenna applications because, in addition to the aforementioned attractive qualities, they inherently possess larger bandwidths and occupy less space which makes them promising components for microwave applications [1–10]. In this frame, various designs of multi-band slot antennas have been proposed [11–19]. Moreover, several techniques have been proposed in the literature to enhance the parameters of conventional microstrip slot antennas and realize the multi-band function, these include using new feeding techniques [20], Defected Ground Structures (DGS) [10,21–23], parasitic elements [24,25], Metamaterial [26,27], etc. Dual or multi-band characteristics are in great demand for various applications such as wireless local area networks (Bluetooth, WLAN, and WiMAX) and Industrial Scientific and Medical (ISM) applications. In modern communication standards a single antenna system that can cover all these bands is considered highly necessary, not only to cover several frequency bands but also to reduce the physical size of the system. A literature survey shows that planar slot antennas are promising candidates to realize broadband or multi-band functions to cover up multi-standard services. Several dual- and/or multi-band slot antenna designs have demonstrated their superior performances for the desired multi-band requirements [28,29]. A compact CPW-fed triple-band antenna for diversity applications is reported in [30], triple-band antennas in [31,32] and quad-band in [33,34] and broad-band circular polarization [35], for example. The dual-band characteristic of these kind of antennas may be achieved by joining small resonant slots to the body of the main one [36] or etching several stubs on the large slots [14]. Slot antennas triple-band characteristic is also possible by etching three folded slots on the ground plane such as the designs reported in [15,16], or several stubs on the slots [17]. Quad-band slot antennas are also possible using the above mentioned techniques [3,19]. Better impedance matching may be achieved by inserting open-ended stubs [37]. Generally speaking, slot antennas have larger bandwidth (BW) than the microstrip antennas because of their lower quality factors due to the bidirectional radiation characteristics [4,35]. Combinations of slots and strips on the feeding side can generate circular polarization [4].

In this paper, we present the technique of introducing small arc-shaped slots in the antenna structure to realize a new high gain miniaturized dual-band radiating antenna. The antenna design consists of two main semi-circular slots. Two adjunct small arc-shaped slots are asymmetrically and partially etched round these two main slots to realize the dual-band function and obtain a small sized-structure. The antenna is fed by a stubbed microstrip line for a better impedance matching in the operating bands. The proposed slot antenna covers the desirable 1.8 GHz and 2.4 GHz bands for hand-held devices and medical applications.

2. Proposed Antenna Geometry and Summarized Results

The geometry of the slot antenna design is shown in Figure 1. In this design, two main semi-circular slots are etched on a copper plane. A feeding L-shaped 50 Ω microstrip line is printed on the other side of the substrate as this approach enables ease of impedance matching and other useful features [38]. The feed line is stubbed for a better impedance matching. Two adjunct arc-shaped slots are asymmetrically etched partially round the two slots. Two angular parameters: $\theta_{11} + \theta_{12}$ and $\theta_{21} + \theta_{22}$ are attributed to the two arc-shaped slots to control their locations, lengths and widths with the aid of the parameters r_1, φ_1 and r_2, φ_2, respectively. The whole structure is designed and realized on an FR4 $30 \times 28.5 \times 0.8$ mm^3 substrate with a relative permittivity $\varepsilon_{rs} = 4.4$ and loss tangent of 0.017.

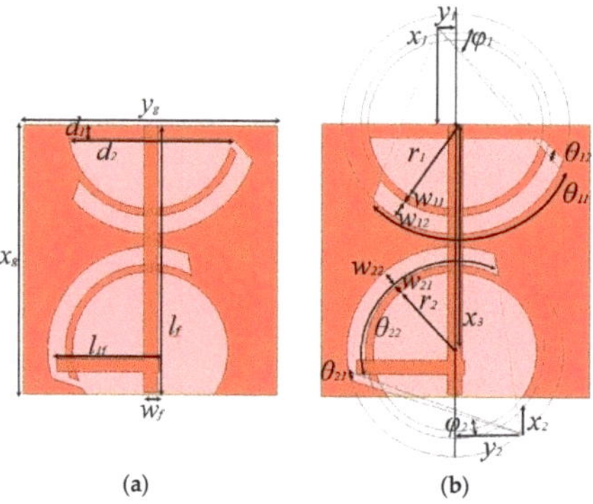

Figure 1. Design parameters of the proposed antenna, (**a**) overall optimized dimensions and (**b**) slots parameters.

The antenna simulations were carried out using High Frequency Structure Simulator (HFSS) software. The final antenna dimensions are: $x_g = 30$, $y_g = 28.5$, $d_1 = 1.5$, $d_2 = 18.45$, $l_f = 29.75$, $w_f = 1.5$, $l_{1f} = 11.5$, $\theta_{11} = 60°$, $\theta_{12} = 2.5°$, $\theta_{22} = 60°$, $\theta_{21} = 2.5°$, $\varphi_1 = 47.23°$, $\varphi_2 = 17.5°$, $r_1 = 9.3$, $w_{11} = 0.8$, $w_{12} = 1.95$, $r_2 = 8.75$, $w_{21} = 0.8$, $w_{22} = 1.9$, $x_1 = 10.59$, $y_1 = 1.87$, $x_2 = 4.03$, $y_2 = 7.20$, $x_3 = 25$ (all dimensions are in mm). All the parameters are clearly represented in Figure 1.

2.1. Effect of the Main Slots and Adjunct Arc-Shaped Slots

The objective of this work is to realize a miniaturized dual-band antenna that covers the bands 1.8 GHz and 2.4 GHz for mobile and medical applications, and solutions with stable radiation patterns. In order to examine the effect of introducing the semi-circular slots and arcs in the radiating structure, the different steps of the antenna design procedure and the comparative study results are presented by Figures 2–4. The first antenna (Figure 2a) is a structure with simple partial circular slots etched in the ground plane and the second one (Figure 2b) is the final proposed structure, where two additional arc-shaped slots are cut round the two main slots. Figure 3a,b shows the S_{11} coefficient and gain, respectively, for the two structures. The first antenna shows a better return loss coefficient at about 2.2 GHz with a gain varying between 1.04 dBi and 4.9 dBi in the frequency range 1.75–4 GHz. The final structure (Figure 2b) manifests itself as a dual-band antenna that operates at the desired 1.8 GHz and 2.4 GHz bands with simulated gains of 3.43 and 4.47 dBi, respectively.

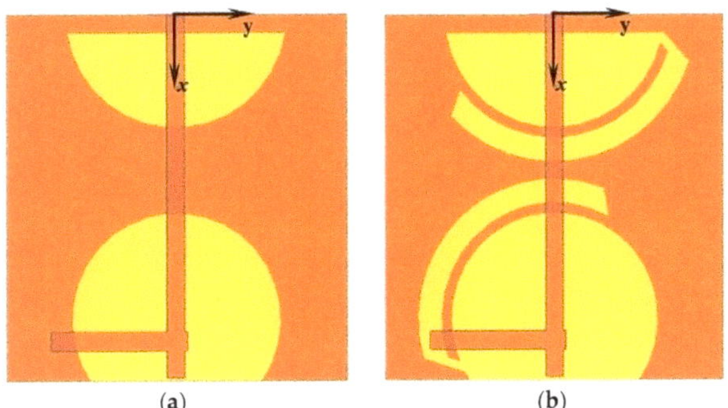

Figure 2. Design steps of the proposed antenna (**a**) slot antenna without adjunct arc-shaped slots and (**b**) slot antenna with adjunct arc-shaped slots.

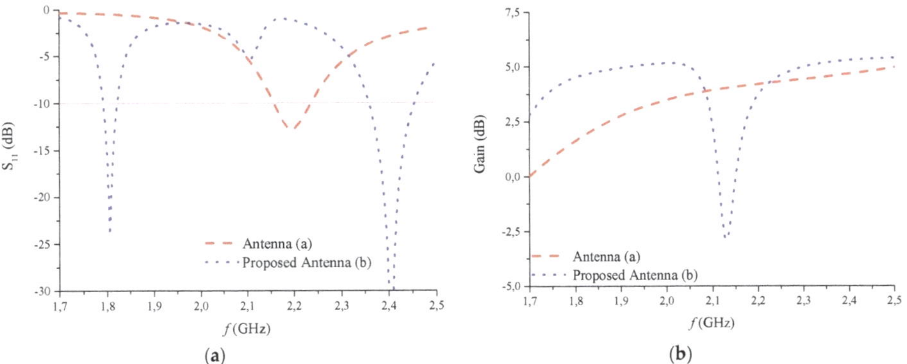

Figure 3. (**a**) Simulated S_{11} of antennas; (**b**) Simulated gain of the two antennas depicted in Figure 2.

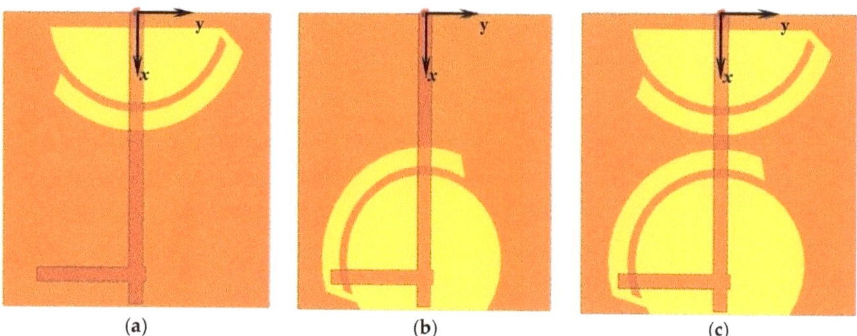

Figure 4. Comparison of three antennas configurations: (**a**) first partial slot antenna; (**b**) second partial slot antenna; and (**c**) final proposed antenna.

The second study aims to investigate the response of the circular slots with arcs according to Figure 4 configuration steps. Three different antennas are designed. In the first structure (Figure 4a), only the slot in the first side of the antenna near the sub miniature version A (SMA) port is considered. In the second structure (Figure 4b), we only consider the slot at the other side of the structure and in Figure 4c, we consider both slots forming our final proposed antenna.

Figure 5a,b shows the compared simulation results of S_{11} coefficients and gains. We notice that the two antennas (a) and (b) are basically mono-band and present operating bands centered at 1.95 GHz and 2.0 GHz, respectively. Moreover, antenna (b) exhibits a very poor gain. The effect of the two main slots and adjunct arc-shaped slots is clear on the properties of the antennas.

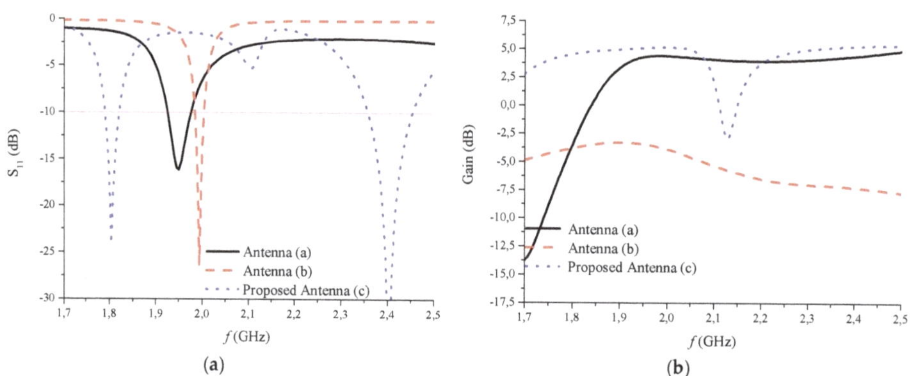

Figure 5. (**a**) Simulated S_{11} of the three antennas depicted in Figure 4; (**b**) Simulated gain of the three antennas depicted in Figure 4.

2.2. Equivalent Circuit Model

In order to analyze the true behavior of the system, co-simulation of the antenna and other RF front-end components is necessary. It is necessary to draw the equivalent circuit model of the antenna, because time domain simulators such as Advanced Design System (ADS) and SPICE are used for majority of the RF front-end design. The equivalent circuit model of the proposed dual-band antenna is shown in Figure 6a. We excited two modes of the antenna using microstrip transmission line in the electromagnetic (EM) model simulations. The two modes of the antenna are represented by the two resistor, inductor and capacitor (RLC) section in the equivalent circuit model [39] as shown in Figure 6a. The coupling associated between two modes of the antenna is represented as the resonant circuit (LC) section [40]. The LC section is responsible for the strong or weak coupling between the modes. The impedance transformer shows the matching of the source with the antenna. The optimized components of the equivalent circuit model is illustrated in Figure 6a. The impedance results of the EM model and equivalent circuit model is compared in Figure 6b. We can see that the results match well in the two resonating modes. We can see that there are several modes in the EM model and only two modes in equivalent circuit model. The reason behind two modes in the equivalent circuit model is that we have designed our equivalent circuit model keeping in mind the two resonating modes only.

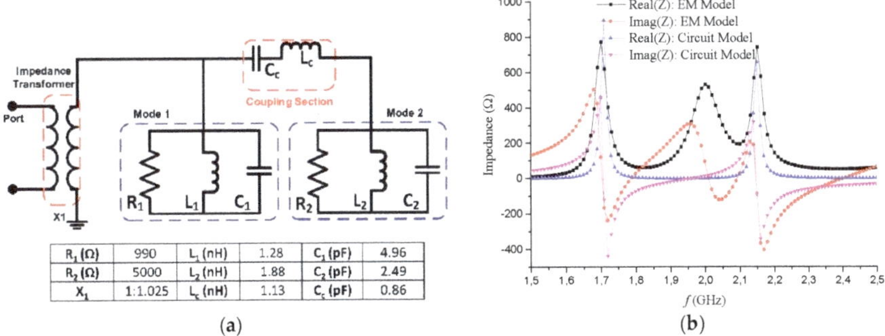

(a) (b)

Figure 6. (a) Equivalent Circuit Model of the dual-band antenna; (b) Impedance comparison of the circuit model and electromagnetic (EM) model.

2.3. Parametric Study

A parametric study was carried out to examine the effects of the main slots as well as the arc-shaped ones on the size reduction coefficient and return loss of the antenna. A direct relationship between the slots parameters and the characteristics of the antenna is noticed. These parameters could control the antenna characteristics by modifying the radial and angular dimensions of the slots.

Note that the position and the size of the two arc-shaped slots are very influential. The characteristics of the antenna may be controlled using the two parameters φ_1 and φ_2 attributed to Arc_1 and Arc_2, respectively. While Arc_1 has a significant effect around 1.8 GHz (Figure 7a,b) and can also realize a tri-band characteristic, Arc_2 presents the same effect but around 2.4 GHz and helps shift the 1.8 GHz band towards higher frequencies.

The length of the two arcs as well as their widths (r_1, r_2, w_{11}, and w_{12}) have almost the same effect around 1.8 GHz, but only Arc_1 can affect the S_{11} around 2.4 GHz (Figure 8a–d). The spacing between the arcs and the ground plane (w_{12} and w_{22}) affects S_{11} in the same way. These remarks are certified by the behavior of the current density, which showed the relationship between the two arcs and the frequencies 1.8 GHz and 2.4 GHz (Figure 8e,f).

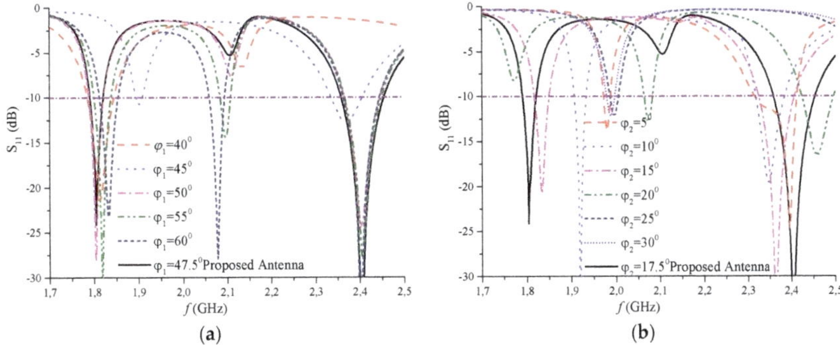

(a) (b)

Figure 7. *Cont.*

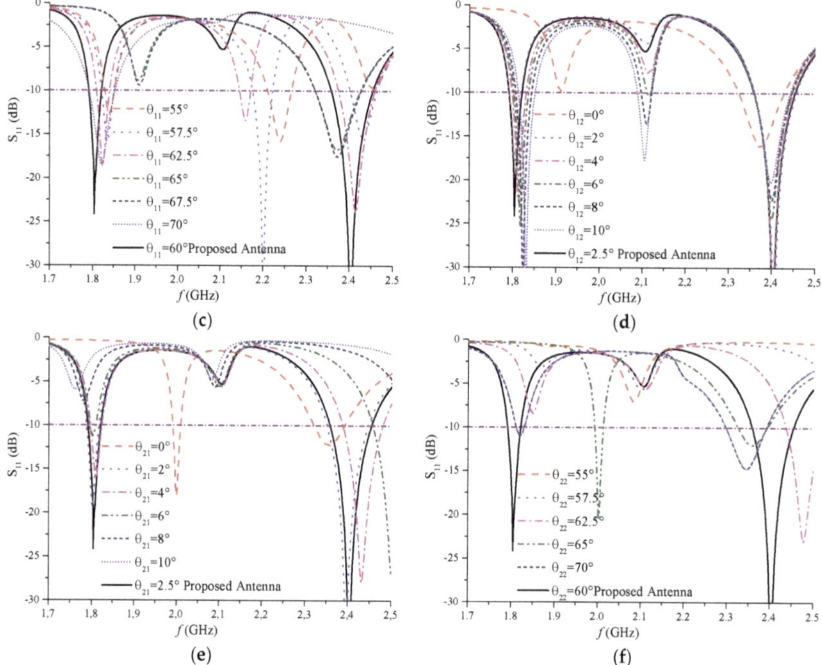

Figure 7. Length and position effects of arc-shaped slots on S_{11}: (**a**) φ_1 angular start of the first arc; (**b**) φ_2 angular start of the second arc; (**c**) θ_{11} angular length of the first arc; (**d**) θ_{12} angular spacing between ground plane and first arc; (**e**) θ_{21} angular spacing between ground plane and second arc; and (**f**) θ_{22} angular length of the second arc.

According to the proposed antenna parametric study, it is easy to reach any frequency in the band 1.8–2.6 GHz by simply modifying the dimensions of the antenna. Moreover, we can realize a single, double or triple band antenna centered at: 1.8, 1.9, 2, 2.2, 2.4, 2.5, and 2.6 GHz for digital communication system (DCS) (1.71–1.88 GHz), personal communication services (PCS) (1.85–1.99 GHz), UMTS (1.92–2.17 GHz), Bluetooth (2.4–2.5 GHz), Wi-Fi (2.4–2.454 GHz), ISM (2.4 GHz), and WLAN (2.4 GHz).

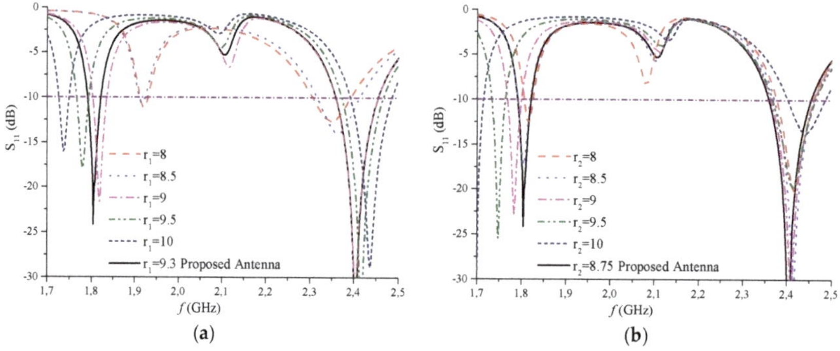

Figure 8. *Cont.*

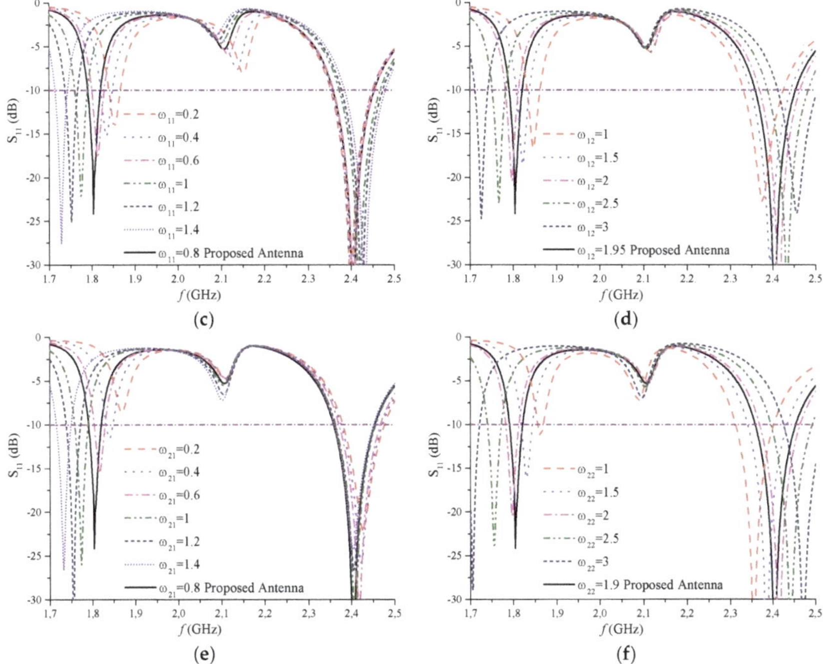

(c) (d) (e) (f)

Figure 8. Effect of arcs thicknesses on S_{11} of the proposed antenna: (**a**) inner radius of the first arc r_1; (**b**) inner radius of the second arc r_2; (**c**) w_{11} thickness of the first arc; (**d**) w_{12} spacing between ground plane and first arc; (**e**) w_{21} spacing between ground plane and second arc; and (**f**) w_{22} thickness of the second arc.

The surface current density at operating frequencies is investigated to better understand the operation mode of the proposed dual-band antenna, as shown in Figure 9a,b. Figure 9a shows that the highest current density is observed along the quarter length of the feed line and the two arc-shaped slots at 1.8 GHz. For 2.4 GHz, the current is mainly concentrated along the microstrip line and the two arc-shaped slots, as shown in Figure 9b. By analyzing the current flux plot at 1.8 and 2.4 GHz, we can conclude that coupling between the arc-shaped slots and the excitation line generates two simultaneous resonance frequencies.

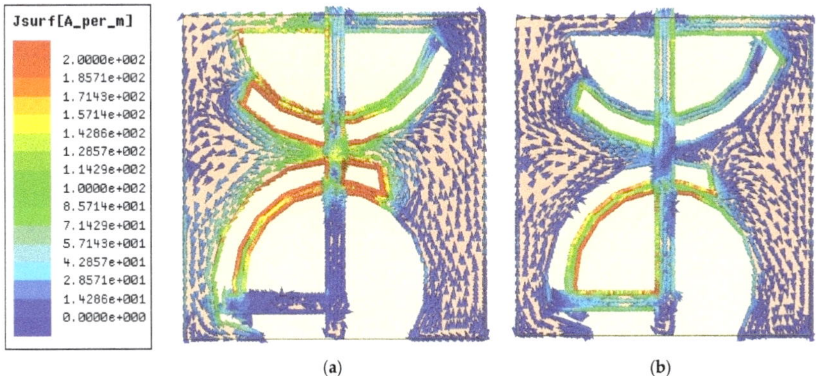

Figure 9. Surface current behavior at: (**a**) 1.8 GHz and (**b**) 2.4 GHz.

Measuring the effective lengths corresponding to the current paths represented by the black lines in Figure 10, we find respectively: L_1 = 57 mm which corresponds to a frequency of f_1 = 2.51 GHz and L_2 = 89 mm which corresponds to a frequency of f_2 = 1.61 GHz.

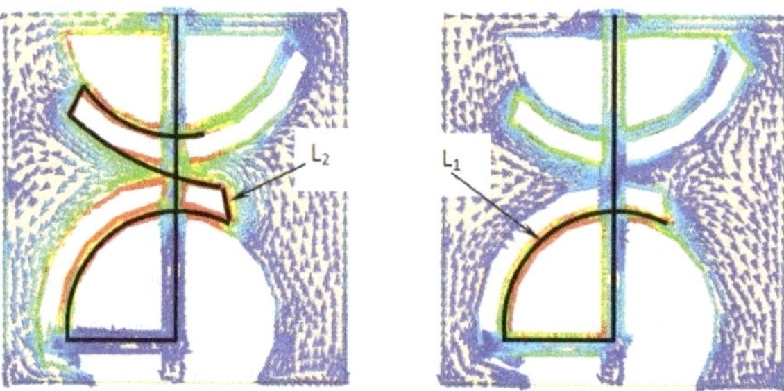

Figure 10. Current paths and antenna effective electric lengths.

According to the radiation pattern plots presented in Figure 11, it is noted that the antenna is omni-directional in the XY plane (H plane) with a weak cross polarization. In the XZ plane (E plane), the radiation pattern is stable with a weak cross polarization as well. Whereas, for the YZ plane, the cross polarization is slightly higher.

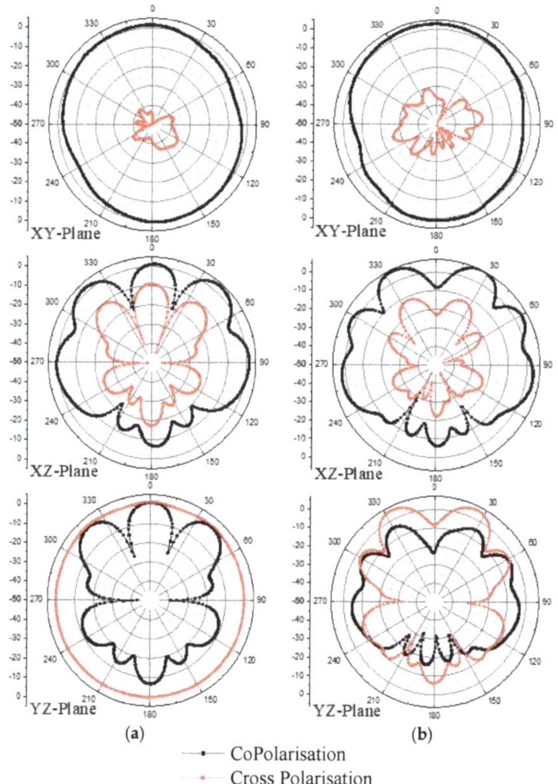

Figure 11. Normalized radiation pattern at: (**a**) 1.8 and (**b**) 2.4 GHz.

To validate the simulated results, an antenna prototype is realized as shown in Figure 12a,b and measurements were performed using an HP8510C vector network analyzer.

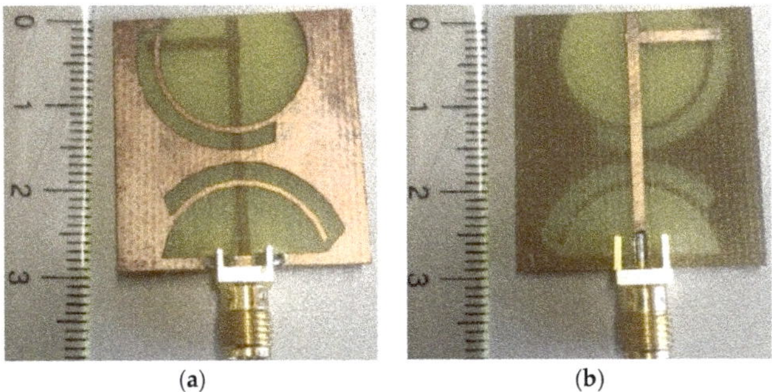

(a)   (b)

Figure 12. Photograph of the realized prototype: (**a**) top view and (**b**) bottom view.

In general, and from Figure 13a,b, measurements of the dual-band antenna S_{11} coefficient and gain coincide well with simulations. For S_{11}, a slight shift of less than 11 MHz between simulations and measurements is observed. For the gain, the obtained results of simulations and measurements for the two frequency bands are: 4.5 and 4.34 dB for 1.8 GHz and 5.29 and 5.09 dB for 2.4 GHz, respectively. The measured gains of the prototype were carried out in a far-field anechoic chamber using a calibrated EMCO type 3115 broadband horn as the reference antenna. The reference fixed antenna was a broadband horn (EMCO type 3115) positioned at 4 m from the antenna under test. We have scanned for three orthogonal planes. For each plane a co-polar and cross-polar gain components were measured.

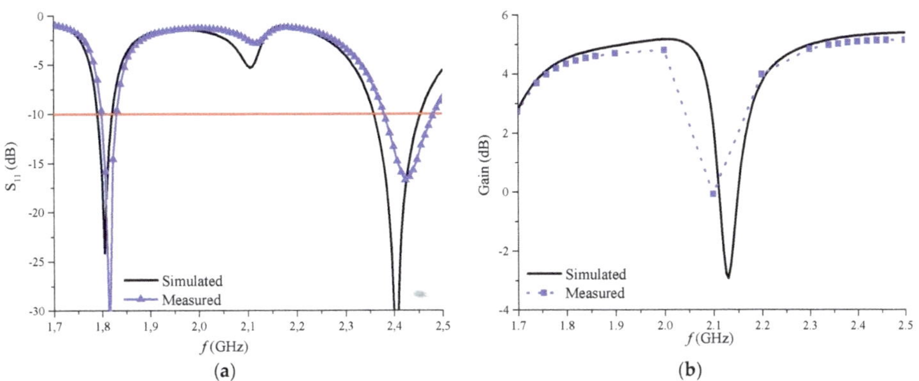

(a)   (b)

Figure 13. Comparison between simulated and measured results of: (**a**) S_{11} and (**b**) gain of the proposed dual-band antenna.

2.4. Specific Absorption Rate (SAR) Analysis

As we have seen in the parametric analysis section, that the antenna can be used as a single, double or triple band antenna centered at: 1.8, 1.9, 2, 2.2, 2.4, 2.5, and 2.6 GHz, like Digital Cellular System DCS (1.71–1.88 GHz), PCS (1.85–1.99 GHz), UMTS (1.92–2.17 GHz), Bluetooth (2.4–2.5 GHz), and Wi-Fi (2.4–2.454 GHz) spectrum Industrial Scientific and Medical radio frequency band (ISM–2.4 GHz), Wireless Local Area Network (WLAN-2.4). The above mentioned bands are commonly used in hand-held modern devices. It is useful to study the impact of the electromagnetic radiations on the human body at these frequencies.

In this work, we only studied the SAR for 1.8 and 2.4 GHz. We modeled a three layer human model using properties of skin, fat and muscle [41,42] as shown in Figure 14a. We placed the antenna at a distance of 5 mm from the skin to avoid direct contact with the skin layer. As can be seen from the Figure 14b, the proposed antenna has 10-g peak SAR values of 1.49 and 1.33 W/kg at 1.8 and 2.4 GHz, respectively for input power of 1 W. According to IEEE C95.1-2005, the SAR values should not exceed 2 W/kg for 10-g sample and 1.6 W/kg for 1-g sample [43]. From the peak SAR values, it is clear that the proposed antenna is safe to be used in the any hand-held device with input power not more than 1.34 W.

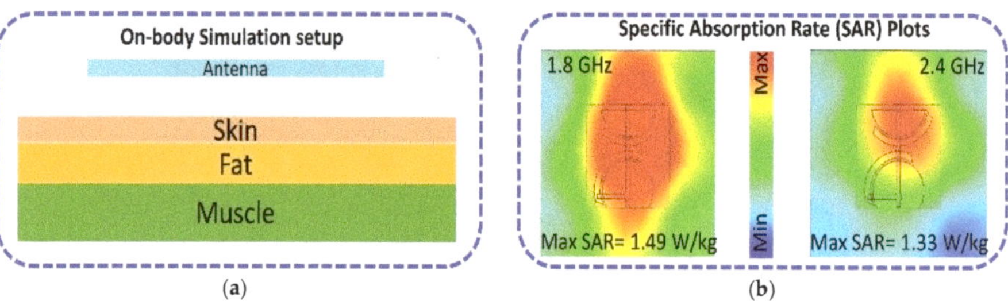

Figure 14. On-body analysis of the proposed antenna: (**a**) simulation setup for on-body analysis (**b**) 10-g SAR distributions at 1.8 and 2.4 GHz.

Let us conclude this part with a comparison of the size, operating bands and gain of our antenna with those realized or simulated available in the literature. It is clear from Table 1 that our proposed structure presents a good compromise between miniaturization and the significant gain obtained compared to the reported studies. Our antenna structure is well miniaturized and has the best ratio gain/size. By comparing with higher gain antenna cases, close to ours, presented in [2,5,6,17,23,44], our proposed antenna size is reduced by more than three times. However, for reported antennas with size close to ours [3,9,10,13,18,26,30,45,46], the proposed antenna presents a higher gain.

Table 1. Antenna dimensions and properties comparison with published data.

Reference, Year	Total Area (mm^2)	Centered Operating Bands (GHz)	Peak Gain in dBi
[2], 2001	90 × 75	1.8, 2.4/2.6	6/4.2/4.5
[5], 2008	75 × 75	2.4/5	4.5/6.2
[17], 2010	75 × 75	2.4–3.0/3.25–3.68/4.9–6.2	3.86/3.52/4.32
[23], 2014	70 × 40	2.4/5.5	1.99/3.71
[12], 2014	30 × 30	2.4/5.2/5.8	not reported
[28], 2014	40 × 50	3.5/5.2	2.84/0.16
[3], 2015	48 × 18	1.6/2.45/3.6/5.5	3.05/3.5/4.2/4.5
[30], 2015	26 × 25	2.5/3.5/5.5	1.73/1.86/2.18
[44], 2015	55 × 50	2.54/3.55/5.7	5.71/6.16/6.48
[45], 2015	31 × 14	2.5/3.5/5.5	nearly 2.9/3.1/4.5
[29], 2016	30 × 15	2.45/3.19–6.44	0.2/2.9
[27], 2016	38 × 38	2.4/3.5/5.8	1.52/1.6/1.5
[46], 2017	22 × 40	2.45/3.49/5.13/5.81	1.72/2/2.08/2.96
[33], 2018	56 × 44	3.1/5.52/7.31/9.72	1.35/1.0/1.07/1.75
[9], 2018	45 × 17	0.868/2.4	1.18/2.1
[10], 2018	20 × 21	913–934/1.5–1.59/2.43–2.50	0.32/1.2/1.5
[13], 2018	28 × 30	1.6/2.5/5.8/9.5	2.9/2.4/3.1/1.8
[18], 2018	32 × 32	3.5/5.9/6.7/8.5/9.8	1.2/1.6/2.1/2.5/2.7
[34], 2018	57.2 × 31.2	0.8/2.45/3.5/5.5	-8.12/-1.31/1.46/3.66
[36], 2018	40 × 40	2.16–3.42	2–3.2
[4], 2019	40 × 45	2.-2.6/3.21-3.5/3.8-6.38	2.2/4.3/6.3
[6], 2019	110 × 89	1.8/2.4	6.31/7.8
[11], 2019	220 × 220	0.9–6.1	Between 3.5–7
[26], 2019	35 × 30	2.3/2.7/3.5/3.8/4.3/5.6	2.3/2.7/3.5/3.8/4.3/5.6
[32], 2019	16.45 × 16	1.89/3.5/5.5	0.136/2.12/3.55
[26], 2019	35 × 30	2.3/2.7/3.5/3.8/4.3/5.6	2.2/2.5/3.4/3.2/3.5/4
Our proposed antenna	30 × 28.5	1.8/2.4	4.39/5.02

3. Conclusions

In this work, we proposed a new slot antenna design for dual-band operation. Two adjunct arc-shaped slots are added to two main semi-circular slots, all etched on the ground plane to realize an appropriate adjustment of the desired frequency bands. Thus, two resonant frequencies centered at 1.8 and 2.4 GHz have been achieved resulting in a dual-band antenna. A radiation peak gain of 4.39/5.09 dB has been obtained. The key design advantages, including size reduction, simple configuration, adequate gain, and dual-band feature make the proposed antenna an excellent candidate with potential features for current and future applications.

Author Contributions: Design and concept, C.Z. and D.S.; methodology, C.Z. and I.E.; investigation, A.I., W.F.A.M. and C.Z.; resources, I.E. and J.R.; writing—original draft preparation, C.Z.; writing—review and editing, R.A.-A., I.E., J.K., and A.I.; validation, visualization, R.A.-A.; supervision, J.R.; project administration, J.R.

Funding: This project has received funding from the European Union's Horizon 2020 research and innovation program under grant agreement H2020-MSCA-ITN-2016 SECRET-722424. This work is also funded by the FCT/MEC through national funds and when applicable co-financed by the ERDF, under the PT2020 Partnership Agreement under the UID/EEA/50008/2019 project.

Acknowledgments: This work is supported by the European Union's Horizon 2020 Research and Innovation program under grant agreement H2020-MSCA-ITN-2016-SECRET-722424. This work is part of the POSITION-II project funded by the ECSEL joint Undertaking under grant number Ecsel-7831132-Postitio-II-2017-IA, www. position-2.eu.

Conflicts of Interest: The authors declare no conflict of interest.

References

1. Tang, M.C.; Shi, T.; Ziolkowski, R.W. Planar ultrawideband antennas with improved realized gain performance. *IEEE Trans. Antennas Propag.* **2015**, *64*, 61–69. [CrossRef]
2. Chen, W.-S.; Wong, K.-L. Dual-frequency operation of a coplanar waveguide-fed dual-slot loop antenna. *Microw. Opt. Technol. Lett.* **2001**, *30*, 38–40. [CrossRef]
3. Cao, Y.F.; Cheung, S.W.; Yuk, T.I. A multi-band slot antenna for GPS/WiMAX/WLAN systems. *IEEE Trans. Antennas Propag.* **2015**, *63*, 952–958. [CrossRef]
4. Kunwar, A.; Gautam, A.K.; Kanaujia, B.K.; Rambabu, K. Circularly polarized D-shaped slot antenna for wireless applications. *Int. J. RF Microw. Comput.-Aided Eng.* **2019**, *29*, e21498. [CrossRef]
5. Sze, J.-Y.; Hsu, C.-I.G.; Hsu, S.-C. Dual-broadband multistandard printed slot antenna with a composite back-patch. *Microw. Antennas Propag.* **2008**, *2*, 205–209. [CrossRef]
6. Hassan, N.; Zakaria, Z.W.; Sam, Y.; Hanapiah, I.N.M.; Mohamad, A.N.; Roslan, A.F.; Aziz, M.Z.A. A Design of dual-band microstrip patch antenna with right-angle triangular aperture slot for energy transfer application. *Int. J. RF Microw. Comput.-Aided Eng.* **2019**, *29*, e21666. [CrossRef]
7. Zebiri, C.; Lashab, M.; Sayad, D.; Elfergani, I.E.; Ali, A.; Khambashi, M.A.; Abd-Alhameed, R. Bandwidth Enhancement of rectangular dielectric resonator antenna using circular and sector slot coupled technique. In Proceedings of the 12th European Conference on Antennas and Propagation (EuCAP 2018), London, UK, 9–13 April 2018.
8. Elfergani, I.; Hussaini, A.S.; Rodriguez, J.; Abd-Alhameed, R. (Eds.) *Antenna Fundamentals for Legacy Mobile Applications and Beyond*, 1st ed.; Springer: Cham, Switzerland, 2018. [CrossRef]
9. Upadhyaya, T.; Desai, A.; Patel, R. Design of printed monopole antenna for wireless energy meter and smart applications. *Prog. Electromagn. Res. Lett.* **2018**, *77*, 27–33. [CrossRef]
10. Patel, R.; Desai, A.; Upadhyaya, T. An electrically small antenna using defected ground structure for RFID, GPS and IEEE 802.11 a/b/g/s applications. *Prog. Electromagn. Res. Lett.* **2018**, *75*, 75–81. [CrossRef]
11. Li, K.; Dong, T.; Xia, Z. Wideband Printed Wide-Slot Antenna with Fork-Shaped Stub. *Electronics* **2019**, *8*, 347. [CrossRef]
12. Ojaroudi, N.; Ghadimi, N. Design of CPW-fed slot antenna for MIMO system applications. *Microw. Opt. Technol. Lett.* **2014**, *56*, 1278–1281. [CrossRef]
13. Tanweer, A.; Muzammil, M.; Biradar, R.C. A multi-band reconfigurable slot antenna for wireless applications. *AEU-Int. J. Electron. Commun.* **2018**, *84*, 273–280.

14. Chiang, M.J.; Wang, S.; Hsu, C.C. Compact multifrequency slot antenna design incorporating embedded arc-strip. *IEEE Antennas Wirel. Propag. Lett.* **2012**, *11*, 834–837. [CrossRef]

15. Saghati, A.P.; Azarmanesh, M.; Zaker, R. A novel switchable singleand multifrequency triple-slot antenna for 2.4-GHz bluetooth, 3.5-GHz WiMax, and 5.8-GHz WLAN. *IEEE Antennas Wirel. Propag. Lett.* **2010**, *9*, 534–537. [CrossRef]

16. Lu, J.H.; Huang, B.J. Planar compact slot antenna with multi-band operation for IEEE 802.16 m application. *IEEE Trans. Antennas Propag.* **2013**, *61*, 1411. [CrossRef]

17. Dang, L.; Lei, Z.Y.; Xie, Y.J.; Ning, G.L.; Fan, J. A compact microstrip slot triple-band antenna for WLAN/WiMAX applications. *IEEE Antennas Wirel. Propag. Lett.* **2010**, *9*, 1178–1181. [CrossRef]

18. Tanweer, A.; Prasad, K.D.; Biradar, R.C. A miniaturized slotted multi-band antenna for wireless applications. *J. Comput. Electron.* **2018**, *17*, 1–15.

19. Bod, M.; Hassani, H.R.; Taheri, M.M. Compact UWB printed slot antenna with extra bluetooth, GSM, and GPS bands. *IEEE Antennas Wirel. Propag. Lett.* **2012**, *11*, 531–534. [CrossRef]

20. Khandelwal, M.K.; Kanaujia, B.K.; Kumar, S. Defected ground structure: Fundamentals, analysis, and applications in modern wireless trends. *Int. J. Antennas Propag.* **2017**. [CrossRef]

21. Jaiswal, A.; Sarin, R.K.; Raj, B.; Sukhija, S. A novel circular slotted microstrip-fed patch antenna with three triangle shape defected ground structure for multi-band applications. *Adv. Electromagn.* **2018**, *7*, 56–63. [CrossRef]

22. Tahar, F.; Barakat, A.; Saad, R.; Yoshitomi, K.; Pokharel, R.K. Dual-Band Defected Ground Structures Wireless Power Transfer System with Independent External and Inter-Resonator Coupling. *IEEE Trans. Circuits Syst. II Express Briefs* **2017**, *64*, 1372–1376. [CrossRef]

23. Lin, W.P.; Yang, D.H.; De Lin, Z. Compact dual-band planar inverted-e-shaped antenna using defected ground structure. *Int. J. Antennas Propag.* **2014**. [CrossRef]

24. Zbitou, J.; Errkik, A. *Emerging Innovations in Microwave and Antenna Engineering*; IGI Global: Hershey, PA, USA, 2018.

25. Yunus, M.; Sinaga, T.J.; Fitri, I.; Wismiana, E.; Munir, A. Bowtie-shaped DGS for reducing coupling between elements of planar array antenna. In Proceedings of the 2017 International Symposium on Electronics and Smart Devices (ISESD), Yogyakarta, Indonesia, 17–19 October 2017; pp. 226–229.

26. Pokkunuri, P.; Madhav, B.T.P.; Sai, G.K.; Venkateswararao, M.; Ganesh, B.; Tarakaram, N.; Teja, D. Metamaterial Inspired Reconfigurable Fractal Monopole Antenna for Multi-band Applications. *Int. J. Intell. Eng. Syst.* **2019**, *12*, 53–61.

27. Saraswat, R.K.; Kumar, M. Miniaturized slotted ground UWB antenna loaded with metamaterial for WLAN and WiMAX applications. *Prog. Electromagn. Res.* **2016**, *65*, 65–80. [CrossRef]

28. Hung, T.; Liu, J.; Wei, C. Dual-band circularly polarized aperture-coupled stack antenna with fractal patch for WLAN and WiMAX applications. *Int. J. RF Microw. Comput. Aided Eng.* **2014**, *24*, 130–138. [CrossRef]

29. Desde, I.; Bozdag, G.; Kustepeli, A. Multi-band CPW fed MIMO antenna for bluetooth, WLAN, and WiMAX. *Microw. Opt. Technol. Lett.* **2016**, *58*, 1023–1026. [CrossRef]

30. Wu, R.Z.; Wang, P.; Zheng, Q.; Li, R. Compact CPW-fed triple-band antenna for diversity applications. *Electron. Lett.* **2015**, *51*, 735–736. [CrossRef]

31. Zhai, H.; Liu, L.; Ma, Z.; Liang, C. A printed monopole antenna for triple-band WLAN/WiMAX applications. *Int. J. Antennas Propag.* **2015**. [CrossRef]

32. Kumar, C.V.A.; Paul, B.; Mohanan, P. Compact Triband Dual F-Shaped Antenna for DCS/WiMAX/WLAN Applications. *Prog. Electromagn. Res.* **2018**, *78*, 97–104. [CrossRef]

33. Tanweer, A.; Saadh, M.; Biradar, R.C. A fractal quad-band antenna loaded with L-shaped slot and metamaterial for wireless applications. *Int. J. Microw. Wirel. Technol.* **2018**, *10*, 826–834.

34. Chu, H.B.; Shirai, H. A compact metamaterial quad-band antenna based on asymmetric E-CRLH unit cell. *Prog. Electromagn. Res. C* **2018**, *81*, 171–179. [CrossRef]

35. Wu, J.; Ren, X.; Li, Z.; Yin, Y.Z. Modified square slot antennas for broadband circular polarization. *Prog. Electromagn. Res.* **2013**, *38*, 1–14. [CrossRef]

36. Gangwar, S.P.; Gangwar, K.; Kumar, A. A compact modified hexagonal slot antenna for wideband applications. *Electromagnetics* **2018**, *38*, 339–351. [CrossRef]

37. Naser-Moghadasi, M.; Sadeghzadeh, R.A.; Asadpor, L.; Soltani, S.; Virdee, B.S. Improved band-notch technique for ultra-wideband antenna. *IET Microw. Antennas Propag.* **2010**, *4*, 1886–1891. [CrossRef]

38. Zebiri, C.-E.; Lashab, M.; Sayad, D.; Elfergani, I.T.E.; Sayidmarie, K.H.; Benabdelaziz, F.; Abd-Alhameed, R.A.; Rodriguez, J.; Noras, J.M. Offset Aperture-Coupled Double-Cylinder Dielectric Resonator Antenna with Extended-Wideband. *IEEE Trans. Antennas Propag.* **2017**, *65*, 5617–5622. [CrossRef]

39. Iqbal, A.; Bouazizi, A.; Kundu, S.; Elfergani, I.; Rodriguez, J. Dielectric resonator antenna with top loaded parasitic strip elements for dual-band operation. *Microw. Opt. Technol. Lett.* **2019**, *61*, 2134–2140. [CrossRef]

40. Iqbal, A.; A Saraereh, O.; Bouazizi, A.; Basir, A. Metamaterial-based highly isolated MIMO antenna for portable wireless applications. *Electronics* **2018**, *7*, 267. [CrossRef]

41. Bouazizi, A.; Zaibi, G.; Iqbal, A.; Basir, A.; Samet, M.; Kachouri, A. A dual-band case-printed planar inverted-F antenna design with independent resonance control for wearable short range telemetric systems. *Int. J. RF Microw. Comput.-Aided Eng.* **2019**, *29*, e21781. [CrossRef]

42. Iqbal, A.; Saraereh, O.A. Design and analysis of flexible cylindrical dielectric resonator antenna for body centric WiMAX and WLAN applications. In Proceedings of the 2016 Loughborough Antennas & Propagation Conference (LAPC), Loughborough, UK, 14–15 November 2016.

43. Basir, A.; Bouazizi, A.; Zada, M.; Iqbal, A.; Ullah, S.; Naeem, U. A dual-band implantable antenna with wide-band characteristics at MICS and ISM bands. *Microw. Opt. Technol. Lett.* **2018**, *60*, 2944–2949. [CrossRef]

44. Ahsan, M.R.; Islam, M.T.; Ullah, M.H. Computational and experimental analysis of high gain antenna for WLAN/WiMAX applications. *J. Comput. Electron.* **2015**, *14*, 634–641. [CrossRef]

45. Sundar, P.S.; Sarat, K.; Ramakrishna, T.V. Novel Miniatured Wide Band Annular Slot Monopole Antenna. *Far East J. Electron. Commun.* **2015**, *14*, 149–159. [CrossRef]

46. Shah, S.A.A.; Khan, M.F.; Ullah, S.; Basir, A.; Ali, U.; Naeem, U. Design and measurement of planar monopole antennas for multi-band wireless applications. *IETE J. Res.* **2017**, *63*, 194–204. [CrossRef]

 *electronics*

Article

Complex Bianisotropy Effect on the Propagation Constant of a Shielded Multilayered Coplanar Waveguide Using Improved Full Generalized Exponential Matrix Technique

Djamel Sayad [1], Chemseddine Zebiri [2,3], Issa Elfergani [4,*], Jonathan Rodriguez [5], Hasan Abobaker [6], Atta Ullah [3], Raed Abd-Alhameed [3], Ifiok Otung [5] and Fatiha Benabdelaziz [7]

[1] Department of Electrical Engineering, University 20 Aout 1955-Skikda, Skikda 21000, Algeria; d.sayad@univ-skikda.dz
[2] Department of Electronics, University of Ferhat Abbas, Sétif -1-, Sétif 19000, Algeria; czebiri@univ-setif.dz
[3] School of Electrical Engineering and Computer Science, University of Bradford, Bradford BD7 1DP, UK; A.Ullah5@bradford.ac.uk (A.U.); R.A.A.Abd@bradford.ac.uk (R.A.-A.)
[4] Instituto de Telecomunicações, Campus Universitário de Santiago, 3810-193 Aveiro, Portugal
[5] Faculty of Computing, Engineering and Science, University of South Wales, Pontypridd CF37 1DL, UK; jonathan.rodriguez@southwales.ac.uk (J.R.); ifiok.otung@southwales.ac.uk (I.O.)
[6] Department of Computer and Informatics, Faculty of Electrical Engineering and Informatics, Technical University of Kosice, Letná 1/9, 040 01 Košice, Slovakia; atayeib@yahoo.com
[7] Department of Electronics, University Mentouri, Constantine 1, Constantine 25000, Algeria; benabdelaziz2003@yahoo.fr
[*] Correspondence: i.t.e.elfergani@av.it.pt; Tel.: +351-234-377-900

Received: 23 December 2019; Accepted: 28 January 2020; Published: 2 February 2020

Abstract: A theoretical study of the electromagnetic propagation in a complex medium suspended multilayer coplanar waveguide (CPW) is presented. The study is based on the generalized exponential matrix technique (GEMT) combined with Galerkin's spectral method of moments applied to a CPW printed on a bianisotropic medium. The analytical formulation is based on a Full-GEMT, a method that avoids usual procedures of heavy and tedious mathematical expressions in the development of calculations and uses matrix-based mathematical expressions instead. These particularities are exploited to develop a mathematical model for the characterization of wave propagation in a three-layer shielded suspended CPW structure. This study is based on the development of mathematical formulations in full compact matrix-based expressions resulting in Green's functions in a matrix form. The implemented method incorporates a new accelerating procedure developed in the GEMT which provides an initial value used to speed up searching for the exact solution in the principal computation code. This helped us to obtain accurate solutions with tolerable computing time. Good agreements have been achieved with the literature in terms of accuracy and rapid convergence. The results for different cases of bianisotropy have been investigated, and particularly, the effect on the dispersion characteristics is presented and compared with the isotropic case.

Keywords: chiral; Tellegen; multilayer CPW structure; dispersion characteristics; full-GEMT; Muller's method; complex propagation constant; acceleration procedure

---

1. Introduction

To establish strong foundations for the development of modern and mass-market applications in the field of telecommunications, microwave designers have to develop further efficient devices, which aim to meet the specific needs of modern telecommunication systems, particularly the 5G technology. In addition to providing far better levels of reliability and performances by offering new services, the new technology should be fully consistent with traditional services including 2G, 3G, 4G, Wi-Fi, and other relevant wireless systems, which requires suitable and high performance microwave devices in terms of miniaturization and ultra-wideband characteristics that are proven to be the most challenging issues of all time.

An important class of existing microwave devices is that exploiting the particular properties of bianisotropic media [1] for the development of special and innovative devices that may respond to the needs of modern technologies [2]. In this class of promising materials, we may mention, for example, non-reciprocal, gyrotropic, ferrites, chirals, metamaterials, metasurfaces, etc. [1,3]. Over the last three decades, the electromagnetism of bianisotropic media have gained a great deal of interest from scientists and researchers within the frame of artificial media with new and exciting properties [4]. However, practical exploitation of these phenomena did not develop on a large scale; many physics and engineering problems needed to be solved. Recently, as the science of materials has tremendously advanced, the concept of bianisotropic media has substantially reemerged as a field of importance in microwaves and optics technology [5–7].

The particular properties of bianisotropic media arise from a coupling between the electric and magnetic fields that can be explicitly described by general constitutive relations. Due to their diversity, they have found many potential applications from microwaves to optical frequencies such as polarization transformers, directional couplers, antenna and transmission line substrates, antenna radomes, radar systems, chiral waveguides and others [8–13].

The electromagnetic properties of bianisotropic media should be analyzed to perceive their exotic characteristics. Several studies have been conducted to characterize the electromagnetic behavior of bianisotropic media [14–18], ferrites [19], metamaterials [20], chiral [21,22], nonreciprocal [23] for simple and complex dielectric based microwave planar structures using numerical and analytical methods [8,14,17,20,24–35]. In [8] and [35], the method of lines is used to analyze planar transmission lines with conductor losses and to analyze integrated optical waveguide structures, respectively. In [17] and [20], the transmission line matrix (TLM) method is used for modeling dispersive chiral media and the analysis of dispersion in metamaterials. In [25], a fast computation of planar microstrip lines using the generalized equivalent circuit method of moments is presented. The finite difference technique and the iterated moment method are utilized for the analysis of dielectric [33] and optical waveguides [34]. In [27], a complex image method based on genetic algorithm (GA) is proposed to calculate the Green's functions of a coplanar waveguide structure. Anisotropic based multilayers, microstrip and waveguide structures are treated in [28,29,31,32]. Recently, in [14] and [24], Karma et al. studied microstrip transmission lines with anisotropic and uniaxial anisotropic substrates using the discrete mode matching method.

To extract the effective constitutive parameters of bianisotropic materials, various techniques such as stepwise method, S-parameters method, resonator method, coaxial probe method, free-space characterization method, rectangular waveguide measurements and recursive algorithms have been employed [36–39].

By knowing the intrinsic physical properties of complex media, designers can predict the response of microwave components for the development of inventive devices. However, the complexity of mathematical modeling of bianisotropic media is a real challenge in the characterization of microwave components. This has recently become a central area of research in microwaves and optics.

This paper introduces a mathematical modeling of complex media characterized by full 3×3 bianisotropic tensors of permittivity, permeability and magneto-electric parameters. The objective of this work is to sufficiently develop predictive mathematical models to judiciously characterize the propagation of electromagnetic waves in a suspended shielded bianisotropic three-layer CPW structure using a Full-GEM technique [28,40,41]. Three primary considerations have been exploited to accomplish new and promising results. These include: the consideration of the most general reciprocal and non-reciprocal chiral and achiral (Tellegen) complex media using the Full-GEMT, estimation of the effective permittivity constant to be used as an initial value to search for the exact solution, and the use of Muller's method for the extraction of the complex solution of the associated propagation constant in both chiral and achiral media.

The efficient spectral Galerkin-based method of moments (SGMoM) is extensively used to analyze microwave planar structures [18,22,25,27,29,30,41–48]. In our recent work [42], we presented an analytical modeling of a shielded microstrip line based on an anisotropic medium with full 3×3 permittivity and permeability tensors. To reduce calculations, some conditions on the permittivity and permeability constants were considered ensuring the decoupling of the TE and TM modes. On the other hand, the computational time drastically increases with the required accuracy because of the slow convergence of integrals and series summations making this approach time intensive [43]. For accelerated convergence and efficient computation, several techniques have been used [44–47]. The time intensive part in the SGMoM is the evaluation of the matrix elements and the determinant calculation, as the matrix size is large in most cases. In [48], the SGMoM calculation of the propagation constant (β) for a shielded microstrip line is accelerated using asymptotic expansions for the Bessel's and the Green's functions with the aid of super convergent series in the approximation of the summation of the leading terms. In [49], for the multilayered shielded microstrip analysis, series summation calculation are accelerated using the Levin's transformation in the spectral method. In [28], the Green's function-based volume integral equation computation is accelerated using the fast Fourier transform technique. In [44,45], the impedance matrix elements based on Sommerfeld-type infinite double integral Green functions evaluation is accelerated by converting the infinite double integral of the impedance-matrix elements into a finite one-dimensional integral by using the asymptotic Green's functions and triangular basis functions with edge condition. In [47], integral-equation formulation in the SGMoM is accelerated by extracting suitable half-space parts of the kernels which leads to an exponentially decaying integrand functions. The integrals of the extracted parts are expressed as combinations of proper integrals and fast converging improper integrals. In [50], an efficient quasi-static analysis is presented, which can be used for speeding up full-wave SGMoM computations as well. Accelerated versions of full-wave spectral domain approach are also reported in [51–54].

In this paper, we propose a novel approach for the numerical acceleration of the SGMoM for the analysis of bianisotropic medium-based microstrip structures by accelerating and fixing problems of the convergence of the series summation in the elements of the Galerkin's matrix based on Green's functions.

The herein considered complex medium is bianisotropic with non-zero magneto-electric tensors ($\xi \neq 0$ and $\eta \neq 0$ at the same time). Our previous study, presented in [41], did not treat the case of bianisotropic media with both magneto-electric tensors; only one tensor was considered non zero. The relative simplicity of this case of medium does not require a more complex resolution technique or longer calculation time. Note that this technique failed to provide accurate solutions for general complex bianisotropic media due to the round-off errors and the highly oscillatory fields behavior. A new procedure is implemented to improve the technique and expand it to support the general case of bianisotropy for a CPW structure. Due to the complexity of the considered bianisotropic medium, the resolution method has required a more efficient technique to overcome the drawbacks in terms of non-convergence or considerable calculation time for a tolerable accuracy. An improvement is achieved by introducing an intermediate calculation procedure based on the GEMT to retrieve an approximated initial value of the relative effective permittivity of the three layer-structure as a function of ε_r, μ_r, ξ and η: the bianisotropic layer constitutive parameters. This value is used for searching for the exact solutions of the normalized complex constant of propagation ($(\beta/\kappa_0)^2$ and $(\alpha/\kappa_0)^2$).

2. Exponential Matrix Technique Formulation

The general CPW geometry and the appropriate coordinate system with the z-axis as the direction of propagation are shown in Figure 1. The considered structure is based on a complex bianisotropic medium (region 1) characterized by full 3×3-magneto-electric tensors expressing the cross coupling between electric and magnetic fields.

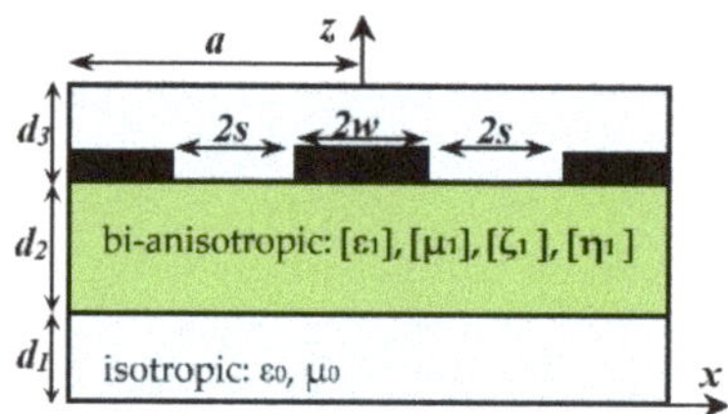

Figure 1. Geometry of the shielded suspended 3-layer CPW structure.

Bianisotropic materials, in their general form, are characterized by the following constitutive relations [30,41,45].

$$\vec{D} = \varepsilon_0[\varepsilon]\vec{E} + \sqrt{\varepsilon_0\mu_0}[\xi]\vec{H}$$
$$\vec{B} = \mu_0[\mu]\vec{H} + \sqrt{\varepsilon_0\mu_0}[\eta]\vec{E}$$

(1)

The tensors of the relative permittivity $[\varepsilon]$, relative permeability $[\mu]$ and magneto-electric elements $[\xi]$ and $[\eta]$ are represented in the Cartesian coordinate system as follows:

$$\psi = \begin{bmatrix} \psi_{xx} & \psi_{xy} & \psi_{xz} \\ \psi_{yx} & \psi_{yy} & \psi_{yz} \\ \psi_{zx} & \psi_{zy} & \psi_{zz} \end{bmatrix},$$

(2)

where ψ stands for $[\varepsilon]$, $[\mu]$, $[\xi]$, or $[\eta]$.

Starting from Maxwell's equations and using the GEMT in the spectral domain, we come to four coupled first-order differential equations for the transverse electromagnetic field components as functions of their derivatives [40,41] given in the Fourier domain:

$$\frac{\partial\left[\widetilde{f}^{(i)}(\alpha,\beta,z)\right]}{\partial z} = \left[P^{(i)}\right]_{4\times4}\left[\widetilde{f}^{(i)}(\alpha,\beta,z)\right],$$

(3)

α and β are the Fourier variables corresponding to the space domain wavenumbers κ_x and κ_y, with

$$\left[\widetilde{f}^{(i)}(\alpha,\beta,z)\right] = \begin{bmatrix} \widetilde{E}_x^{(i)}(\alpha,\beta,z) \\ \widetilde{E}_y^{(i)}(\alpha,\beta,z) \\ \widetilde{H}_x^{(i)}(\alpha,\beta,z) \\ \widetilde{H}_y^{(i)}(\alpha,\beta,z) \end{bmatrix},$$

(4)

and

$$[P]_{4\times4} = j\kappa_0\left\{[M] + [N][Q][R]\right\},$$

(5)

where

$$[M] = \begin{bmatrix} -\eta_{yx} & -\eta_{yy} & -{}_0\mu_{yx} & -{}_0\mu_{yy} \\ \eta_{xx} & \eta_{xy} & {}_0\mu_{xx} & {}_0\mu_{xy} \\ Y_0\varepsilon_{yx} & Y_0\varepsilon_{yy} & \xi_{yx} & \xi_{yy} \\ -Y_0\varepsilon_{xx} & -Y_0\varepsilon_{xy} & -\xi_{xx} & -\xi_{xy} \end{bmatrix} = \begin{bmatrix} [\eta_T] & {}_0[\mu_T] \\ -Y_0[\varepsilon_T] & -[\xi_T] \end{bmatrix},$$

(6a)

$$[N] = \begin{bmatrix} -\left(\eta_{yz}+\kappa_x^n\right) & -{}_0\mu_{yz} \\ \left(\eta_{xz}-\kappa_y^n\right) & {}_0\mu_{xz} \\ Y_0\varepsilon_{yz} & \left(\xi_{yz}-\kappa_x^n\right) \\ -Y_0\varepsilon_{xz} & -\left(\xi_{xz}+\kappa_y^n\right) \end{bmatrix},$$

(6b)

$$[Q] = \frac{-1}{\left(\varepsilon_{zz}\mu_{zz}-\xi_{zz}\eta_{zz}\right)}\begin{bmatrix} {}_0\mu_{zz} & \xi_{zz} \\ -\eta_{zz} & -Y_0\varepsilon_{zz} \end{bmatrix},$$

(6c)

$$[R] = \begin{bmatrix} Y_0 \varepsilon_{xz} & Y_0 \varepsilon_{yz} & \left(\xi_{xz} - \kappa_y^n\right) & \left(\xi_{yz} + \kappa_x^n\right) \\ -\left(\eta_{xz} + \kappa_y^n\right) & -\left(\eta_{yz} - \kappa_x^n\right) & -{}_0\mu_{xz} & -{}_0\mu_{yz} \end{bmatrix},$$

(6d)

$$\kappa_x^n = \frac{\kappa_x}{\kappa_0},$$

(6e)

$$\kappa_y^n = \frac{\kappa_y}{\kappa_0},$$

(6f)

$$_0 = \frac{1}{Y_0} = \sqrt{\frac{\mu_0}{\varepsilon_0}},$$

(6g)

with

$$[\eta_T] = \begin{bmatrix} -\eta_{yx} & -\eta_{yy} \\ \eta_{xx} & \eta_{xy} \end{bmatrix}, \ [\varepsilon_T] = \begin{bmatrix} -\varepsilon_{yx} & -\varepsilon_{yy} \\ \varepsilon_{xx} & \varepsilon_{xy} \end{bmatrix}, \ [\xi_T] = \begin{bmatrix} -\xi_{yx} & -\xi_{yy} \\ \xi_{xx} & \xi_{xy} \end{bmatrix}, \ [\mu_T] = \begin{bmatrix} -\mu_{yx} & -\mu_{yy} \\ \mu_{xx} & \mu_{xy} \end{bmatrix}.$$

(6h)

where κ_0 is the free space wavenumber and ω is the angular frequency.

This study is essentially based on the development of mathematical formulations in compact matrix-based forms; this is deemed as a promising approach, since it avoids excessive and complex calculation developments. This can dramatically reduce the complexity of wave propagation modeling in complex media.

The matrix $[P]$ (Equation (5)) is the first foundation for this technique associated with the studied bianisotropic-medium based CPW structure. Its elements are given as functions of the constitutive tensors elements. In previous works [30,45], for particular cases of media calculations were explicitly developed, which is not obvious with heavy mathematical calculations case that characterize the herein studied complex bianisotropic structure.

Equation (3) admits a general solution of the form:

$$\begin{bmatrix} E_x(z) \\ E_y(z) \\ H_x(z) \\ H_y(z) \end{bmatrix} = T(\kappa_x, \kappa_y, z) \begin{bmatrix} E_x(0) \\ E_y(0) \\ H_x(0) \\ H_y(0) \end{bmatrix},$$

(7)

with

$$T(\kappa_x, \kappa_y, z) = \exp([P] \cdot z),$$

(8)

and

$$\tilde{f}^{(i)}(\alpha, \beta, z(i)) = T(\kappa_x, \kappa_y, z)\tilde{f}^{(i)}(\alpha, \beta, z(i-1)).$$

(9)

The 4×4 transfer matrix $T(\kappa_x, \kappa_y; z)$ is calculated in the formulation of the GEMT by means of the Cayley Hamilton theorem for the determination of the complex function roots [40]. It is expressed in the following polynomial form:

$$T(z) = a_0[I] + a_1[P] + a_2[P]^2 + a_3[P]^3,$$

(10)

where a_j are scalar expansion coefficients, determined by solving the Vandermode linear algebraic system [40], and $[I]$ is a 4×4 identity matrix. The transfer matrix is easily obtained by multiplying the different transfer matrices related to the different layers of the structure, this constitutes the main advantage of this new technique. By imposing the appropriate boundary conditions between the heterogeneous medium layers, the appropriate Green's tensor which models the CPW structure is derived. Details can be found in [41]. This technique exhibits a compact matrix form with the advantage of being easily inserted in the calculation code.

2.1. Implementation of the Acceleration Procedure

To overcome the drawbacks of the resolution method used in [41], when applied for a general complex bianisotropic medium, the resolution method has to be improved in terms of convergence and accuracy for a tolerable computing time. A new procedure is introduced in the calculation technique. This latter is based on the GEMT technique (detailed calculations can be found in [40]) used to retrieve an approximated value for the effective relative permittivity of the whole inhomogeneous structure to be used as an initial value to search for the exact solution of the propagation constant. This value is evaluated for each frequency point by extracting the eigenvalues of matrix $[P]$ by resolving Equation (12) for $\kappa_x = 0$ and $\kappa_y = 0$. By applying this procedure, analytical expressions of the effective relative permittivity maybe obtained as functions of the constitutive parameters of the bianisotropic layer ε_r, μ_r, ξ and η. The application of this procedure allows the acceleration of the Matlab® [55] calculation code and provides a better solution accuracy.

The expansion coefficients a_i $(i = 0, 1, 2, 3)$ in Equation (10) are determined by solving the Vandermode linear algebraic system:

$$
\begin{bmatrix}
1 & \lambda_0 & \lambda_0^2 & \lambda_0^3 \\
1 & \lambda_1 & \lambda_1^2 & \lambda_1^3 \\
1 & \lambda_2 & \lambda_2^2 & \lambda_2^3 \\
1 & \lambda_3 & \lambda_3^2 & \lambda_3^3
\end{bmatrix}
\begin{bmatrix}
a_0(z) \\
a_1(z) \\
a_2(z) \\
a_3(z)
\end{bmatrix}
=
\begin{bmatrix}
\exp(\lambda_0 z) \\
\exp(\lambda_1 z) \\
\exp(\lambda_2 z) \\
\exp(\lambda_3 z)
\end{bmatrix},
\tag{11}
$$

λ_i $(i = 0, 1, 2, 3)$: are eigenvalues of $[P]$ which correspond to propagating waves [56] defined by:

$$
\det(\lambda[I] - [P]) = \lambda^4 + \alpha_1 \lambda^3 + \alpha_2 \lambda^2 + \alpha_3 \lambda + \alpha_4 = 0.
\tag{12}
$$

The coefficients α_i $(i = 0, 1, 2, 3)$ are given in terms of the matrix $[P]$, explicit expressions may be found in [40].

2.2. Derivation of the Initial Value Expression of the Effective Relative Permittivity

a. Isotropic case

The initial effective permittivity value expression is calculated in terms of the medium constitutive parameters ε_r, μ_r, ξ and η using the total transfer matrix (Equation (10)) for $\kappa_x = 0$ and $\kappa_y = 0$, which permits the extraction of an approximated value. As an example of calculations, we present the derived analytical expressions of the initial effective permittivity of the isotropic and some bianisotropic cases. For the isotropic case, the derived matrix $[P]$ is given by:

$$
[P] =
\begin{bmatrix}
0 & 0 & 0 & -\mu_r Z_0 \kappa_0 \\
0 & 0 & \mu_r Z_0 \kappa_0 & 0 \\
0 & \varepsilon_r \kappa_0 / Z_0 & 0 & 0 \\
-\varepsilon_r \kappa_0 / Z_0 & 0 & 0 & 0
\end{bmatrix}.
\tag{13a}
$$

In this case, the normalized eigenvalues of matrix $[P]$, with respect to κ_0, are:

$$
\lambda_{n0} = \sqrt{\varepsilon_r \mu_r}, \quad \lambda_{n1} = -\sqrt{\varepsilon_r \mu_r}, \quad \lambda_{n2} = \sqrt{\varepsilon_r \mu_r} \text{ and } \lambda_{n3} = -\sqrt{\varepsilon_r \mu_r};
\tag{13b}
$$

functions of the relative permittivity and permeability of the isotropic medium. For the fundamental propagating mode, the numerical maximal value is taken as the initial value.

$$
\varepsilon_{reff0} = \lambda_{n0}^2 = \varepsilon_r \mu_r
\tag{13c}
$$

b. Uniaxial anisotropic case

In this example, we consider the following special case of biaxial anisotropy

$$[\varepsilon] = \begin{bmatrix} \varepsilon_x & 0 & 0 \\ 0 & \varepsilon_y & 0 \\ 0 & 0 & \varepsilon_z \end{bmatrix}, [\mu] = \begin{bmatrix} \mu_x & 0 & 0 \\ 0 & \mu_y & 0 \\ 0 & 0 & \mu_z \end{bmatrix}, [\xi] = 0, [\eta] = 0, \tag{14a}$$

the derived matrix $[P]$ is:

$$[P] = \begin{bmatrix} 0 & 0 & 0 & -\mu_y Z_0 \kappa_0 \\ 0 & 0 & \mu_x Z_0 \kappa_0 & 0 \\ 0 & \varepsilon_y \kappa_0/Z_0 & 0 & 0 \\ -\varepsilon_x \kappa_0/Z_0 & 0 & 0 & 0 \end{bmatrix}, \tag{14b}$$

where the four normalized eigenvalues are found to be

$$\lambda_{n0} = \sqrt{\tfrac{\mu_x}{\mu_z}\left(\varepsilon_y\mu_z - 1\right)}, \; \lambda_{n1} = -\sqrt{\tfrac{\mu_x}{\mu_z}\left(\varepsilon_y\mu_z - 1\right)},$$
$$\lambda_{n2} = \sqrt{\tfrac{\varepsilon_x}{\varepsilon_z}\left(\varepsilon_z\mu_y - 1\right)}, \; \lambda_{n3} = -\sqrt{\tfrac{\varepsilon_x}{\varepsilon_z}\left(\varepsilon_z\mu_y - 1\right)} \tag{14c}$$

and

$$\varepsilon_{reff0} = \max\left(\lambda_{ni}^2\right) = \max\left(\sqrt{\tfrac{\mu_x}{\mu_z}\left(\varepsilon_y\mu_z - 1\right)}, \; \sqrt{\tfrac{\varepsilon_x}{\varepsilon_z}\left(\varepsilon_z\mu_y - 1\right)}\right) \tag{14d}$$

c. Diagonal bianisotropy case

As an example, we take the following case:

$$[\varepsilon] = \varepsilon_r[I], \, [\mu] = \mu_r[I], \, [\xi] = \begin{bmatrix} \xi_{xx} & 0 & 0 \\ 0 & 0 & 0 \\ 0 & 0 & 0 \end{bmatrix}, \, [\eta] = \begin{bmatrix} \eta_{xx} & 0 & 0 \\ 0 & 0 & 0 \\ 0 & 0 & 0 \end{bmatrix}, \tag{15a}$$

where $[I]$ is a 3×3 identity matrix. The derived $[P]$ is:

$$[P] = \begin{bmatrix} 0 & 0 & 0 & -\mu_r Z_0 \kappa_0 \\ \eta_{xx}\kappa_0 & 0 & \mu_r Z_0 \kappa_0 & 0 \\ 0 & \varepsilon_r \kappa_0/Z_0 & 0 & 0 \\ -\varepsilon_r \kappa_0/Z_0 & 0 & -\xi_{xx}\kappa_0 & 0 \end{bmatrix}, \tag{15b}$$

and the four normalized eigenvalues of $[P]$ are:

$$\lambda_{n0} = \sqrt{\varepsilon_r\mu_r + \sqrt{\varepsilon_r\mu_r\xi_{xx}\eta_{xx}}}, \; \lambda_{n1} = -\sqrt{\varepsilon_r\mu_r + \sqrt{\varepsilon_r\mu_r\xi_{xx}\eta_{xx}}},$$
$$\lambda_{n2} = \sqrt{\varepsilon_r\mu_r - \sqrt{\varepsilon_r\mu_r\xi_{xx}\eta_{xx}}}, \; \lambda_{n3} = -\sqrt{\varepsilon_r\mu_r - \sqrt{\varepsilon_r\mu_r\xi_{xx}\eta_{xx}}} \tag{15c}$$

The initial value depends on the constitutive parameters ε_r, μ_r, ξ_{xx} and η_{xx}.

$$\varepsilon_{reff0} = \max\left(\lambda_{ni}^2\right) = \varepsilon_r\mu_r + \sqrt{\varepsilon_r\mu_r\xi_{xx}\eta_{xx}} \tag{15d}$$

d. Gyrotropic bianisotropy case

Considering the following medium case:

$$[\varepsilon] = \varepsilon_r[I], \, [\mu] = \mu_r[I], \, [\xi] = \begin{bmatrix} 0 & 0 & \xi_{xz} \\ 0 & 0 & 0 \\ \xi_{zx} & 0 & 0 \end{bmatrix}, \, [\eta] = \begin{bmatrix} 0 & 0 & \eta_{xz} \\ 0 & 0 & 0 \\ \eta_{zx} & 0 & 0 \end{bmatrix}. \tag{16a}$$

The derived corresponding $[P]$ is:

$$[P] = \begin{bmatrix} 0 & 0 & 0 & -\mu_r Z_0 \kappa_0 \\ 0 & 0 & -\frac{\kappa_0 Z_0}{\varepsilon_r}(\varepsilon_r\mu_r - \xi_{zx}\eta_{xz}) & 0 \\ 0 & \varepsilon_r \kappa_0/Z_0 & 0 & 0 \\ -\frac{\kappa_0}{\mu_r Z_0}(\varepsilon_r\mu_r - \xi_{xz}\eta_{zx}) & 0 & 0 & 0 \end{bmatrix}, \tag{16b}$$

which gives as solutions:

$$\begin{aligned} \lambda_{n0} &= \sqrt{\varepsilon_r\mu_r - \xi_{zx}\eta_{xz}}, \ \lambda_{n1} = -\sqrt{\varepsilon_r\mu_r - \xi_{zx}\eta_{xz}} \\ \lambda_{n2} &= \sqrt{\varepsilon_r\mu_r - \xi_{xz}\eta_{zx}}, \ \lambda_{n3} = -\sqrt{\varepsilon_r\mu_r - \xi_{xz}\eta_{zx}} \end{aligned}, \tag{16c}$$

and

$$\varepsilon_{reff0} = \max(\lambda_{ni}^2) = \max(\varepsilon_r\mu_r - \xi_{xz}\eta_{zx}, \ \varepsilon_r\mu_r - \xi_{zx}\eta_{xz}). \tag{16d}$$

It is medium-case dependent.

Notice that if ξ_{xz}, ξ_{zx}, η_{xz} and η_{zx} are taken so that $\xi_{xz}\eta_{zx} \neq \xi_{zx}\eta_{xz}$, two different solutions are obtained corresponding to bifurcating modes [21]. This may constitute an independent issue, which is outside the scope of this work. This shows the efficiency of the procedure, without which, the calculation code would diverge to adjacent solutions or give spurious ones [57], since the electromagnetic fields of bianisotropic media are highly oscillatory [56]. The result presented by Equation (15d) shows that the use of this approach has not only made it feasible to get the optimal initial value, in some cases, but also to more accurately infer the appropriate approximation, mainly in the presence of the bifurcating modes phenomenon, in which two neighboring modes are excited.

In order to show the benefits of using the new procedure, two examples of the studied cases of bianisotropy are considered. Case1: $\xi_{zx,1} = \xi_{xz,1} = -0.75\sqrt{\varepsilon_r}$, $\eta_{zx,1} = -\xi_{xz,1}$ and $\eta_{zx,1} = \eta_{xz,1}$, and Case2: $\eta_{xz,8} = -\xi_{xz,8} = -j\sqrt{\varepsilon_r}$, $\xi_{zx,8} = -\xi_{xz,8}$ and $\eta_{zx,8} = \eta_{xz,8}$. By the introduction of the new procedure, the computing time for Case1, for a frequency point, is reduced by about 33% from 1.937 s to 1.302 s. In Case2, with the aid of the procedure, we get a solution in 1.442 s while without the procedure, the technique failed to find a solution and gave a spurious value instead, after a long execution time. This is due to the oscillating behavior of the series summations of the manipulated complex Galerkin's matrix.

3. Method of Solution

By applying the boundary conditions, the expressions of the electric and magnetic tangential components are evaluated at the interface air-dielectric in terms of the tangential current densities on the strips $\widetilde{j}_x$ and $\widetilde{j}_y$. A matrix of the Green's tensor elements $G_{ij}(\alpha_n, \beta)$ for the CPW structure is achieved. It is arranged in the following system of equations

$$\begin{bmatrix} \widetilde{j}_x \\ \widetilde{j}_y \end{bmatrix} = \frac{1}{\Delta_G}\begin{bmatrix} G_{22}(\alpha_n,\beta) & -G_{12}(\alpha_n,\beta) \\ -G_{21}(\alpha_n,\beta) & G_{11}(\alpha_n,\beta) \end{bmatrix}\begin{bmatrix} \widetilde{E}_x \\ \widetilde{E}_y \end{bmatrix}, \tag{17a}$$

with

$$\Delta_G = G_{11}G_{22} - G_{12}G_{21}. \tag{17b}$$

and α_n: the discrete Fourier transform variable with n the Fourier number of terms $n = 1, 2, 3, \ldots, N$.

For the resolution of the problem, the SGMoM is applied, the spectral electric field components are expanded in terms of trigonometric basis function sets [58,59]. A homogeneous system of linear equations arranged in a compact matrix form is derived [58]:

$$[M(\beta)]\,[a_1\,a_2\,\ldots a_M\,b_1\,b_2\ldots b_N]^T = [0], \tag{18}$$

where

$$[M(\beta)] = \begin{bmatrix} M_{q',p}^{1,1} & M_{q',q}^{1,2} \\ M_{p',p}^{2,1} & M_{p',q}^{1,1} \end{bmatrix}, \tag{19}$$

and

$$M_{q',p}^{1,1}(\beta) = \sum_n \frac{1}{\Delta_G} G_{22}(\alpha_n, \beta)\, \widetilde{J}_{x,p}\widetilde{J}_{y,q'}^*; \tag{20a}$$

$$M_{q',q}^{1,2}(\beta) = -\sum_n \frac{1}{\Delta_G} G_{12}(\alpha_n, \beta)\, \widetilde{J}_{y,q}\widetilde{J}_{y,q'}^*; \tag{20b}$$

$$M_{p',p}^{2,1}(\beta) = -\sum_n \frac{1}{\Delta_G} G_{12}(\alpha_n, \beta)\, \widetilde{J}_{x,p}\widetilde{J}_{x,p'}^*; \tag{20c}$$

$$M_{p',q}^{2,2}(\beta) = \sum_n \frac{1}{\Delta_G} G_{11}(\alpha_n, \beta)\, \widetilde{J}_{y,q}\widetilde{J}_{x,p'}^*. \tag{20d}$$

The system admits nontrivial solutions when *det* $[M(\beta)] = 0$ [41,45,58,60], from which the frequency dependent propagation constant can be determined. For lossy media, a complex constant solution is expected.

Using the new technique, original results for the dispersion characteristics of complex bianisotropic chiral and achiral media are obtained through the ratio $(\beta/\kappa_0)^2$, discussed and compared with the isotropic case ($\xi = \eta = 0$) using the technique in [41].

Due to the great number of possible medium cases, we have restricted our analysis to highlighting the main results of achiral media that have been less addressed in the literature. Accurate solutions of the determinant roots in Equation (18) are obtained within a tolerance of 10^{-12}.

4. Results and Discussions

In order to validate our calculations and test the efficiency of the proposed method, three magnetic anisotropic cases have initially been considered. Numerical results have been computed and compared with available literature [61,62] (Figure 2) and good agreements are observed. On the other hand, a rapid convergence has been achieved with a reduced Fourier number (N = 500) and basis functions (K = 8) against (N = 3000) used by Khodja et al. [61] for the same number of basis functions.

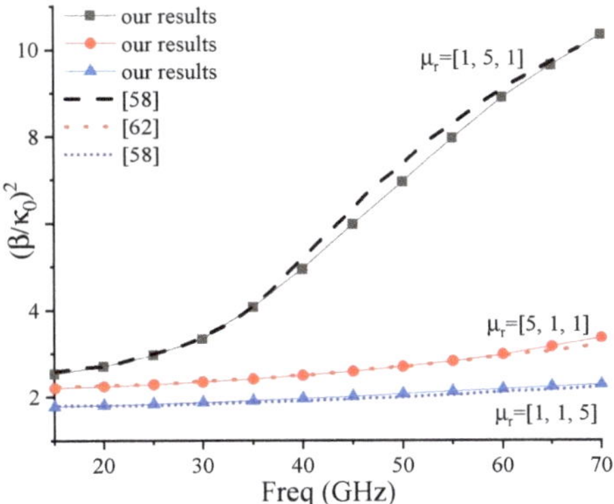

Figure 2. $(\beta/\kappa_0)^2$ for the dominant mode of a shielded CPW with magnetic anisotropy, ($2a$ = 3.556 mm, $2w = 2s$ = 0.7112 mm, $d1$ = 2.8448 mm, $d2$ = 0.7112 mm, $d3$ = 3.556 mm and ε_r = 3).

In this study, we consider a suspended three-layer CPW structure implanted on a complex bianisotropic dielectric material (Figure 1) with the following geometrical dimensions a = 10 mm, $d1$ = 4.5 mm, $d2$ = 1 mm, $d3$ = 4.5 mm, w = 1 mm, s = 1 mm, ε_r = 2.53, μ_r = 1. Different sub-figures are differentiated by the included legends where only the sign of the constitutive element changes respectively to the previous case in the same figure.

In order to examine the effect of the magneto-electric parameters on the dispersion characteristics, we first start with the examination of the axial bianisotropy effect. The magneto-electric elements, whether they are real, imaginary, positive or negative directly affect the phase constant as well as the attenuation factor.

The obtained results, treat two principal cases of diagonal bianisotropic medium: achiral and chiral. In each case, the magneto-electric pair (ξ_{ij}, η_{ij}) is considered non-zero, so that the new original results of the achiral medium case can be validated and compared with the chiral case. In addition, in this parametric study, we examined the effects of the gyrotropic elements of the magneto-electric tensors on the complex propagation coefficient. Results are grouped in figures according to the constitutive parameters effects.

4.1. Effect of Diagonal Bianisotropy

For diagonal bianisotropy three cases of magneto-electric elements are considered:

1. $\xi_{ii,1} = a\sqrt{\varepsilon_r},\ \eta_{ii} = \xi_{ii}$;
2. $\xi_{ii,2} = a\sqrt{\varepsilon_r},\ \eta_{ii} = -\xi_{ii}$;
3. $\xi_{ii,3} = ja\sqrt{\varepsilon_r},\ \eta_{ii} = -\xi_{ii}$.

where $(i = x, y, z)$ and $a = (-1, -0.75, -0.5, -0.25, 0.25, 0.5, 0.75, 1)$. Two main cases are distinguished: achiral $(\xi_{ii} = a\sqrt{\varepsilon_r})$ and chiral $(\xi_{ii} = ja\sqrt{\varepsilon_r})$.

In Figure 3, the effect of element ξ_{yy}, for achiral and chiral media cases is presented. An identical effect is observed on the ratio $(\beta/\kappa_0)^2$ (Figure 3a), with reciprocal effect (curves superposition for $\xi = \pm a\sqrt{\varepsilon_r}$), for both chiral and achiral cases. These media cases show low losses for achirality (case (i)) with $\eta_{yy} = \xi_{yy}$ and almost zero losses for chirality with $\eta_{yy} = -\xi_{yy}$ (case (ii)) (Figure 3b). It can be concluded that the effect of a reciprocal chirality is almost the same as for an achiral medium with equal magneto-electric elements. In these cases $(ii = yy)$ propagating modes are excited in the guiding structure, however, for achiral with $\eta_{ii} = \xi_{ii}$ and chiral with $\eta_{ii} = -\xi_{ii}$, no solutions are obtained for $ii = xx$ and $ii = zz$, hence, the medium does not support any propagating modes.

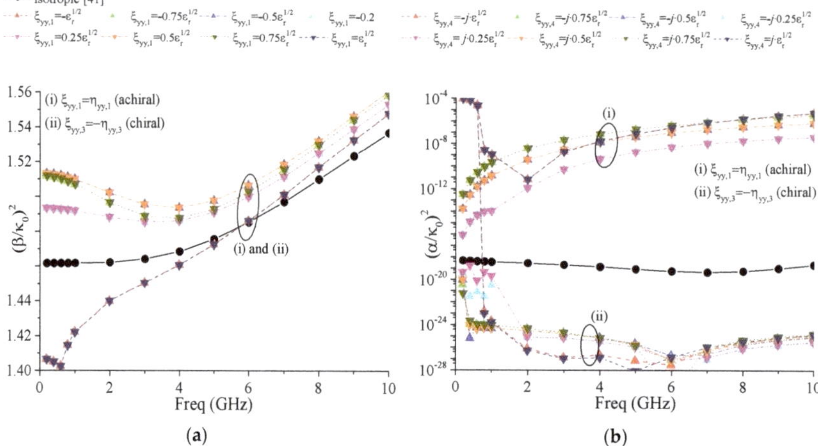

Figure 3. Effect of diagonal bianisotropy ((i) ξ_{yy} real with $\eta_{yy} = \xi_{yy}$ and (ii) ξ_{yy} imaginary with $\eta_{yy} = -\xi_{yy}$) on (**a**): $(\beta/\kappa_0)^2$ and (**b**): $(\alpha/\kappa_0)^2$.

Figure 4 illustrates the effect of diagonal bianisotropic media for the case achiral with $\eta_{ii} = -\xi_{ii}$. Unlike the previous case, the fundamental propagating mode is excited for all diagonal magneto-electric elements ($ii = xx, yy, zz$). However, each of the elements has its own effect. According to Figure 4, for ($ii = xx$), a non-reciprocity for the achiral case (Figure 4a) is observed on $(\beta/\kappa_0)^2$. Higher losses are observed for $a \geq 0.5$ (Figure 4b). The effect of element $\xi_{yy,2}$ is presented in Figure 4c and d. These cases are non-reciprocal and exhibit relatively lower losses. The effect on $(\beta/\kappa_0)^2$ is almost identical for both cases $\xi_{yy,2}$ and $\xi_{zz,2}$ (Figure 4c,e).

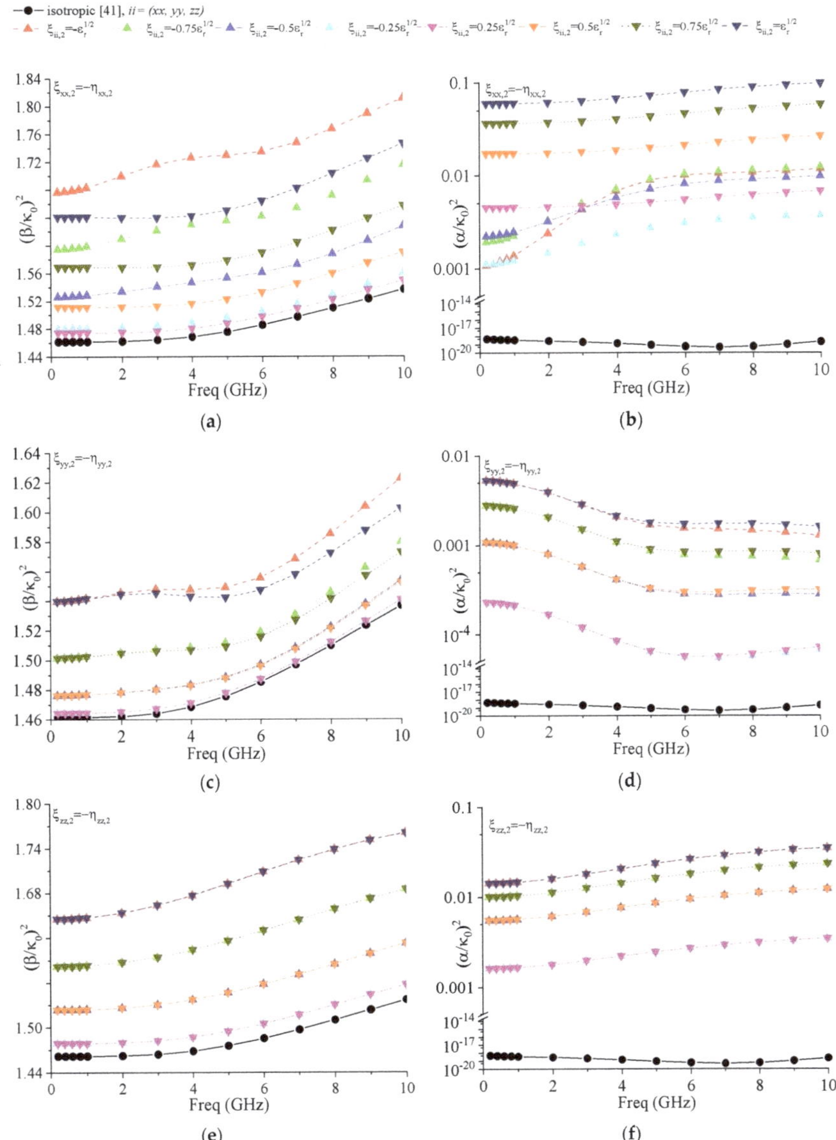

Figure 4. Effect of diagonal bianisotropy elements, ξ_{xx} on: (**a**) $(\beta/\kappa_0)^2$ and (**b**) $(\alpha/\kappa_0)^2$; ξ_{yy} on: (**c**) $(\beta/\kappa_0)^2$ and (**d**) $(\alpha/\kappa_0)^2$; ξ_{zz} on: (**e**) $(\beta/\kappa_0)^2$ and (**f**) $(\alpha/\kappa_0)^2$ with $\eta_{xx} = -\xi_{xx}$.

4.2. Effect of Gyrotropic Bianisotropy

For the gyrotropic elements, five achiral cases are considered:

1. $\xi_{ij,1} = a\sqrt{\varepsilon_r}, \eta_{ij} = \xi_{ij}, \xi_{ji} = \xi_{ij}, \eta_{ji} = \eta_{ij};$
2. $\xi_{ij,2} = a\sqrt{\varepsilon_r}, \eta_{ij} = \xi_{ij}, \xi_{ji} = \xi_{ij}, \eta_{ji} = -\eta_{ij};$
3. $\xi_{ij,3} = a\sqrt{\varepsilon_r}, \eta_{ij} = \xi_{ij}, \xi_{ji} = -\xi_{ij}, \eta_{ji} = \eta_{ij};$
4. $\xi_{ij,4} = a\sqrt{\varepsilon_r}, \eta_{ij} = -\xi_{ij}, \xi_{ji} = -\xi_{ij}, \eta_{ji} = \eta_{ij};$
5. $\xi_{ij,5} = a\sqrt{\varepsilon_r}, \eta_{ij} = -\xi_{ij}, \xi_{ji} = \xi_{ij}, \eta_{ji} = -\eta_{ij};$

In Figure 5a–d, five cases were grouped, each differs from the other by a single change in sign of the magneto-electric element. It can be seen that the combination of the constitutive parameters shows non-reciprocity and a distinct effect on $(\beta/\kappa_0)^2$ and $(\alpha/\kappa_0)^2$ parameters. In Figure 5a,b, for $\left|\xi_{xy,1}\right|$ and $\left|\xi_{xy,2}\right| \leq 0.25\sqrt{\varepsilon_r}$, we observe a weak effect on the ratio $(\beta/\kappa_0)^2$ compared to the isotropic case with lower losses for $\xi_{xy,2}$ (case (ii)). The sign change between $\eta_{xy,1} = \eta_{yx,1}$ (case (i)) and $\eta_{xy,2} = -\eta_{yx,2}$ (case (ii)) keeps the same effect on $(\beta/\kappa_0)^2$ (Figure 5a). Only the ratio $(\alpha/\kappa_0)^2$ is affected with the appearance of non-reciprocity (Figure 5b). As shown in Figure 5c,d, for the combinations $\xi_{xy,3}$, $\xi_{xy,4}$ and $\xi_{xy,5}$, a sign change has only a weak effect on $(\beta/\kappa_0)^2$ and $(\alpha/\kappa_0)^2$.

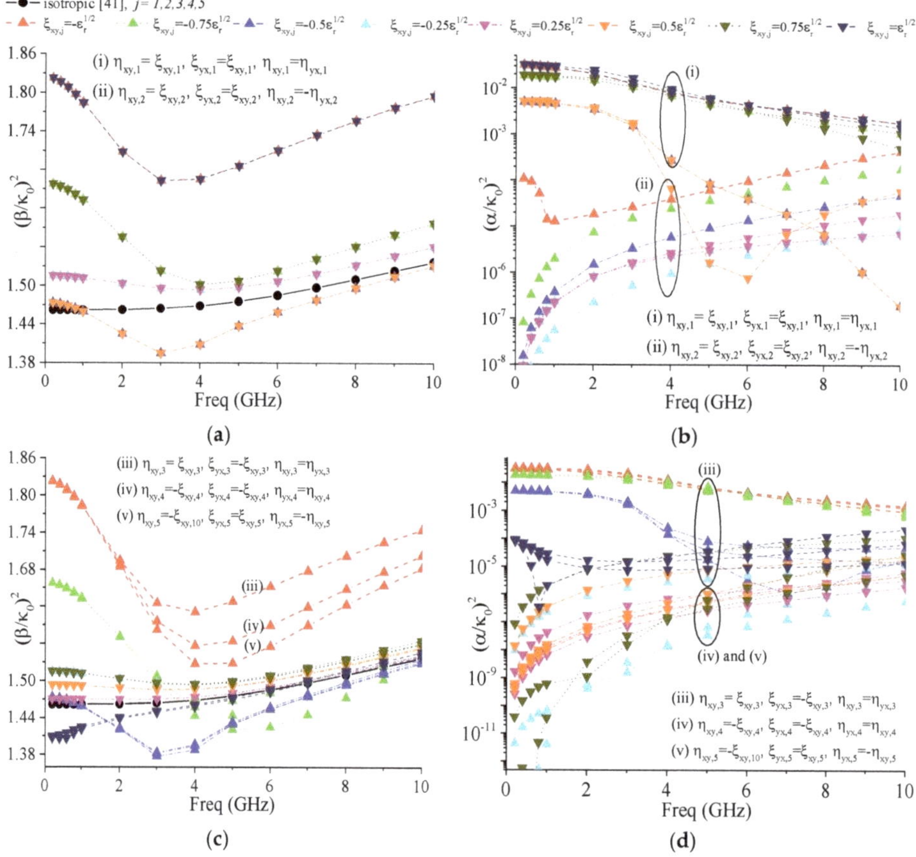

Figure 5. Effect of gyrotropic achiral bianisotropy for different ξ_{xy}, ξ_{yx}, η_{xy} and η_{yx} combinations on: (a) $(\beta/\kappa_0)^2$, (b) $(\alpha/\kappa_0)^2$, (c) $(\beta/\kappa_0)^2$ and (d) $(\alpha/\kappa_0)^2$.

In Figure 6 are presented results of the element ξ_{xz} combinations. The remarkable effect is that the ratio $(\beta/\kappa_0)^2$ is almost constant with respect to frequency and decreases with $\xi_{xz,1}$ to reach the unity for $\left|\xi_{xz,1}\right| = 0.75\sqrt{\varepsilon_r}$ and zero for $\left|\xi_{xz,1}\right| = \sqrt{\varepsilon_r}$ (Figure 6a), all with negligible losses (Figure 6b,d).

The elements, $\xi_{xz,2}$ and $\xi_{xz,3}$ have the same effect on both $(\beta/\kappa_0)^2$ and $(\alpha/\kappa_0)^2$ (Figure 6c,d). In this case, only 3 combinations: $\xi_{xz,1}$, $\xi_{xz,2}$ and $\xi_{xz,3}$ support propagating modes, with the appearance of the non-reciprocal effect. For the two other cases $\xi_{xz,4}$ and $\xi_{xz,5}$, no solutions are obtained. The $(\beta/\kappa_0)^2$ and $(\alpha/\kappa_0)^2$ variations of both cases (ii) and (iii) are completely different from case (i). The exchange of sign between $\xi_{xz,3} = -\xi_{zx,3}$ and $\eta_{xz,2} = -\eta_{zx,2}$ preserved the same effect ((ii) and (iii) curves are superposed), as shown in Figure 6c,d.

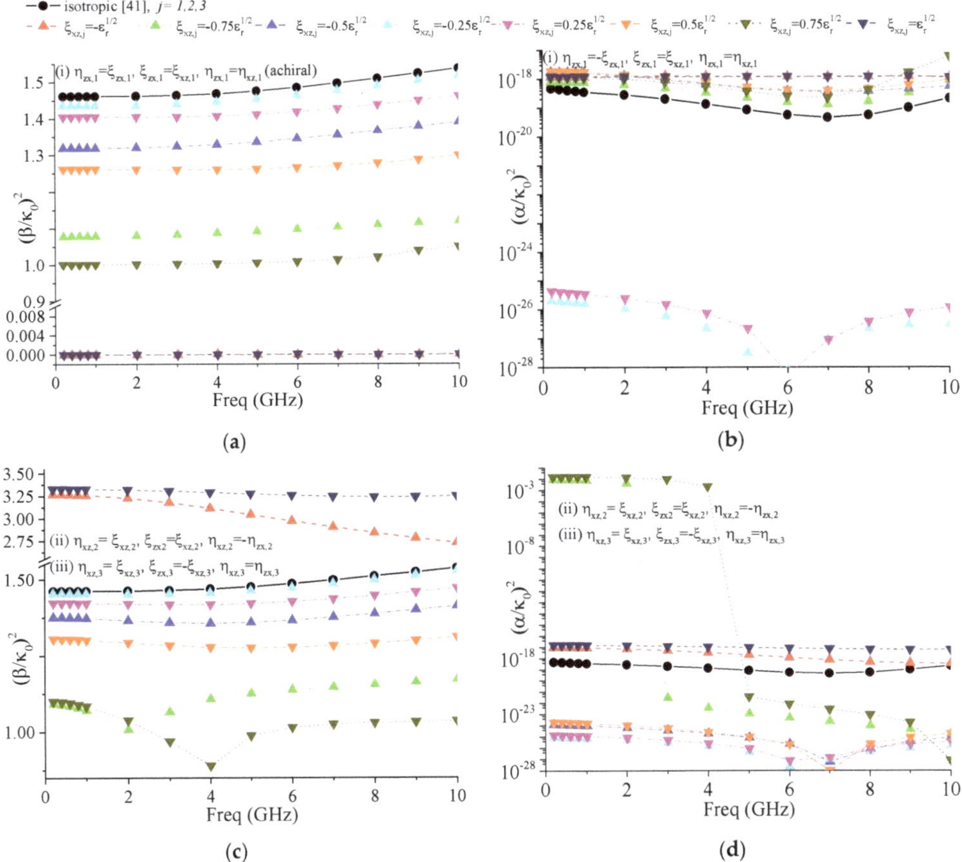

Figure 6. Effect of gyrotropic achiral bianisotropy for different ξ_{xz}, ξ_{zx}, η_{xz} and η_{zx} combinations on: (a) $(\beta/\kappa_0)^2$, (b) $(\alpha/\kappa_0)^2$, (c) $(\beta/\kappa_0)^2$ and (d) $(\alpha/\kappa_0)^2$.

Figure 7a,b presents the effect of sign exchange between the magneto-electric element of achiral medium cases. The positive sign $\xi_{yz,2} = \eta_{yz,2}$ and $\xi_{yz,3} = \eta_{yz,3}$ reveals a non-reciprocity on $(\beta/\kappa_0)^2$, while it shows a significant effect on $(\alpha/\kappa_0)^2$ and reduced losses compared to isotropic case.

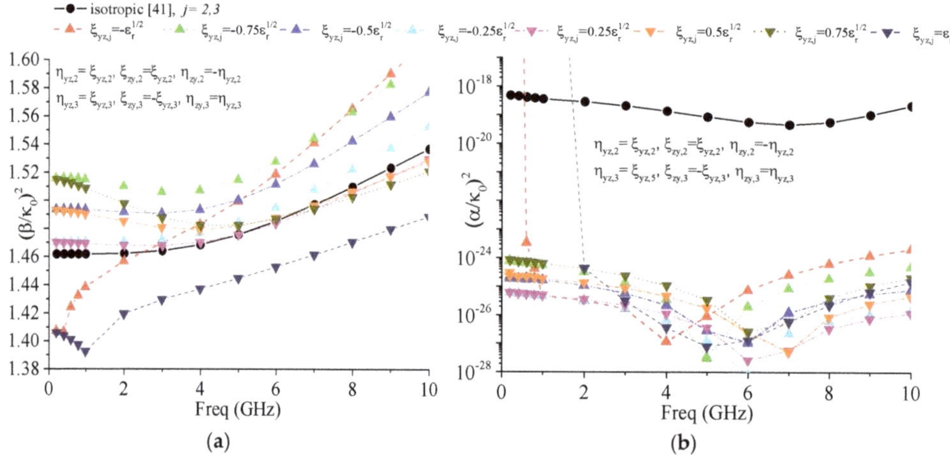

Figure 7. Effect of gyrotropic achiral bianisotropy for different ξ_{yz}, ξ_{zy}, η_{yz} and η_{zy} combinations on: (**a**) $(\beta/\kappa_0)^2$ and (**b**) $(\alpha/\kappa_0)^2$.

5. Conclusions

In this work, an analytical modeling of a three-layer CPW structure implanted on a complex medium is presented. This study is based on the Full-GEMT developed in a matrix form for the characterization of the bianisotropic-substrate CPW structure. This resulted in compact matrix form expressions of the Green's tensor. The implemented resolution method includes a new accelerating procedure developed in the GEMT that contributed to accomplishing accurate solutions with improved computing time. The computing time, for one frequency point, has been reduced by 33% for some calculation cases and more for others. A satisfactory calculation convergence is achieved with a significantly reduced Fourier number compared to literature. The presented technique can dramatically reduce the complexity of wave propagation characterization, in highly complex media, in terms of mathematical modelling and computational solution method.

According to our calculations, cases of complex media have been achieved, where the ratio $(\beta/\kappa_0)^2$ is close to unity such as for cases $\xi_{xz,1}$, $\xi_{xz,2}$. and $\xi_{xz,3}$. This characteristic is vital for the realization of media with permittivity close to unity for better use in the design of radiating antenna structures.

For cases $\xi_{xy,1}$ and $\xi_{xy,2}$, where there is only one element that changes sign between one case and another, the results of $(\beta/\kappa_0)^2$ are similar (reversed cases with similar effects), however, $(\alpha/\kappa_0)^2$ presents a different variation either in form or in magnitude.

It is noted that for the case of achiral medium when $\xi = \eta$, $\xi_{ij} = \xi_{ji}$ and $\eta_{ij} = -\eta_{ji}$ ($\xi_{xy,2}$), the medium is reciprocal and the effect is well reversed (non-reciprocal) when $\xi_{ij} = -\xi_{ji}$ ($\xi_{xy,3}$). For the cases $\xi_{xy,3}$ and $\xi_{xy,5}$, one had to have inverted cases, whereas, it is not the case. This can be explained by the properties imposed by the geometry of the studied structure.

An original result that should be taken into consideration is the losses, which show changes with each sign change between the magneto-electric elements. On the other hand, it is found that losses in achiral media are of the same magnitude as those in non-reciprocal chiral media, which must be taken into account. Furthermore, it is worth noting that the transverse elements ξ_{xz} are the most influential on the phase constant in bianisotropic a CPW structure, and this may be attributed to the geometry of the studied structure.

Investigation of some achiral media cases has shown new results such as the notion of achiral media with a relative permittivity approaching unity with reduced losses. This new finding could serve as a valuable concept from which designers of unusual synthetic materials may benefit to enhance the media intrinsic properties for future innovative applications use.

Finally, the technique discussed in this paper may be extended to deal with propagation of bifurcated modes and enclosed multilayer microstrip structures.

Author Contributions: Design and concept, D.S. and C.Z.; methodology, C.Z. and I.E.; investigation, I.O., A.U.; resources, I.E. and J.R.; writing—original draft preparation, C.Z. and D.S.; writing—review and editing, I.E., A.U., H.A. and R.A.-A.; supervision, R.A.-A.; project administration, J.R., formal analysis H.A. and F.B. All authors have read and agreed to the published version of the manuscript.

Funding: This project has received funding from the European Union's Horizon 2020 research and innovation program under grant agreement H2020-MSCA-ITN-2016 SECRET-722424. This work is also funded by the FCT/MEC through national funds and when applicable co-financed by the ERDF, under the PT2020 Partnership Agreement under the UID/EEA/50008/2019 project.

Acknowledgments: This work is supported by the European Union's Horizon 2020 Research and Innovation program under grant agreement H2020-MSCA-ITN-2016-SECRET-722424. This work is also funded by the FCT/MEC through national funds and when applicable co-financed by the ERDF, under the PT2020 Partnership Agreement under the UID/EEA/50008/2019 project.

Conflicts of Interest: The authors declare no conflict of interest.

References

1. Capolino, F. *Theory and Phenomena of Metamaterials*; CRC Press, Taylor & Francis Group: Boca Raton, FL, USA, 2009.
2. Asadchy, V.S.; Díaz-Rubio, A.; Tretyakov, S.A. Bianisotropic metasurfaces: Physics and applications. *Nanophotonics* **2018**, *7*, 1069–1094. [CrossRef]
3. Ra'di, Y.; Grbic, A. Magnet-free nonreciprocal bianisotropic metasurfaces. *Phys. Rev. B* **2016**, *94*, 195432. [CrossRef]
4. Sihvola, A.; Semchenko, I.; Khakhomov, S. View on the history of electromagnetics of metamaterials: Evolution of the congress series of complex media. *Photonics Nanostructures-Fundam. Appl.* **2014**, *12*, 279–283. [CrossRef]
5. Novitsky, A.; Shalin, A.S.; Lavrinenko, A.V. Spherically symmetric inhomogeneous bianisotropic media: Wave propagation and light scattering. *Phys. Rev. A* **2017**, *95*, 053818. [CrossRef]
6. Wang, N.; Wang, G.P. Effective medium theory with closed-form expressions for bi-anisotropic optical metamaterials. *Opt. Express* **2019**, *27*, 23739–23750. [CrossRef] [PubMed]
7. Sihvola, A.; Lindell, I.V. Bianisotropic materials and PEMC. In *Theory and Phenomena of Metamaterials*; CRC Press: Boca Raton, FL, USA, 2017; pp. 26-1–26-7.
8. Kesari, V.; Keshari, J.P. Hybrid-mode analysis of circular waveguide with chiral dielectric lining for dispersion characteristics for potential application in broadbanding a gyro-traveling-wave tube. *J. Electromagn. Waves Appl.* **2019**, *33*, 204–214. [CrossRef]
9. Crowgey, B.R.; Tuncer, O.; Tang, J.; Rothwell, E.J.; Shanker, B.; Kempel, L.C.; Havrilla, M.J. Characterization of Biaxial Anisotropic Material Using a Reduced Aperture Waveguide. *IEEE Trans. Instrum. Meas.* **2013**, *62*, 2739–2750. [CrossRef]
10. Wu, B.; Wang, W.; Pacheco, J.; Chen, X.; Lu, J.; Grzegorczyk, T.; Kong, J.A.; Kao, P.; Theophelakes, P.A.; Hogan, M.J. Anisotropic metamaterials as antenna substrate to enhance directivity. *Microw. Opt. Technol. Lett.* **2006**, *48*, 680–683. [CrossRef]
11. Bodnar, D.G.; Bassett, H.L. Analysis of an anisotropic dielectric radome. *IEEE Trans. Antennas Propag.* **1975**, *23*, 841–846. [CrossRef]
12. Zebiri, C.; Sayad, D.; Elfergani, I.; Iqbal, A.; Mshwat, W.F.; Kosha, J.; Rodriguez, J.; Abd-Alhameed, R. A compact semi-circular and arc-shaped slot antenna for heterogeneous RF front-ends. *Electronics* **2019**, *8*, 1123. [CrossRef]
13. Meshram, M.R.; Agrawal, N.K.; Sinha, B.; Misra, P.S. Characterization of M-type barium hexagonal ferrite-based wide band microwave absorber. *J. Magn. Magn. Mater.* **2004**, *271*, 207–214. [CrossRef]
14. Kamra, V.; Dreher, A. Efficient analysis of multiple microstrip transmission lines with anisotropic substrates. *IEEE Microw. Wirel. Compon. Lett.* **2018**, *28*, 636–638. [CrossRef]

15. Buzov, A.L.; Buzova, M.A.; Klyuev, D.S.; Mishin, D.V.; Neshcheret, A.M. Calculating the Input Impedance of a Microstrip Antenna with a Substrate of a Chiral Metamaterial. *J. Commun. Technol. Electron.* **2018**, *63*, 1259–1264. [CrossRef]

16. Klyuev, D.S.; Minkin, M.A.; Mishin, D.V.; Neshcheret, A.M.; Tabakov, D.P. Characteristics of Radiation from a Microstrip Antenna on a Substrate Made of a Chiral Metamaterial. *Radiophys. Quantum Electron.* **2018**, *61*, 445–455. [CrossRef]

17. Zhou, Z.; Keller, S.M. The Application of Least-Squares Finite-Element Method to Simulate Wave Propagation in Bianisotropic Media. *IEEE Trans. Antennas Propag.* **2019**, *67*, 2574–2582. [CrossRef]

18. Zebiri, C.; Lashab, M.; Benabdelaziz, F. Effect of anisotropic magneto-chirality on the characteristics of a microstrip resonator. *IET Microw. Antennas Propag.* **2010**, *4*, 446–452. [CrossRef]

19. Balbastre, J.V.; Nuño, L. Modelling the propagation of electromagnetic waves across complex metamaterials in closed structures. *J. Comput. Appl. Math.* **2019**, *352*, 40–49. [CrossRef]

20. Xiong, Y.; Russer, J.A.; Che, W.; Shen, G.; Han, Y.; Russer, P. Dispersion analysis of a fishnet metamaterial based on the rotated transmission-line matrix method. *IET Microw. Antennas Propag.* **2015**, *9*, 1345–1353. [CrossRef]

21. Aib, S.; Benabdelaziz, F.; Zebiri, C.; Sayad, D. Propagation in diagonal anisotropic chirowaveguides. *Adv. Optoelectron.* **2017**, *2017*, 9524046. [CrossRef]

22. Zebiri, C.; Daoudi, S.; Benabdelaziz, F.; Lashab, M.; Sayad, D.; Ali, N.T.; Abd-Alhameed, R.A. Gyro-chirality effect of bianisotropic substrate on the operational of rectangular microstrip patch antenna. *Int. J. Appl. Electromagn. Mech.* **2016**, *51*, 249–260. [CrossRef]

23. Alù, A.; Krishnaswamy, H. Artificial nonreciprocal photonic materials at GHz-to-THz frequencies. *MRS Bull.* **2018**, *43*, 436–442. [CrossRef]

24. Kamra, V.; Dreher, A. Discrete mode matching method for the analysis of microstrip lines on uniaxial anisotropic substrates. In Proceedings of the International Applied Computational Electromagnetics Society Symposium ACES, Firenze, Italy, 26–30 March 2017; pp. 1–2.

25. Oueslati, N.; Aguili, T. An Improved MoM-GEC Method for Fast and Accurate Computation of Transmission Planar Structures in Waveguides: Application to Planar Microstrip Lines. *Prog. Electromagn. Res.* **2016**, *48*, 9–24. [CrossRef]

26. Ardakani, H.H.; Mehrdadian, A.; Forooraghi, K. Analysis of graphene-based microstrip structures. *IEEE Access* **2017**, *5*, 20887–20897. [CrossRef]

27. Han, D.; Lee, C.; Kahng, S. Formulation of the Green's Functions for Coplanar Waveguide Microwave Devices as Genetic Algorithm-Based Complex Images. *J. Electr. Eng. Technol.* **2017**, *12*, 1600–1604.

28. Hu, Y.; Fang, Y.; Wang, D.; Zhan, Q.; Zhang, R.; Liu, Q.H. The scattering of electromagnetic fields from anisotropic objects embedded in anisotropic multilayers. *IEEE Trans. Antennas Propag.* **2019**, *67*, 7561–7568. [CrossRef]

29. Daoudi, S.; Benabdelaziz, F.; Zebiri, C.; Sayad, D.; Abdussalam, F.M.; Abd-Alhameed, R. Dispersion characteristics of a gyro-chiro-ferrite shielded multilayered microstrip line using the generalized exponential matrix technique. In Proceedings of the ITA Internet Technologies and Applications, Wrexham, UK, 12–15 September 2017; pp. 293–298.

30. Zebiri, C.; Lashab, M.; Benabdelaziz, F. Rectangular microstrip antenna with uniaxial bianisotropic chiral substrate-superstrate. *IET Microw. Antennas Propag.* **2011**, *5*, 17–29. [CrossRef]

31. Kamra, V.; Dreher, A. Analysis of Anisotropic Inhomogeneous Dielectric Waveguides With Discrete Mode Matching Method. In Proceedings of the IMS International Microwave Symposium, Boston, MA, USA, 2–8 June 2019; pp. 24–27.

32. Kamra, V.; Dreher, A. Multilayered Transmission Lines on Quasi-planar Substrates With Anisotropic Medium. *Adv. Radio Sci.* **2019**, *17*, 77–82. [CrossRef]

33. Horikis, T.P. Dielectric waveguides of arbitrary cross sectional shape. *Appl. Math. Model.* **2013**, *37*, 5080–5091. [CrossRef]

34. She, S.X. Iterated-moment method for the analysis of optical waveguides of arbitrary cross section. *JOSA A* **1989**, *6*, 1031–1037. [CrossRef]

35. Yang, W.D.; Pregla, R. The method of lines for analysis of integrated optical waveguide structures with arbitrary curved interfaces. *J. Lightwave Technol.* **1996**, *14*, 879–884. [CrossRef]

36. Hasar, U.C.; Barroso, J.J.; Sabah, C.; Kaya, Y.; Ertugrul, M. Stepwise technique for accurate and unique retrieval of electromagnetic properties of bianisotropic metamaterials. *J. Opt. Soc. Am. B* **2013**, *30*, 1058–1068. [CrossRef]

37. Hasar, U.C.; Barroso, J.J.; Bute, M.; Muratoglu, A.; Ertugrul, M. Boundary effects on the determination of electromagnetic properties of bianisotropic metamaterials from scattering parameters. *IEEE Trans. Antennas Propag.* **2016**, *64*, 3459–3469. [CrossRef]

38. Hasar, U.C.; Muratoglu, A.; Bute, M.; Barroso, J.J.; Ertugrul, M. Effective Constitutive Parameters Retrieval Method for Bianisotropic Metamaterials Using Waveguide Measurements. *IEEE Trans. Microw. Theory Tech.* **2017**, *65*, 1488–1497. [CrossRef]

39. Hasar, U.C.; Buldu, G.; Kaya, Y.; Ozturk, G. Determination of Effective Constitutive Parameters of Inhomogeneous Metamaterials With Bi-anisotropy. *IEEE Trans. Microw. Theory Tech.* **2018**, *66*, 3734–3744. [CrossRef]

40. Tsalamengas, J.L. Interaction of electromagnetic waves with general bianisotropic slabs. *IEEE Trans. Microw. Theory Tech.* **1992**, *40*, 1870–1878. [CrossRef]

41. Daoudi, S.; Benabdelaziz, F.; Zebiri, C.; Sayad, D. Generalized Exponential Matrix Technique Application for the Evaluation of the Dispersion Characteristics of a Chiro-Ferrite shielded Multilayered Microstrip Line. *Prog. Electromagn. Res. M* **2017**, *61*, 1–14. [CrossRef]

42. Sayad, D.; Zebiri, C.; Daoudi, S.; Benabdelaziz, F. Analysis of the Effect of a Gyrotropic Anisotropy on the Phase Constant and Characteristic Impedance of a Shielded Microstrip Line. *Adv. Electromagn.* **2019**, *8*, 15–22.

43. Bianconi, G.; Mittra, R. Efficient Numerical Techniques for Analyzing Microstrip Circuits and Antennas Etched on Layered Media via the Characteristic Basis Function Method. In *Computational Electromagnetics*; Springer: New York, NY, USA, 2014; pp. 111–148.

44. Park, S.-O.; Balanis, C.A. Analytical technique to evaluate the asymptotic part of the impedance matrix of Sommerfeld-type integrals. *IEEE Trans. Antennas Propag.* **1997**, *45*, 798–805. [CrossRef]

45. Sayad, D.; Benabdelaziz, F.; Zebiri, C.; Daoudi, S.; Abd-Alhameed, R.A. Spectral domain analysis of gyrotropic anisotropy chiral effect on the input impedance of a printed dipole antenna. *Prog. Electromagn. Res. M* **2016**, *51*, 1–8. [CrossRef]

46. Jain, S.; Song, J. Accelerated spectral domain approach for shielded microstrip lines by approximating summation with super convergent series. In Proceedings of the Digests of the 14th Biennial IEEE Conference on Electromagnetic Field Computation, Chicago, IL, USA, 9–12 May 2010; p. 1.

47. Lucido, M. A new high-efficient spectral-domain analysis of single and multiple coupled microstrip lines in planarly layered media. *IEEE Trans. Microw. Theory Tech.* **2012**, *60*, 2025–2034. [CrossRef]

48. Jain, S.; Song, J.; Kamgaing, T.; Mekonnen, Y.S. Acceleration of spectral domain approach for generalized multilayered shielded microstrip interconnects using two fast convergent series. *IEEE Trans. Compon. Packag. Manuf. Technol.* **2013**, *3*, 401–410. [CrossRef]

49. Xu, H.; Jain, S.; Song, J.; Kamgaing, T.; Mekonnen, Y.S. Acceleration of spectral domain immitance approach for generalized multilayered shielded microstrips using the Levin's transformation. *IEEE Antennas Wirel. Propag. Lett.* **2014**, *14*, 92–95. [CrossRef]

50. Medina, F.; Horno, M. Quasianalytical static solution of the boxed microstrip line embedded in a layered medium. *IEEE Trans. Microw. Theory Tech.* **1992**, *40*, 1748–1756. [CrossRef]

51. Tsalamengas, J.L.; Fikioris, G. Rapidly converging spectral-domain analysis of rectangularly shielded layered microstrip lines. *IEEE Trans. Microw. Theory Tech.* **2003**, *51*, 1729–1734. [CrossRef]

52. Railton, C.J.; McGeehan, J.P. A rigorous and computationally efficient analysis of microstrip for use as an electro-optic modulator. *IEEE Trans. Microw. Theory Tech.* **1989**, *37*, 1099–1104. [CrossRef]

53. Tao, J.-W.; Angenieux, G.; Flechet, B. Full-wave description of propagation and losses in quasi-planar transmission lines by quasi-analytical solution. *IEEE Trans. Microw. Theory Tech.* **1994**, *42*, 1246–1253. [CrossRef]

54. Cano, G.; Medina, F.; Horno, M. Efficient spectral domain analysis of generalized microstrip lines in stratified media including thin, anisotropic and lossy substrates. *IEEE Trans. Microw. Theory Tech.* **1992**, *40*, 217–227. [CrossRef]

55. *MATLAB 9.4 (R2018a)*, The MathWorks, Inc.: Natick, MA, USA, 2018.

56. Yang, H.Y. A numerical method of evaluating electromagnetic fields in a generalized anisotropic medium. *IEEE Trans. Microw. Theory Tech.* **1995**, *43*, 1626–1628. [CrossRef]

57. Mariotte, F.; Pelet, P.; Engheta, N. A review of recent study of guided waves in chiral media. *Prog. Electromagn. Res.* **1994**, *9*, 311–350.

58. Krowne, C.M. Electromagnetic propagation and field behavior in highly anisotropic media. *Adv. Imaging Electron Phys.* **1995**, *92*, 79–214.

59. Nguyen, C. *Analysis Methods for RF, Microwave, and Millimeter-Wave Planar Transmission Line Structures*; John Wiley & Sons: New York, NY, USA, 2003.

60. Mirshekar-Syhakal, D. *Spectral Domain Method for Microwave Integrated Circuits*; Wiley: New York, NY, USA, 1990.

61. Maze-Merceur, G.; Tedjini, S.; Bonnefoy, J.L. Analysis of a CPW on electric and magnetic biaxial substrate. *IEEE Trans. Microw. Theory Tech.* **1993**, *41*, 457–461. [CrossRef]

62. Khodja, A.; Yagoub, M.C.E.; Touhami, R.; Baudrand, H. Practical Recurrence Formulation for Composite Substrates: Application to Coplanar Structures with Bi-Anisotropic Dielectrics. In Proceedings of the 18th Mediterranean Microwave Symposium (MMS), Istanbul, Turkey, 31 October–2 November 2018; pp. 341–344.

Article

Low-Profile and Closely Spaced Four-Element MIMO Antenna for Wireless Body Area Networks

Issa Elfergani [1,*], Amjad Iqbal [2,3,*], Chemseddine Zebiri [4], Abdul Basir [5], Jonathan Rodriguez [1,6], Maryam Sajedin [1], Artur de Oliveira Pereira [1], Widad Mshwat [7], Raed Abd-Alhameed [7,8] and Sadiq Ullah [5]

[1] Mobile Systems Group, Instituto de Telecomunicações, 3810-193 Aveiro, Portugal; Jonathan@av.it.pt (J.R.); maryam.sajedin@av.it.pt (M.S.); aopereira@av.it.pt (A.d.O.P.)
[2] Centre For Wireless Technology, Faculty of Engineering, Multimedia University, Cyberjaya 63100, Malaysia
[3] Department of Electrical Engineering, CECOS University of IT and Emerging Sciences, Peshawar 25000, Pakistan
[4] Department of Electronics, University of Ferhat Abbas, Setif -1-, 19000 Setif, Algeria; czebiri@univ-setif.dz
[5] Department of Telecommunication Engineering, UET Mardan 23200, Pakistan; engrobasir@gmail.com (A.B.); sadiqullah@uetmardan.edu.pk (S.U.)
[6] Faculty of Computing, Engineering and Science, University of South Wales, Pontypridd CF37 1DL, UK
[7] School of Engineering and Informatics, University of Bradford, Bradford BD7 1DP, UK; W.F.A.A.Mshwat@bradford.ac.uk (W.M.); R.A.A.Abd@bradford.ac.uk (R.A.-A.)
[8] Department of Communication and Informatics Engineering, Basra University College of Science and Technology, Basra 61004, Iraq
* Correspondence: i.t.e.elfergani@av.it.pt (I.E.); amjad730@gmail.com (A.I.); Tel.: +351-920-019-023 (I.E.); +601-128-784-475 (A.I.)

Received: 10 December 2019; Accepted: 27 January 2020; Published: 4 February 2020

Abstract: A compact four-element multiple-input multiple output (MIMO) antenna is proposed for medical applications operating at a 2.4 GHz ISM band. The proposed MIMO design occupies an overall volume of 26 mm × 26 mm × 0.8 mm. This antenna exhibits a good impedance matching at the operating frequency of the ISM band, whose performance attributes include: isolation around 25 dB, envelope correlation coefficient (ECC) less than 0.02, average channel capacity loss (CCL) less than 0.3 bits/s/Hz and diversity gain (DG) of around 10 dB. The average peak realized gain of the four-element MIMO antenna is 2.4 dBi with more than 77 % radiation efficiency at the frequency of interest (ISM 2.4 GHz). The compact volume and adequate bandwidth, as well as the good achieved gain, make this antenna a strong candidate for bio-medical wearable applications.

Keywords: MIMO antenna; ISM 2.4 GHz; isolation; envelope correlation coefficient (ECC); channel capacity loss (CCL)

1. Introduction

Due to the tremendous demand for high data rate and large channel capacity, developing multiple-input multiple-output (MIMO) antenna systems is given more attention by industry and academia. The key goals of the MIMO system is to accomplish multiband operation, enhanced bandwidth, improved gain/efficiency, good channel capacity and diversity performance [1].

To achieve a high quality data transmission, the multi-input multi-output (MIMO) antenna technology has been promoted and has become an eye-catching choice for antenna designers. However, a good MIMO antenna should meet several important factors, such as ensuring high isolation between elements, while obtaining a good impedance characteristics and highly efficient individual elements [2,3]. Therefore, the MIMO technology has been extensively considered and exploited to improve the channel capacity and signal quality.

In this context, the isolation between the closely spaced elements largely affects the antenna system performance [4,5]. Subsequently, one of the most challenging tasks in the MIMO paradigm is to accomplish high isolation and low correlation between different elements, in which this will contribute to increasing the channel capacity and consequently obtain both a higher gain as well as good spectral efficiency.

The conventional methods for mitigating the mutual coupling is to position antenna elements with adequate spacing (typically half wavelength distance). Although this approach can solve the coupling issue, it needs such a huge space of deployment, which is not a convenient approach to be used within hand held wireless devices due to the limited space. Thus, for the goal of having higher functionality and reliability, numerous approaches have been lately used to reduce inter-element coupling and hence improve isolation. The achievable data rate throughput in wireless communication would be multiplied and increased within the multiple-input multiple-output (MIMO) system [6,7]. The MIMO approach helps in optimizing the use of the transmission spectrum and power [8,9]; by exploiting MIMO synthesis, additional paths may be utilized in order to improve the link capacity [10,11].

The industrial, scientific and medical (ISM) band of 2.4 GHz is broadly considered as an exempted license band, with minimal free access regulation. This band has been largely exploited in several applications such as cordless phones, medical applications and Wi-Fi standards of 802.11 a,b,g,n.The 802.11n standard was defined in October 2009 and has since then become broadly used in many applications. This standard utilizes ISM frequency bands of 2.4 GHz or 5.8 GHz and is interoperable with 802.11b, and uses 802.11g-like modulation, but with MIMO. There have been various investigations on antenna design at the 2.4 GHz ISM band frequency as in [12–18].

However, such antennas are not wearable antennas for use in real applications. Thus, several investigations on the wearable antenna designs have been carried out. These include different kinds of antenna shapes and structures [19–30], such as conformal dipole and monopole antennas [19,20], wideband implantable antennas [21,22], circularly polarized antennas [23,24], and capsule antennas [25,26]. The main goal of such investigations was to meet wearable antenna requirements in accomplishing robust communication ability for wearable devices to detect major vital signs and bio-electric activity.

On the other hand, most wearable antennas in [19–30] are single-input–single-output (SISO) designs. Once their structure and characteristics are accomplished, size dimensions as well as bandwidth are static. For further improvements, such as enhancing the transmission rate data and avoiding the multipath interferences phenomena, an introduction of an additional antenna characteristic should be taken into account. Special consideration and attentions have been given to multiple-input multiple-output (MIMO) technology, with the potentiality of improving channel capacity, but without need for additional frequency spectrum or power.

In MIMO wireless systems, multiple antenna elements are adopted to work as transmitter and receiver, improving the data transmission rates of high-resolution images from the capsule to outside base station/gateway over single antenna topology in multipath channels [31]. Recently, planar monopole antennas with a pair of elements were proposed, verifying that antennas operate at ISM 2.4 GHz along with MIMO technique that can be applied for wireless devices, realizing high data rate in limited range and space [32–38].

However, none of the above-mentioned antenna structures have been placed on the human body to check the impact on performance, limiting studies targeting medical applications. Thus, this work proposes a four-element monopole compact antenna operating over the ISM 2.4 GHz band, which is then placed on the human body for medical investigations. To claim the novelty or/and advantages of the present work, a comparison table of previous similar published antennas has been included. By reviewing all ISM 2.4 GHz band MIMO antennas in Table 1, one can observe that the proposed antenna has achieved a smaller size compared to all works in [32–38], higher isolation in contrast to antennas in [32,33,37], improved ECC with comparison to refs [30,35], and better power gains over the designs in [33,36,38]. Moreover, the proposed antenna has been embedded on the body and its performance compared with free space antennas.

Table 1. Comparison of the performance of the published scientific and medical (ISM) band multiple-input multiple-output (MIMO) antennas.

Ref.	Frequency (GHz)	Ports	Size (mm^3)	Isolation (dB)	ECC	Isolation (dB)	Peak Gain (dBi)	Efficiency (%)
[32]	2.4–2.48	4	18.5 × 18.5 × 1.27	−15.99	0.0025	NG*	−15.18	NG
[33]	2.4	2	22 × 12 × 6	−10	0.015	8.85	NG	NG
[17]	2.45	4	100 × 50 × 0.8	−10	0.1	NG	−0.8	29
[16]	2.31–2.51	2	65.25 × 65.25 × 1.52	−17	0.01	NG	3	87
[34]	1.44, 2.3, 4.2	4	44 × 44 × 0.8	−24.8	NG	NG	NG	NG
[35]	2.4	2	76.8 × 57.8 × 1.6	−30	0.1	9.9	NG	NG
[36]	2.4	4	186 × 188 × 1.6	−30	0.5	7.95	NG	85
[37]	2.4	2	38.1 × 38.1 × 2	−12	0.01	NG	1.6	NG
[18]	2.45	4	34 × 18 × 1.6	−15	0.12	NG	1.26	78
[38]	2.4	2	30 × 44 × 1.6	−35	0.02	NG	1.4	85
Present work	2.4	4	26 × 26 × 0.8	−25	0.02	9.9	2.4	77

NG* = Not given.

2. Antenna Design And Concept

The proposed design consists of four p-shaped monopole radiators as shown in Figure 1. The proposed ISM 2.4 GHz MIMO antenna is printed on FR-4 dielectric substrate (permittivity of 4.3, loss tangent of 0.025, and thickness of 0.8 mm). The proposed geometry of the four-element MIMO monopole antenna is considered as one of the smallest designs in wearable antennas with a total size of 26 × 26 × 0.8 mm^3. As can be seen in Figure 1, the proposed antenna is fed by a single 50 Ω microstrip line designed and printed on FR-4 substrate. Initially, a single-element antennas was investigated. The dimensions of the antenna were optimized at the desired frequency (2.4 GHz). The fabricated prototype of the single-element antenna is portrayed in Figure 2a. The simulated and measured reflection coefficients of the antenna are shown in Figure 2b. We see that the simulated and measured results have the same resonant frequency. The bandwidth of the fabricated prototype is smaller than the simulated one but still wider than the targeted ISM band.

The four p-shaped elements are symmetrically placed with a separation distance of 4.5 mm (0.06λ). The very small ground plane is printed on the lower side of the substrate, where this appellation is used in [39–41]. The ground plane was defected into four rectangle parts printed underneath each radiating element. The schematic views of the considered radiating element (front side) and partial ground (back side) are shown in Figure 1, along with the optimized dimensions. The present design is simulated with the aid of CST [42]. The detailed geometrical parameters and their sizes are listed in Table 2.

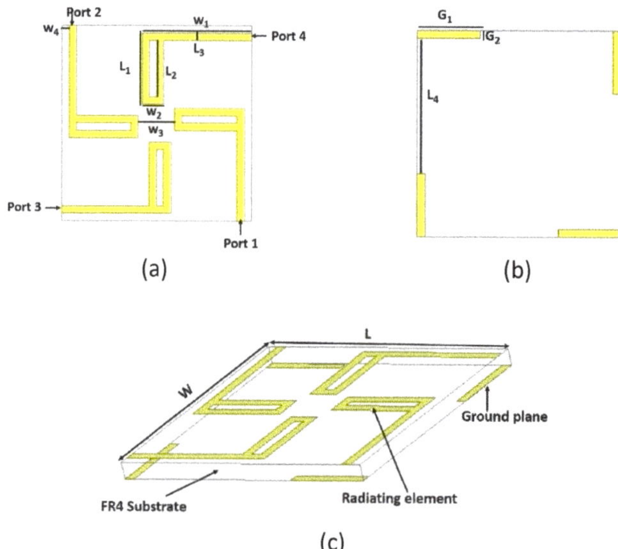

Figure 1. Multiple-input multiple-output (MIMO) antenna structure; (**a**) Top view; (**b**) ground view; (**c**) 3D view.

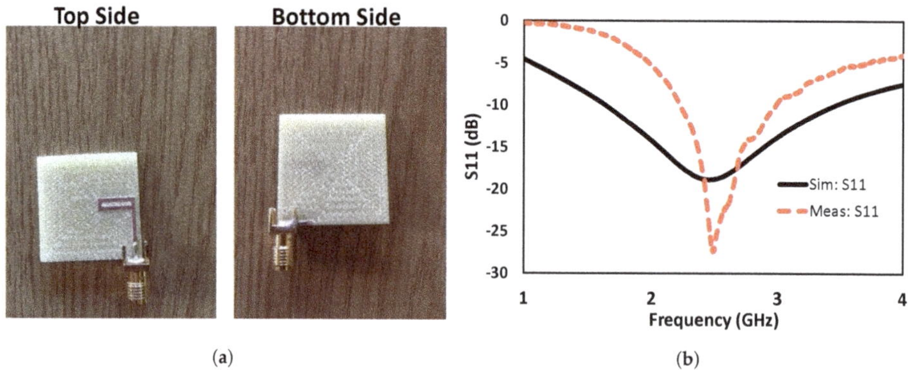

Figure 2. (**a**) Fabricated prototype of single-element antenna, and (**b**) simulated and measured reflection coefficient of the single-element antenna.

Table 2. The dimensions of the proposed MIMO antenna.

Parameters	Value (mm)	Parameters	Value (mm)
W_1	15	L_3	1
W_2	3	L_4	17
W_3	4.5	G_1	7
W_4	1	G_2	1
L_1	9.8	L	26
L_2	7	W	26

3. Equivalent Circuit Model

In order to evaluate the true behavior of the proposed four elements MIMO antenna, an equivalent lumped circuit model is generated using advance design system (ADS). The equivalent circuit model is illustrated in Figure 3. Each monopole radiator of the MIMO antenna system is represented by the parallel RLC resonant circuit [43,44]. Each monopole is excited by a 50 Ω terminal. The associated coupling between the MIMO elements is represented by an LC circuit [45,46]. The equivalent circuit is fine tuned to extract the optimized circuit values. The optimized circuit parameters are listed in Figure 3. The s-parameters of the circuit and EM model are compared in Figure 4. We can see that the plotted curves present satisfactory agreement.

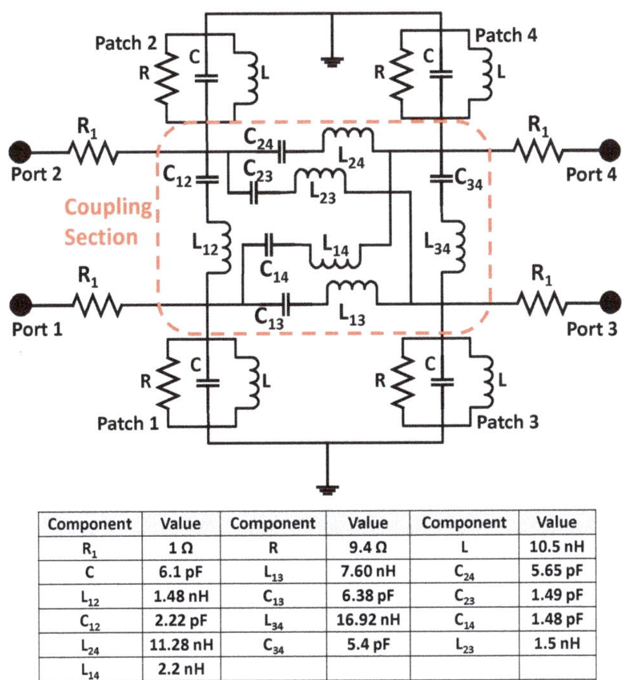

Component	Value	Component	Value	Component	Value
R_1	1 Ω	R	9.4 Ω	L	10.5 nH
C	6.1 pF	L_{13}	7.60 nH	C_{24}	5.65 pF
L_{12}	1.48 nH	C_{13}	6.38 pF	C_{23}	1.49 pF
C_{12}	2.22 pF	L_{34}	16.92 nH	C_{14}	1.48 pF
L_{24}	11.28 nH	C_{34}	5.4 pF	L_{23}	1.5 nH
L_{14}	2.2 nH				

Figure 3. Equivalent circuit model of the MIMO antenna.

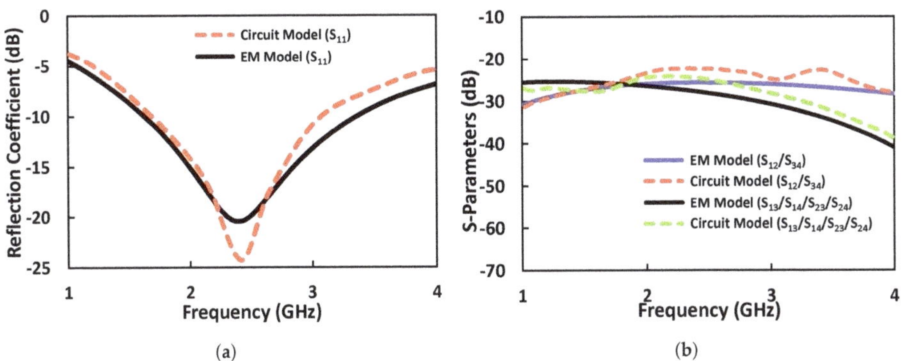

Figure 4. S-parameters comparison of the electromagnetic (EM) model and equivalent circuit model (**a**) Reflection coefficient, and (**b**) Transmission coefficient.

4. Simulation and Measurement Results

The computed and measured s-parameters of the antenna are demonstrated in Figure 5. It is observed that the proposed MIMO antenna has simulated −10 dB impedance bandwidth at the 2.4 GHz ISM band, where it shows isolations between the adjacent and diagonal ports of around −25 dB.

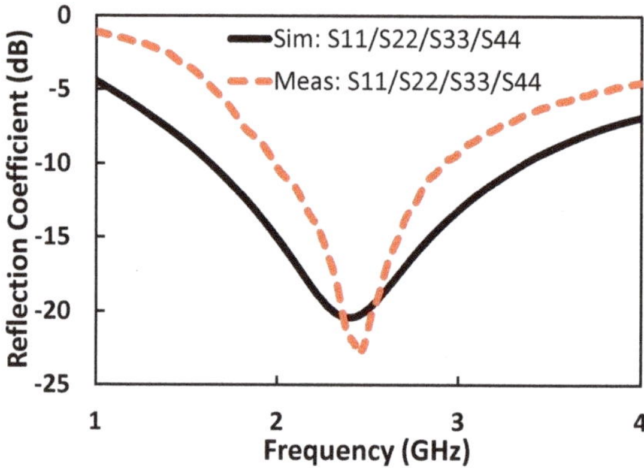

Figure 5. Simulated and measured s-parameters of the proposed MIMO antenna.

For proof-of-concept, the four-element MIMO antenna model is fabricated as depicted in Figure 6. The MIMO antenna was printed over FR-4 substrate with thickness of 0.8 mm. Four rectangle copper parts were printed underneath each radiating element to mimic the partial ground plane of the antenna. Four SMA connectors were exploited to feed the four-element radiators as illustrated in Figure 6. The measured s-parameters of the proposed MIMO antenna in Figure 5 show a good agreement with the simulated outcomes, with an insignificant move in the resonant frequency, which can be attributed to several factors such as the fabrication process and error and port/cable coupling losses.

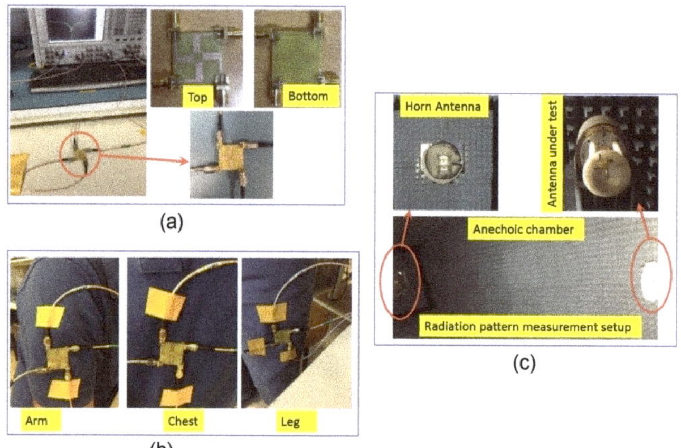

Figure 6. The prototype of the proposed P-shaped MIMO antenna, (**a**) reflection coefficient measurement in free space, (**b**) on-body reflection coefficient measurement, and (**c**) radiation pattern measurement setup.

To further investigate the MIMO antenna diversity performance, the isolation between the elements of the proposed antenna were studied and analysed as it is directly related to the minimum coupling between individual antenna elements. The simulated and measured isolation outcomes between the MIMO antenna elements are indicated in Figure 7. The isolation values vary from 18 to 23 dB in the impedance band of the proposed antenna. The results show good performance for MIMO applications around 2.4 GHz with isolation greater than −25 dB.

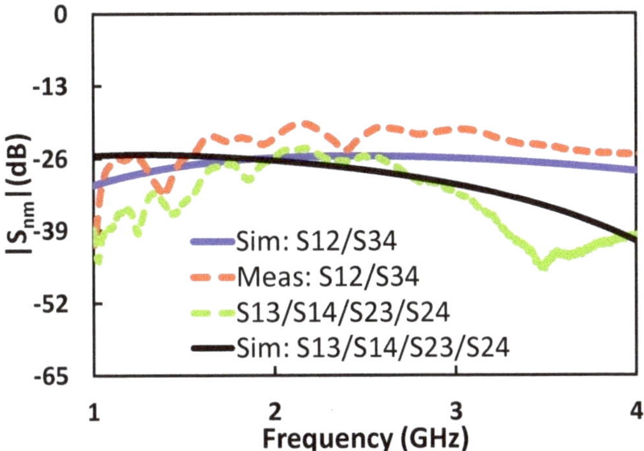

Figure 7. The isolation performance of the proposed MIMO antenna.

5. Current Distribution

To validate the outcome of the mutual coupling between the antenna elements of the proposed antenna obtained in Figures 4, 5 and 7, we can refer to the antenna surface current as indicated in Figure 8. Generally, to understand the influence of the coupling between the antenna elements using this approach, one of the four ports should be excited, whereas the other three ports remain terminated with 50 Ω load as shown in Figure 8. The space between the ports is considered as an important factor in order to figure out how much the antennas are being isolated. Generally speaking, when the antenna elements are largely separated, this will lead to a high isolation.

In the case of Figure 8a, only port 1 is excited. As observable from Figure 8a, the higher value of current concentrates on the feed arm, as there is no current linked to the adjacent ports. Therefore, its effect can be seen in terms of the S12 (or S34) isolation parameter. When port 2 is excited and the other three ports are terminated with 50 Ω load, the current only exist on this port while being negligible on the remaining three. The same case is illustrated when port 3 and port 4 are only excited as indicated in Figure 8c,d. From Figure 8, it can be observed that the surface current is mainly focused over the excited port. Therefore, the measured and simulated isolations for the proposed MIMO antenna that were shown in Figures 5 and 7 are more than 10 dB.

Two planes, i.e., E-plane and H-plane at 2.4 GHz are taken in order to examine the radiation patterns of the proposed antenna as illustrated in Figure 9, when one port is excited and the others are matched to a 50 Ω load. As it is noted, the E and H plane simulated field patterns of the present MIMO antenna are shown at the ISM 2.4 GHz resonant frequencies, which are generated by the CST software package. The measured radiation is taken and carried out inside an anechoic chamber in the presence of a standard horn antenna (as transmitter) and the power meter (to record the received field). In the measurement process, any one port is chosen to work as a receiver, while the other ports are loaded with 50 Ω load, for the reason of avoiding any signal pick-up. Such a procedure is repeated with every port of the proposed MIMO design. The measured and CST simulated radiation patterns are compared in Figure 9 in the case of the two planes. Thanks to the symmetric geometry of the proposed antenna that helps in accomplishing a stable radiation pattern, we consider an appropriate design for use in communication systems. However,certain irregularities between the simulated and measured patterns are observed and this may be attributed to the fabrication process errors, and the losses of the ports/cables.

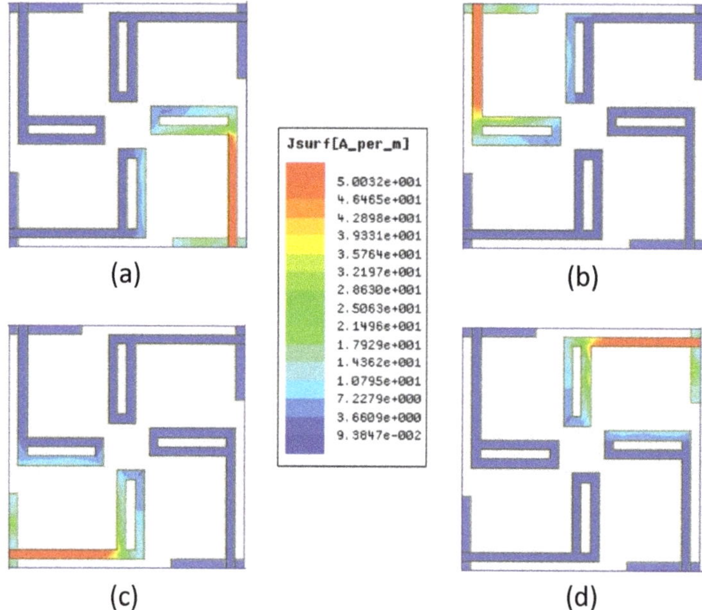

Figure 8. Surface current distribution at 2.4 GHz, (**a**) port 1 excited, (**b**) port 2 excited, (**c**) port 3 excited, (**d**) port 4 excited.

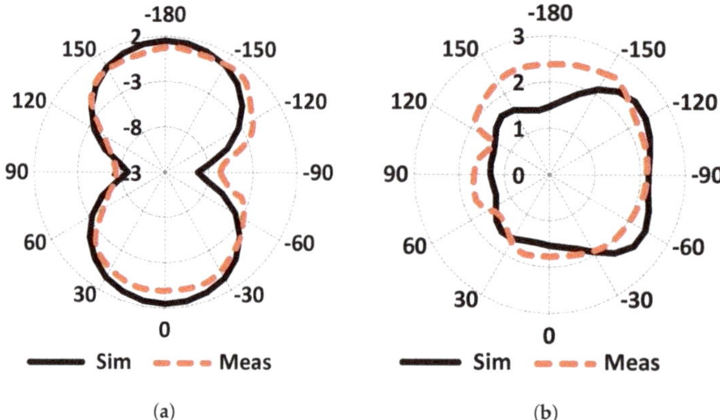

Figure 9. Simulated and measured radiation patterns of the proposed MIMO design, (**a**) E-plane, (**b**) H-plane.

6. Antenna ON-Body Investigation

The human body can absorb electromagnetic (EM) waves (termed as specific absorption rate, SAR), generated by antenna operated in its vicinity. If this absorption of EM waves exceeds some standard limits, the temperature of the tissues will rise [47]. These limits are standardized by the electromagnetic field (EMF) regulation authorities (ICNIRP, IEEE, and FCC). Considering public safety concerns, we have calculated the specific absorption rate (SAR) according to IEEE C95.1-2005 as shown in Figure 10. The antenna was placed on the chest, arm, and leg of a real realistic Ella model while maintaining some gap to avoid direct contact with tissue as shown in Figure 10. Figure 10a–c show the SAR distributions on the chest, arm, and leg of the Ella model. We selected small areas for SAR calculation instead of full body to reduce simulation time. The peak 10-g SAR values 20 W/kg, 8.36 W/kg, and 7.17 W/kg at 1 W input power to each array element on the chest, arm, and leg, respectively. The peak SAR values exceed the standard limits; however, the devices operating near the body use little power. The standard maximum allowable power (EIR) for such devices is set in the mW range [48–50]. To restrict the SAR value under the limit for our proposed antenna, the maximum input powers to the antenna are 100 mW, 239.2 mW, and 278.9 mW for the chest, arm, and leg, respectively, which are much greater than the standard radiated limit (in mW range). Based on the above SAR calculated value, our antenna is safe if the input power is less than than 100 mW, 239.2 mW, and 278.9 mW for the chest, arm, and leg cases, respectively.

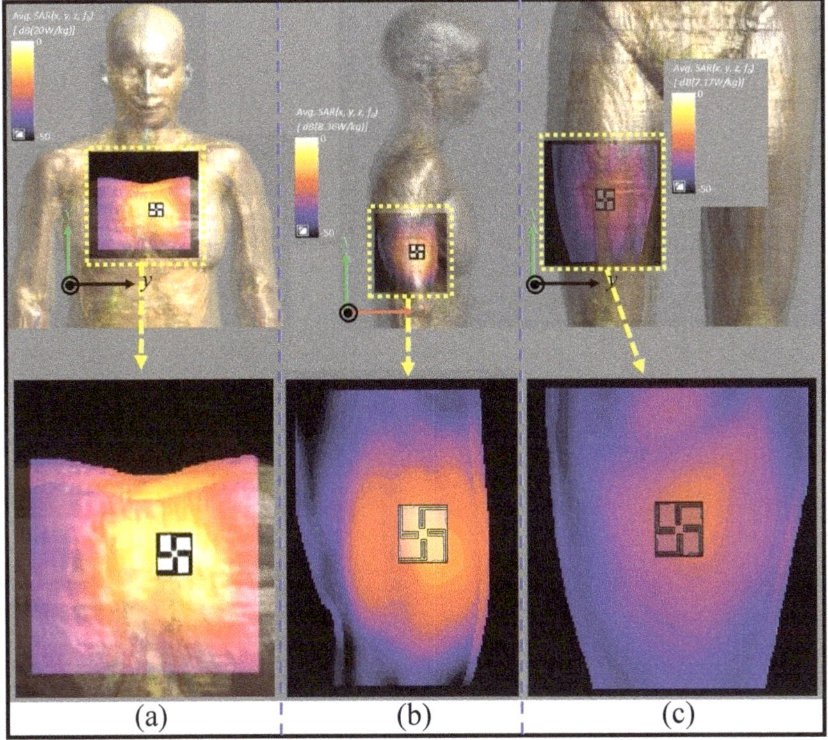

Figure 10. Specific absorption rate (SAR) analysis of the MIMO antenna on (**a**) chest, (**b**) arm, and (**c**) leg.

Additionally, the antenna's performance in terms of reflection coefficient, gain, and radiation pattern is evaluated in the vicinity of the human body. The reflection coefficient of the antenna in free space and in on-body worn scenarios is illustrated in Figure 11a. We can see that the resonant frequency shifts a bit towards the lower frequency side due to body loading effects [51]; however, the antenna still retains a good impedance bandwidth and covers the targeted ISM band (2.4 GHz). Figure 11b,c illustrate the radiation pattern of the antenna in free space and in the vicinity of the human body. We can see that the front-to-back ratio of the antenna in Phi = 0^0 is enhanced due to the fact that skin behaves as an extension of the ground plane [52].

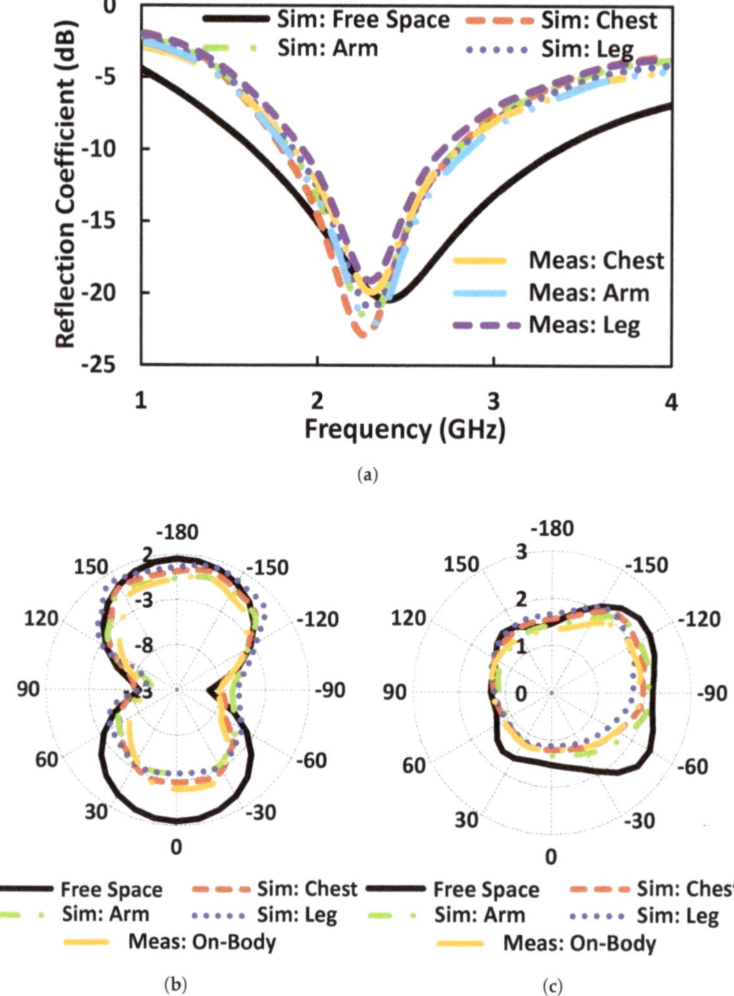

Figure 11. Performance comparison of the antenna in free space and in vicinity of the human body in term of simulated (**a**) reflection coefficient, (**b**) radiation pattern at 0° , and (**c**) radiation pattern at 90°.

The antenna gain is also one of the diversity parameters. The MIMO antenna gain is a far-field parameter and is measured in an Anechoic chamber using the two standard horn antennas and proposed structure. The simulated and measured gains on the free space and on-body paradigms were carried out as depicted in Figure 12. In the case of our design, the minimum value of the computed and measured peak gain is varied between 2.20 dBi and 2.5 dBi. The simulated antenna gain for the free space and on-body analysis at 2.4 GHz resonant frequency is around 2.49 dBi, while the measured gain in the case of free space and on-body worn at the same band is about 2.20 dBi.

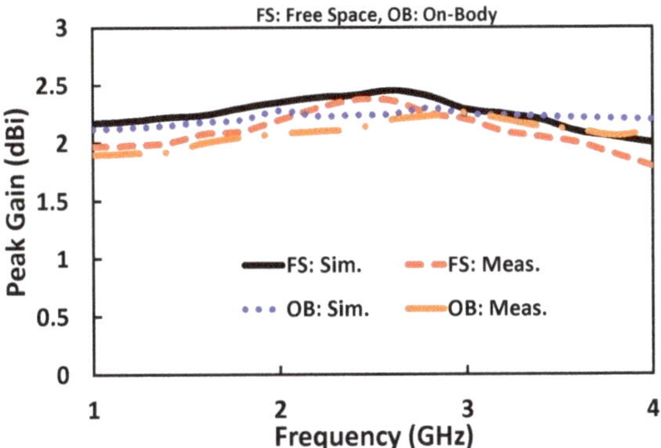

Figure 12. Simulated and measured peak gain of the antenna.

Figure 13a illustrates the outcomes of the computed envelope correlation coefficient (ECC) between the ports of the MIMO antenna system in both scenarios, namely, the free space and on-body worn. ECC of the MIMO system portrays how independent the individual elements of the MIMO system are in terms of radiation pattern. It is always desirable in a MIMO system that each element is independent of the other; however, zero value of ECC is difficult in practical applications. In this case, the three ECC values may be accomplished by substituting the appropriate terms ($S_{11} = S_{22} = S_{33} = S_{44}$, and $S_{12} = S_{21} = S_{34} = S_{43}$, and $S_{13} = S_{31} = S_{24} = S_{24}$, and $S_{14} = S_{41} = S_{23} = S_{32}$) into the following equations, and can be written as shown in Equations (1)–(3):

$$\rho_{e12} = \frac{|S_{11}^* S_{12} + S_{12}^* S_{22} + S_{13}^* S_{32} + S_{14}^* S_{42}|^2}{(1 - (|S_{11}|^2 + |S_{12}|^2) + (|S_{13}|^2 + |S_{14}|^2))^2} \tag{1}$$

$$\rho_{e13} = \frac{|S_{11}^* S_{13} + S_{12}^* S_{23} + S_{13}^* S_{33} + S_{14}^* S_{43}|^2}{(1 - (|S_{11}|^2 + |S_{12}|^2) + (|S_{13}|^2 + |S_{14}|^2))^2} \tag{2}$$

$$\rho_{e13} = \frac{|S_{11}^* S_{14} + S_{12}^* S_{24} + S_{13}^* S_{34} + S_{14}^* S_{44}|^2}{(1 - (|S_{11}|^2 + |S_{12}|^2) + (|S_{13}|^2 + |S_{14}|^2))^2} \tag{3}$$

From Figure 13a, the ECC between the ports is lower than 0.03 for the band of interest. Thus, this value of ECC is reasonable and comparable to the ones achieved in [53], which show a good indication for promising diversity performance.

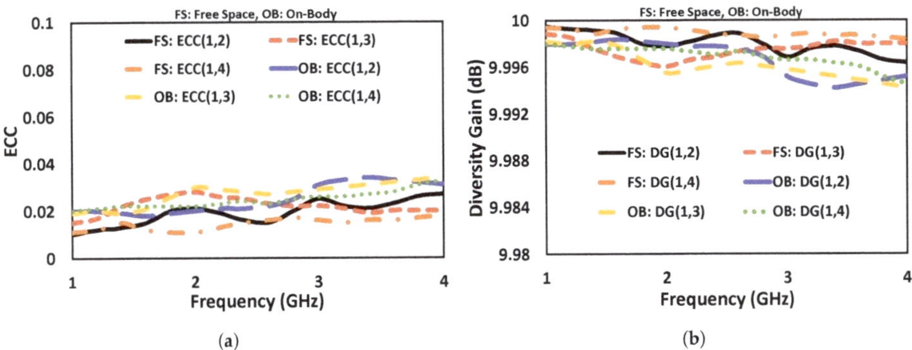

Figure 13. (a) The envelope correlation coefficient (ECC) and (b) diversity gain (DG) of the proposed MIMO antenna.

The diversity gain also is deemed as a paramount diversity parameter [54,55], which may be accomplished in terms of maximum theoretical diversity gain (10 dB) and correlation coefficient exploiting Equation (4)–(6). Diversity gain refers to improvements in signal-to-interference ratio by applying any diversity scheme. As the antenna shows diversity gain with higher values, this means a better isolation is achieved and vice versa. The plots of the diversity gain of the present MIMO antenna in the case of free space and on-body investigations are shown in Figure 13b. From Figure 13b, it is noticed that the values of DG are about 10 dB at the operating band for both paradigms.

$$DG_{12} = 10\sqrt{1 - |\rho_{e12}|^2} \tag{4}$$

$$DG_{13} = 10\sqrt{1 - |\rho_{e13}|^2} \tag{5}$$

$$DG_{14} = 10\sqrt{1 - |\rho_{e14}|^2} \tag{6}$$

The capacity capacity loss (CCL) provides information regarding the maximum limit of the message rate up to which the message can be continuously transmitted without any loss over a communication channel. In theory, the channel capacity can be enhanced by increasing the number of antennas of the MIMO system. Nevertheless, the presence of Rayleigh-fading MIMO channels will induce loss of channel capacity. This loss can be calculated from the correlation matrices given in [55,56]. In the case of a 2 × 2 MIMO system, assuming that only the receiving antenna patterns are correlated and assuming the worst scenario where high signal-to-noise ratio is occurring, the channel capacity loss (CCL) can be evaluated by using the following equation [54–56]:

$$CCL = -log_2 \det(\Psi^R) \tag{7}$$

where Ψ^R is the receiving antenna correlation matrix that is given by:

$$\Psi^R = \begin{bmatrix} \rho_{11} & \rho_{12} & \rho_{13} & \rho_{14} \\ \rho_{21} & \rho_{22} & \rho_{23} & \rho_{24} \\ \rho_{31} & \rho_{32} & \rho_{33} & \rho_{34} \\ \rho_{41} & \rho_{42} & \rho_{43} & \rho_{44} \end{bmatrix} \tag{8}$$

where

$$\rho_{ii} = 1 - |\sum_{n=1}^{N} S_{in}^* S_{ni}|, \; for \; i,j = 1,2,3, or \; 4.$$

and

$$\rho_{ij} = -|\sum_{n=1}^{N} S_{in}^* S_{nj}|, \; for \; i,j = 1,2,3, or \; 4.$$

The CCLs of the proposed radiator for both scenarios are shown in Figure 14a. Figure 14a reveals that the proposed MIMO antenna achieves acceptable CCL values, which vary from 0.2 to 0.3 around the targeted band of ISM 2.4 GHz (ideally, CCL should be less than 0.4 bps/Hz within the entire operating band). Thus, our investigation proves that good impedance matching and isolation between the two antenna elements lead to low capacity loss in the case of free space as well as on-body worn devices.

The efficiency also describes the diversity behaviour of the proposed MIMO antenna. The values of the radiation efficiencies for the 2.4 GHz ISM band in both cases are around 77%, as shown in Figure 14b.

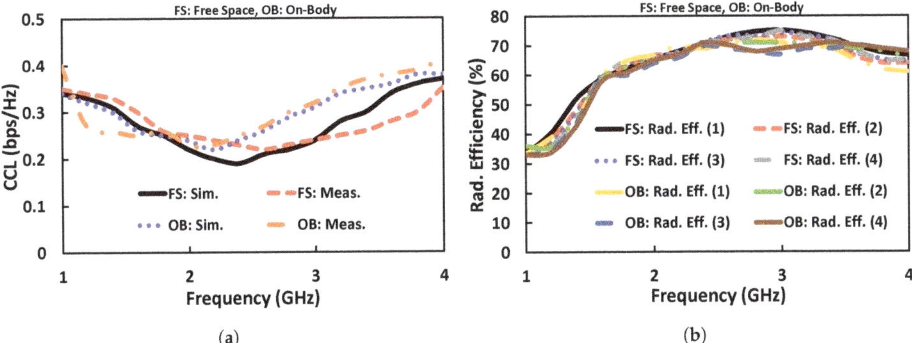

Figure 14. (**a**) The channel capacity loss (CCL) and (**b**) radiation efficiencies of the proposed antenna.

7. Conclusions

A compact four-element MIMO antenna is designed and presented for the ISM band. The P-shaped radiating elements are realized on a low-cost FR-4 substrate. The proposed antenna occupies a compact envelope dimension of $0.20\lambda_o \times 0.20\lambda_o \times 0.006\lambda_o$, where λ_o is the wavelength of the operating frequency. The MIMO antenna design has been simulated and also fabricated. Both the simulations and measurements showed high isolation of around -25 dB at the targeted band of 2.4 GHz . In addition, the envelope correlation coefficient, antenna gain, efficiency and channel capacity loss and other performances are also provided. The proposed antenna has also been placed on a human body, where its performance was investigated and showed acceptable results compared to the free space antenna design. Therefore, the proposed MIMO antenna has been shown to be a strong candidate for use in medical applications, as well as for mobile and satellite communications.

Author Contributions: Design and concept, I.E., A.I. and C.Z.; methodology, C.Z. and I.E.; investigation, A.I., W.M., M.S., S.U.; resources, A.I., A.B., and S.U.; writing—original draft preparation, I.E.; writing—review and editing, R.A.-A., I.E., and A.d.O.P.; validation, A.d.O.P., R.A.-A.; supervision, J.R.; project administration, J.R. All authors have read and agreed to the published version of the manuscript.

Funding: This project has received funding from the European Union's Horizon 2020 research and innovation program under grant agreement H2020-MSCA-ITN-2016 SECRET-722424. This work is also funded by the FCT/MEC through national funds and when applicable co-financed by the ERDF, under the PT2020 Partnership Agreement under the UID/EEA/50008/2019 project.

Acknowledgments: This work is supported by the European Union's Horizon 2020 Research and Innovation program under grant agreement H2020-MSCA-ITN-2016-SECRET-722424. This work is part of the POSITION-II project funded by the ECSEL joint Undertaking under grant number Ecsel-7831132-Postitio-II-2017-IA, www.position-2.eu.

References

1. Elfergani, I.; Hussaini, A.S.; Rodriguez, J.; Abd-Alhameed, R. *Antenna Fundamentals for Legacy Mobile Applications and Beyond*; Springer: New York, NY, USA, 2018.
2. Liao, W.J.; Hsieh, C.Y.; Dai, B.Y.; Hsiao, B.R. Inverted-F/slot integrated dual-band four-antenna system for WLAN access points. *IEEE Antenn. Wirel. Pr.* **2014**, *14*, 847–850. [CrossRef]
3. Iqbal, A.; Saraereh, O.A.; Ahmad, A.W.; Bashir, S. Mutual coupling reduction using F-shaped stubs in UWB-MIMO antenna. *IEEE Access* **2017**, *6*, 2755–2759. [CrossRef]
4. Fletcher, P.; Dean, M.; Nix, A. Mutual coupling in multi-element array antennas and its influence on MIMO channel capacity. *Electron. Lett.* **2003**, *39*, 342–344. [CrossRef]
5. Abdullah, M.; Kiani, S.H.; Iqbal, A. Eight Element Multiple-Input Multiple-Output (MIMO) Antenna for 5G Mobile Applications. *IEEE Access* **2019**, *7*, 134488–134495. [CrossRef]

6. Tse, D.; Viswanath, P. *Fundamentals of Wireless Communication*; Cambridge University Press: Cambridge, UK, 2005.

7. Abdullah, M.; Kiani, S.H.; Abdulrazak, L.F.; Iqbal, A.; Bashir, M.; Khan, S.; Kim, S. High-Performance Multiple-Input Multiple-Output Antenna System For 5G Mobile Terminals. *Electronics* **2019**, *8*, 1090. [CrossRef]

8. Winters, J. On the capacity of radio communication systems with diversity in a Rayleigh fading environment. *IEEE J. Sel. Area. Comm.* **1987**, *5*, 871–878. [CrossRef]

9. Foschini, G.J.; Gans, M.J. On limits of wireless communications in a fading environment when using multiple antennas. *Wireless Pers. Commun.* **1998**, *6*, 311–335. [CrossRef]

10. Garcia-Pardo, C.; Molina-Garcia-Pardo, J.M.; Rodriguez, J.V.; Juan-Llacer, L. MIMO capacity in UWB channels in an office environment for different polarizations. *Prog. Electromagn. Res.* **2013**, *44*, 109–122. [CrossRef]

11. Iqbal, A.; Basir, A.; Smida, A.; Mallat, N.K.; Elfergani, I.; Rodriguez, J.; Kim, S. Electromagnetic bandgap backed millimeter-wave MIMO antenna for wearable applications. *IEEE Access* **2019**, *7*, 111135–111144. [CrossRef]

12. Murmu, S.K.; Misra, I.S. Design of V-shaped microstrip patch antenna at 2.4 GHz. *Microw. Opt. Technol. Lett.* **2011**, *53*, 806–811. [CrossRef]

13. Moosazadeh, M.; Esmati, Z. Small Planar Dual-Band Microstrip-Fed Monopole Antenna for Wireless Local Area Network Applications Using Slotted Conductor-Backed Plane. *Microw. Opt. Technol. Lett.* **2013**, *55*, 2380–2383. [CrossRef]

14. Khan, M.U.; Sharawi, M.S.; Steffes, A.; Aloi, D.N. A 4-element MIMO antenna system loaded with CSRRs and patch antenna elements. In Proceedings of the 7th European Conference on Antennas and Propagation (EuCAP), Gothenburg, Sweden, 8–12 April 2013; pp. 2016–2019.

15. Kang, D.G.; Tak, J.; Choi, J. MIMO antenna with high isolation for WBAN applications. *Int. J. Antennas Propag.* **2015**, *2015*. [CrossRef]

16. Malviya, L.; Panigrahi, R.; Kartikeyan, M. 2×2 MIMO antenna for ISM band application. In Proceedings of the 11th International Conference on Industrial and Information Systems (ICIIS), Roorkee, India, 3–4 December 2016; pp. 794–797.

17. Sharawi, M.S.; Khan, M.U.; Numan, A.B.; Aloi, D.N. A CSRR loaded MIMO antenna system for ISM band operation. *IEEE Trans. Antennas Propag.* **2013**, *61*, 4265–4274. [CrossRef]

18. Yang, L.; Yan, S.; Li, T. Compact printed four-element MIMO antenna system for LTE/ISM operations. *Prog. Electromagn. Res.* **2015**, *54*, 47–53. [CrossRef]

19. Hammoodi, A.I.; Al-Rizzo, H.M.; Isaac, A.A. A wearable dual-band square slot antenna with stub for ISM and WiMAX applications. In Proceedings of the IEEE International Symposium on Antennas and Propagation & USNC/URSI National Radio Science Meeting, Vancouver, BC, Canada, 19–24 July 2015; pp. 732–733.

20. Raihan, R.; Bhuiyan, M.S.A.; Hasan, R.R.; Chowdhury, T.; Farhin, R. A wearable microstrip patch antenna for detecting brain cancer. In Proceedings of the IEEE 2nd International Conference on Signal and Image Processing (ICSIP), Singapore, 4–6 August 2017; pp. 432–436.

21. Ali, T.; Subhash, B.; Pathan, S.; Biradar, R.C. A compact decagonal-shaped UWB monopole planar antenna with truncated ground plane. *Microw. Opt. Technol. Lett.* **2018**, *60*, 2937–2944. [CrossRef]

22. Kwon, K.; Tak, J.; Choi, J. Design of a dual-band antenna for wearable wireless body area network repeater systems. In Proceedings of the 7th European Conference on Antennas and Propagation (EuCAP), Gothenburg, Sweden, 8–12 April 2013; pp. 418–421.

23. Ashyap, A.Y.; Abidin, Z.Z.; Dahlan, S.H.; Majid, H.A.; Shah, S.M.; Kamarudin, M.R.; Alomainy, A. Compact and low-profile textile EBG-based antenna for wearable medical applications. *IEEE Antennas Wirel. Propag. Lett.* **2017**, *16*, 2550–2553. [CrossRef]

24. Yang, H.L.; Yao, W.; Yi, Y.; Huang, X.; Wu, S.; Xiao, B. A dual-band low-profile metasurface-enabled wearable antenna for WLAN devices. *Prog. Electromagn. Res.* **2016**, *61*, 115–125. [CrossRef]

25. Velan, S.; Sundarsingh, E.F.; Kanagasabai, M.; Sarma, A.K.; Raviteja, C.; Sivasamy, R.; Pakkathillam, J.K. Dual-band EBG integrated monopole antenna deploying fractal geometry for wearable applications. *IEEE Antennas Wirel. Propag. Lett.* **2014**, *14*, 249–252. [CrossRef]

26. Jiang, Z.H.; Brocker, D.E.; Sieber, P.E.; Werner, D.H. A compact, low-profile metasurface-enabled antenna for wearable medical body-area network devices. *IEEE Trans. Antennas Propag.* **2014**, *62*, 4021–4030. [CrossRef]

27. Araghi, A.; Khalily, M.; Ghannad, A.A.; Xiao, P.; Tafazolli, R. Compact Dual Band Antenna for Off-Body-Centric Communications. In Proceedings of the 13th European Conference on Antennas and Propagation (EuCAP), Krakow, Poland, 31 March–5 April 2019; pp. 1–5.

28. Sanz-Izquierdo, B.; Miller, J.; Batchelor, J.C.; Sobhy, M. Dual-band wearable metallic button antennas and transmission in body area networks. *IET Microw. Antennas Propag.* **2010**, *4*, 182–190. [CrossRef]

29. Ullah, M.; Islam, M.; Alam, T.; Ashraf, F. Based Flexible Antenna for Wearable Telemedicine Applications at 2.4 GHz ISM Band. *Sensors* **2018**, *18*, 4214. [CrossRef] [PubMed]

30. Sreelakshmy, R.; Ashok Kumar, S.; Shanmuganantham, T. A wearable type embroidered logo antenna at ISM band for military applications. *Microw. Opt. Technol. Lett.* **2017**, *59*, 2159–2163. [CrossRef]

31. Jensen, M.A.; Wallace, J.W. A review of antennas and propagation for MIMO wireless communications. *IEEE Trans. Antennas Propag.* **2004**, *52*, 2810–2824. [CrossRef]

32. Fan, Y.; Huang, J.; Chang, T.; Liu, X. A Miniaturized Four-Element MIMO Antenna With EBG for Implantable Medical Devices. *IEEE J. Electromagn. RF Microw. Med. Biol.* **2018**, *2*, 226–233. [CrossRef]

33. Ahmad, M.S.; Mohyuddin, W.; Choi, H.C.; Kim, K.W. 4×4 MIMO antenna design with folded ground plane for 2.4 GHz WLAN applications. *Microw. Opt. Technol. Lett.* **2018**, *60*, 395–399. [CrossRef]

34. Likhitha, T.; Ashok Kumar, S.; Shanmuganantham, T. Design of Compact Four Port MIMO Antenna Using SRR Ring for High Isolation. In Proceedings of the International Conference on Antenna Testing and Measurement Society (ATMS 2018), Pune, India, 6–7 February 2018; pp. 1–4.

35. Nigam, H.; Kumar, M. A compact MIMO antenna design for 2.4 GHz ISM band frequency applications. *Int. J. Electron. Comput. Sci. Eng.* **2014**, 324–331.

36. Subhanrao Bhadade, R.; Padmakar Mahajan, S. Circularly polarized 4×4 MIMO antenna for WLAN applications. *Electromagnetics* **2019**, *39*, 325–342. [CrossRef]

37. Li, H.; Sun, S.; Wang, B.; Wu, F. Design of compact single-layer textile MIMO antenna for wearable applications. *IEEE Trans. Antennas Propag.* **2018**, *66*, 3136–3141. [CrossRef]

38. Kiem, N.K.; Phuong, H.N.B.; Hieu, Q.N.; Chien, D.N. A novel metamaterial MIMO antenna with high isolation for WLAN applications. *Int. J. Antennas Propag.* **2015**, *2015*, 9. [CrossRef]

39. Zhang, Z.; Jiao, Y.C.; Song, Y.; Zhang, T.L.; Ning, S.M.; Zhang, F.S. A modified CPW-fed monopole antenna with very small ground for multiband WLAN applications. *Microw. Opt. Technol. Let.* **2010**, *52*, 463–466. [CrossRef]

40. Wadkar, S.P.; Hogade, B.; Rathod, S.; Kumar, H.; Kumar, G. Normal Mode Helical Antenna on Small Circular Ground Plane. *IETE J. Res.* **2018**, 1–8. [CrossRef]

41. Sanad, M. Microstrip antennas on very small ground planes for portable communication systems. In Proceedings of the IEEE Antennas and Propagation Society International Symposium and URSI National Radio Science Meeting, Seattle, WA, USA, 20–24 June 1994; Volume 2, pp. 810–813.

42. Studio, C.M. CST Studio Suite 2014. *Comput. Simul. Technol. AG* **2014**.

43. Iqbal, A.; A Saraereh, O.; Bouazizi, A.; Basir, A. Metamaterial-based highly isolated MIMO antenna for portable wireless applications. *Electronics* **2018**, *7*, 267. [CrossRef]

44. Alibakhshikenari, M.; Khalily, M.; Virdee, B.S.; See, C.H.; Abd-Alhameed, R.A.; Limiti, E. Mutual-Coupling Isolation Using Embedded Metamaterial EM Bandgap Decoupling Slab for Densely Packed Array Antennas. *IEEE Access* **2019**, *7*, 51827–51840. [CrossRef]

45. Iqbal, A.; Bouazizi, A.; Kundu, S.; Elfergani, I.; Rodriguez, J. Dielectric resonator antenna with top loaded parasitic strip elements for dual-band operation. *Microw. Opt. Technol. Let.* **2019**, *61*, 2134–2140. [CrossRef]

46. Alibakhshikenari, M.; Virdee, B.S.; See, C.H.; Abd-Alhameed, R.; Ali, A.H.; Falcone, F.; Limiti, E. Study on isolation improvement between closely-packed patch antenna arrays based on fractal metamaterial electromagnetic bandgap structures. *IET Microw. Antennas Propag.* **2018**, *12*, 2241–2247. [CrossRef]

47. Bouazizi, A.; Zaibi, G.; Iqbal, A.; Basir, A.; Samet, M.; Kachouri, A. A dual-band case-printed planar inverted-F antenna design with independent resonance control for wearable short range telemetric systems. *Int. J. RF Microw. Computer-Aided Eng.* **2019**, *29*, e21781. [CrossRef]

48. Iqbal, A.; Smida, A.; Abdulrazak, L.F.; Saraereh, O.A.; Mallat, N.K.; Elfergani, I.; Kim, S. Low-Profile Frequency Reconfigurable Antenna for Heterogeneous Wireless Systems. *Electronics* **2019**, *8*, 976. [CrossRef]

49. Basir, A.; Bouazizi, A.; Zada, M.; Iqbal, A.; Ullah, S.; Naeem, U. A dual-band implantable antenna with wide-band characteristics at MICS and ISM bands. *Microw. Opt. Technol. Let.* **2018**, *60*, 2944–2949. [CrossRef]
50. Zebiri, C.; Sayad, D.; Elfergani, I.; Iqbal, A.; Mshwat, W.F.; Kosha, J.; Rodriguez, J.; Abd-Alhameed, R. A compact semi-circular and arc-shaped slot antenna for heterogeneous RF front-ends. *Electronics* **2019**, *8*, 1123. [CrossRef]
51. Thielens, A.; Benarrouch, R.; Wielandt, S.; Anderson, M.; Moin, A.; Cathelin, A.; Rabaey, J. A Comparative Study of On-Body Radio-Frequency Links in the 420 MHz–2.4 GHz Range. *Sensors* **2018**, *18*, 4165. [CrossRef] [PubMed]
52. Abbasi, M.A.B.; Nikolaou, S.S.; Antoniades, M.A.; Stevanović, M.N.; Vryonides, P. Compact EBG-backed planar monopole for BAN wearable applications. *IEEE Trans. Antennas Propag.* **2016**, *65*, 453–463. [CrossRef]
53. Yang, L.; Li, T.; Yan, S. Highly compact MIMO antenna system for LTE/ISM applications. *Int. J. Antennas Propag.* **2015**, *2015*, 714817. [CrossRef]
54. Malviya, L.; Panigrahi, R.K.; Kartikeyan, M. Four element planar MIMO antenna design for long-term evolution operation. *IETE J. Res.* **2018**, *64*, 367–373. [CrossRef]
55. Sharawi, M.S. *Printed MIMO Antenna Engineering*; Artech House: Norwood, MA, USA, 2014.
56. Singh, H.S.; Pandey, G.K.; Bharti, P.K.; Meshram, M.K. A compact dual-band diversity antenna for WLAN applications with high isolation. *Microw. Opt. Technol. Let.* **2015**, *57*, 906–912. [CrossRef]

 *electronics* 

Article

A New CPW-Fed Diversity Antenna for MIMO 5G Smartphones

Naser Ojaroudi Parchin [1,*], **Haleh Jahanbakhsh Basherlou** [2], **Yasir I. A. Al-Yasir** [1], **Ahmed M. Abdulkhaleq** [1,3], **Mohammad Patwary** [4] **and Raed A. Abd-Alhameed** [1]

[1] Engineering and Informatics, University of Bradford, Bradford BD7 1DP, UK; Y.I.A.Al-Yasir@bradford.ac.uk (Y.I.A.A.-Y.); A.Abd@sarastech.co.uk (A.M.A.); R.A.A.Abd@bradford.ac.uk (R.A.A.-A.)

[2] Bradford College, Bradford BD7 1AY, UK; Hale.Jahanbakhsh@gmail.com

[3] SARAS Technology Limited, Leeds LS12 4NQ, UK

[4] School of Computing and Digital Technology, Birmingham City University, Birmingham B5 5JU, UK; Mohammad.patwary@bcu.ac.uk

* Correspondence: N.OjaroudiParchin@Bradford.ac.uk; Tel.: +447341436156

Received: 31 December 2019; Accepted: 28 January 2020; Published: 4 February 2020

Abstract: In this study, a new coplanar waveguide (CPW)-fed diversity antenna design is introduced for multiple-input–multiple-output (MIMO) smartphone applications. The diversity antenna is composed of a double-fed CPW-fed antenna with a pair of modified T-ring radiators. The antenna is designed to cover the frequency spectrum of commercial sub-6 GHz 5G communication (3.4–3.8 and 3.8–4.2 GHz). It also provides high isolation, better than −16 dB, without an additional decoupling structure. It offers good potential to be deployed in future smartphones. Therefore, the characteristics and performance of an 8-port 5G smartphone antenna were investigated using four pairs of the proposed diversity antennas. Due to the compact size and also the placement of the elements, the presented CPW-fed smartphone antenna array design occupies a very small part of the smartphone board. Its operation band spans from 3.4 to 4.4 GHz. The simulated results agree well with measured results, and the performance of the smartphone antenna design in the presence of a user is given in this paper as well. The proposed MIMO design provides not only sufficient radiation coverage supporting different sides of the mainboard but also polarization diversity.

Keywords: 5G technology; CPW-fed antenna; diversity antenna; future smartphones; MIMO systems

---

1. Introduction

With the creation of standards for and the development of fifth-generation (5G) mobile communication, more and more research has been conducted into related technologies with the hope of achieving a higher transmission rate, lower cost and higher gain [1–3]. Multiple-input–multiple-output (MIMO) technology is a key to realizing a higher transmission rate [4]. By using MIMO technology, multiple independent channels can be achieved on the original spectrum by the diversity method, and multipath fading can be reduced so as to improve the data transmission rate. A MIMO antenna is a significant facility to improve the channel capacity of a MIMO system [5–8]. MIMO systems of 2 × 2 are successfully employed for 4G mobile networks, and a large number of antenna elements are expected to be applied for 5G communications [9,10].

Several kinds of 5G MIMO smartphone antennas have been put forward recently [11–25] (see Table 2 for details). However, these MIMO antenna designs either suffer from a narrow frequency bandwidth or occupy a huge space on a smartphone mainboard. Furthermore, some of the reported designs use uniplanar radiators, which are difficult to fabricate and integrate with the 5G smartphone circuit. In the designs of many MIMO antennas, it is common to avoid placing elements in parallel and to choose instead to place them vertically, which can avoid strong mutual couplings caused by the same polarization mode. In this paper, however, the antenna elements are both perpendicular and parallel to each other to exhibit the diversity function. In addition, the T-shaped strip of the antenna configuration can act as a decoupling structure. Due to compact size and also placement of the antenna, the proposed MIMO design occupies a very small part of the smartphone printed circuit board (PCB). Therefore, the antenna achieves not only low mutual couplings but also small clearance.

The antenna elements of the MIMO design are fed using the coplanar waveguide feeding mechanism in order to operate at sub-6 GHz 5G communication (3.4–3.8 and 3.8–4.2 GHz) [26]. Compared with probe-fed and microstrip-fed antennas, coplanar waveguide (CPW)-fed antennas can easily achieve the wideband impedance matching [27,28]. Therefore, CPW-fed antennas are widely used and becoming increasingly popular in wireless applications owing to their attractive features such as compact size, conformal status, their light weight and ease of fabrication and integration with wireless communication systems [29,30]. The paper is organized as follows: The design and characteristics of the diversity antenna element are represented in Section 2. Section 3 discusses the MIMO performance and radiation characteristics of the proposed 5G smartphone antenna array. Section 4 investigates the radiation behavior of the designed smartphone antenna array in the vicinity of the user. Section 5 gives the conclusion of this paper.

2. The Proposed CPW-Fed Diversity Antenna

The characteristics of the single-element diversity antenna are discussed in this section. Its structure is shown in Figure 1a. It is shown that the schematic of the diversity antenna contains a pair of modified CPW-fed T-ring resonators. As seen, the proposed antenna is designed on one side of the FR4 dielectric. In addition, as seen, SMA (SubMiniature version A) connectors are also embedded in the simulations. Figure 1b depicts the S-parameters of the proposed CPW-fed diversity antenna. As illustrated, the designed antenna provides a wide operation band of 3.2–4.4 GHz, supporting both target bands, including 3.4–3.8 and 3.8–4.2 GHz. It should be noted that the arrow-shaped strip of the design, placed between the elements, can act as a decoupling structure and increase the isolation between the antenna ports. Therefore, the mutual coupling (S_{12}/S_{21}) is successfully reduced. As can be observed, greater than −15 dB (with −20 dB value at the center frequency (4 GHz)) has been achieved for the designed diversity antenna. The characteristics of the antenna are investigated using computer simulation technology (CST) software [31]. The detailed dimensions of the designed CPW-fed diversity antenna are shown in Table 1.

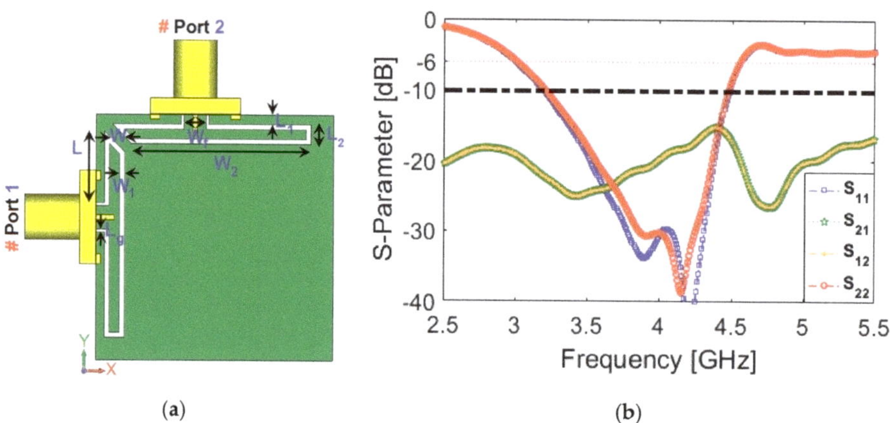

Figure 1. (a) Geometry of the coplanar waveguide (CPW)-fed diversity antenna and (b) its simulated S-parameters.

Table 1. The dimension of the diversity antenna.

Parameter	Value (mm)	Parameter	Value (mm)	Parameter	Value (mm)	Parameter	Value (mm)
W	1.1	W_1	0.5	W_2	19	W_f	2.4
L	7	L_1	1	L_2	2	L_g	0.25

The frequency behavior of the proposed diversity antenna is very flexible. Figure 2 discusses the impedance matching and frequency tuning of the antenna. Figure 2a illustrates the S_{11} and S_{21} characteristics versus different values of W_2. As seen, when its value increases from 17 to 21 mm, the lower and upper operation frequencies of the antenna increase from 3 to 3.4 and 4.5 to 4.8 GHz, respectively. In addition, as can be observed, the S_{21} function of the diversity antenna tunes by changing the value of W_2. Figure 2b investigates the impedance matching function of the antenna for various values of L_2: when its size changes from 2.75 to 1.75 mm, the matching characteristic of the diversity antenna varies from −14 to less than −30 dB. However, unlike Figure 2a, the S_{21} is almost constant with an insignificant variation.

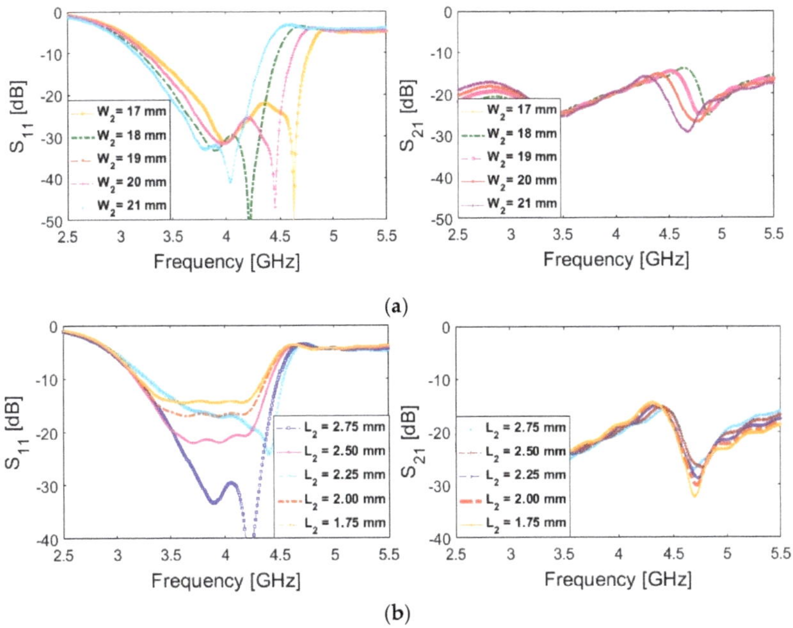

Figure 2. S_{11}/S_{21} results of the diversity antenna for various values of (**a**) W_2 and (**b**) L_2.

In order to have a better illumination about the working mechanism of the CPW-Fed, its simulated current distributions at 3.8 and 4.2 GHz are shown in Figure 3. As shown at 3.8 GHz, the current is mainly distributed near the arrow strip and outer boundary of the modified T-ring resonator, which verifies the role of the strip in creating a new resonance at 3.8 GHz. At 4.2 GHz, the currents are mainly concentrated inside of the modified T-ring slot [31,32]. The radiation patterns of the diversity antenna for each exciting port at 4 GHz (center frequency of the antenna operation band) are plotted in Figure 4. It is shown that well-defined polarization and pattern diversity is obtained for the antenna. The radiation patterns of the antenna are symmetrical, covering the top/bottom sides of the substrate and providing similar radiation behavior with a gain value of 3.6 dB. The fundamental radiation characteristics of the diversity antenna are also given in Figure 5 within the range of 3.4–4.4 GHz (with 0.1 GHz/step). As illustrated in Figure 5, the antenna exhibits high efficiencies over its 1 GHz impedance bandwidth. In addition, the antenna offers sufficient gain and directivity in the range of 3.4–4.4 GHz.

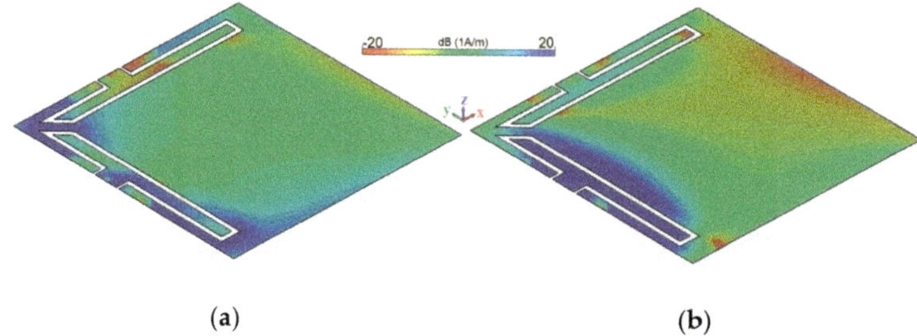

(**a**)   (**b**)

Figure 3. The current densities of the diversity antenna design from port 1 at (**a**) 3.8 and (**b**) 4.2 GHz.

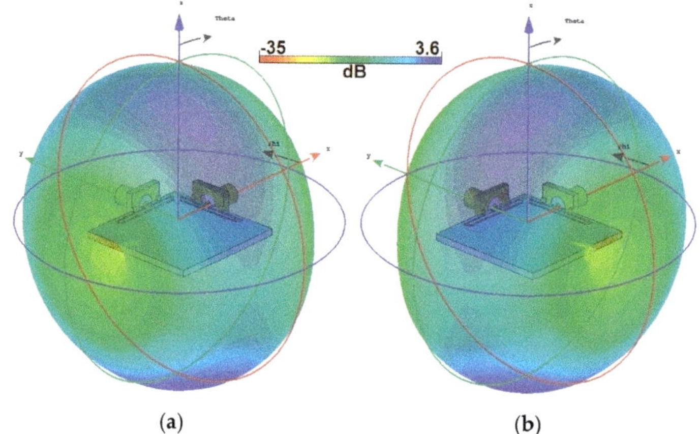

(**a**)   (**b**)

Figure 4. Radiation patterns of the CPW-fed diversity antenna at 4 GHz from (**a**) ports 1 and (**b**) 2.

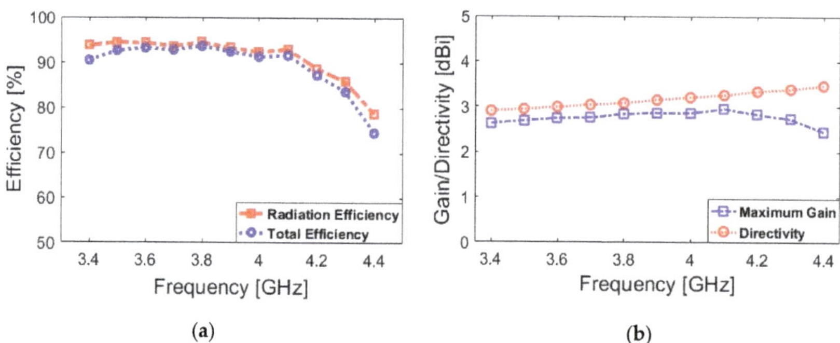

(**a**)   (**b**)

Figure 5. Simulated (**a**) efficiency and (**b**) gain/directivity characteristics of the CPW-fed antenna.

A prototype of the design was fabricated and its S parameters were tested. In order to verify the simulated S-parameter results mentioned above, the single-element diversity antenna was fabricated and measured. A vector network analyzer was used to measure the antenna in our research. The fabricated dual-port antenna and the measured results of S-parameters are given in Figure 6a,b. As illustrated in Figure 6b, the measured results of the diversity antenna based on FR-4 are very close to the simulated results' values within 2.5–5.5 GHz; it provides quite a good impedance bandwidth ($S_{11} <$ −10 dB within 3.2–4.4 GHz), and its mutual coupling (S_{21}) is less than −15 dB.

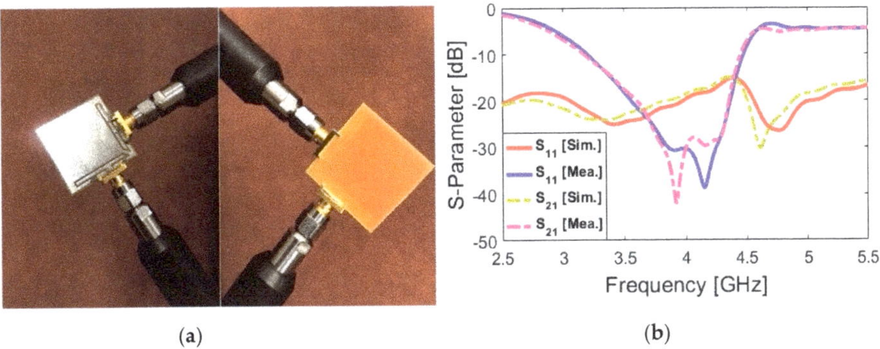

Figure 6. (**a**) Fabricated antenna's top/bottom sides and (**b**) the simulated/measured S-parameters.

The total active reflection coefficient (TARC), envelope correlation coefficient (ECC) and diversity gain (DG) characteristics are important parameters to be considered in diversity/MIMO antennas and can be calculated using the below formulas [33–35].

$$TARC = -\sqrt{\frac{(S_{11} + S_{12})^2 + (S_{21} + S_{22})^2}{2}} \tag{1}$$

$$ECC = \frac{\left|S_{11}^* S_{21} + S_{12}^* S_{22}\right|^2}{\left(1 - |S_{11}|^2 - |S_{12}|^2\right)\left(1 - |S_{21}|^2 - |S_{22}|^2\right)^*} \tag{2}$$

$$DG = 10\sqrt{1 - (ECC)^2} \tag{3}$$

Figure 7 represents the calculated TARC, ECC and DG characteristics for the proposed dual-port diversity antenna. As shown in Figure 7a,b, the TARC and ECC results of this diversity antenna are very low within the band, which means the antenna is competent for diversity reception/transmission in the MIMO channels [33]. In addition, as can be observed from Figure 7c, the DG function of the design is greater than 9.97 dB over the entire band.

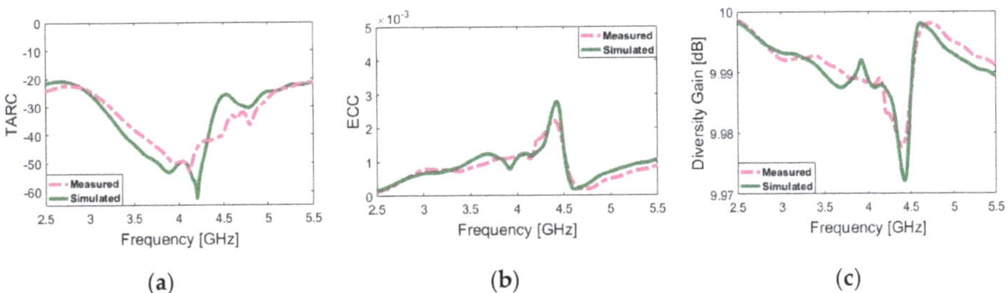

Figure 7. Calculated (**a**) total active reflection coefficient (TARC), (**b**) envelope correlation coefficient (ECC) and (**c**) diversity gain (DG) characteristics of the CPW-fed diversity antenna.

3. Mobile-Phone Antenna Design

Four pairs of the modified CPW-fed diversity antenna mentioned above were placed in different corners of the smartphone board to form an eight-port MIMO antenna with a standard size of $150 \times 75 \times 1.6$ mm^3. Its structure is shown in Figure 8. As can be observed, due to the compact size and also the placement of the CPW-fed ring antenna, the proposed MIMO design occupies a very small part of the board.

Figure 9 shows the S-parameters of the CPW-fed MIMO smartphone antenna. It can be observed from Figure 9a that all antenna elements exhibit good return loss results covering 3.4–4.4 GHz. It should also be noted that due to the effect of the MIMO configuration and also the big ground plane of the smartphone board, the lower operation frequency shifted from 3.2 to 3.4 GHz. However, it still covers the target 5G bands including 3.4–3.8 and 3.8–4.2 GHz. The isolations between ports are shown in Figure 9b. The maximum mutual couplings of the diversity antenna arrays are usually between the closely spaced diversity elements such as Ant. 1 and Ant. 2. Due, however, to similar performances and placements of the antenna pairs in the configuration of the proposed smartphone antenna, it is not necessary to show all S-parameters. As seen from Figure 9b, the isolation levels of the antenna ports are less than 16 dB within 3.4–4.4 GHz. This is mainly due to the strong mutual couplings between the adjacent ports. Figure 10 plots the radiation patterns at the middle frequency (4 GHz) of the first CPW-fed diversity antenna (with ports 1 and 2) mounted onto the smartphone PCB. As seen, the radiation patterns are symmetrical covering the top/bottom sides of the substrate and providing similar radiation behavior with gain vale of 4 dBi.

Figure 8. Schematic of the CPW-fed eight-port 5G smartphone antenna.

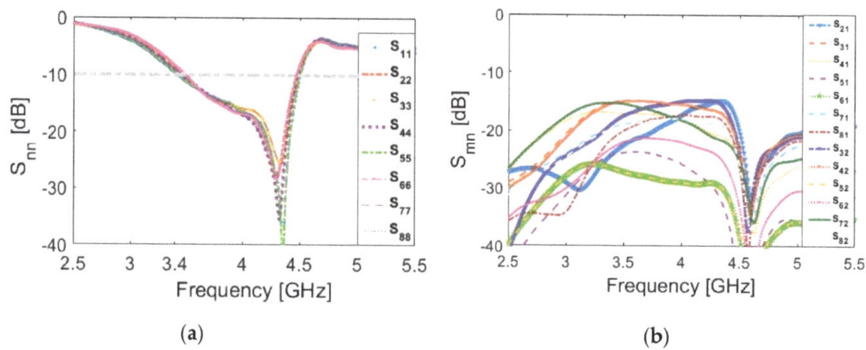

(a) (b)

Figure 9. The simulated (**a**) S_{nn} and (**b**) S_{mn} results.

The 3D patterns of antenna radiations at 4 GHz for each feeding port have been illustrated in Figure 11. It can be observed that the CPW-fed resonators not only can cover different sides of the mobile-phone board but also support different polarizations, which is a unique function of the MIMO design [36,37]. In addition, due to the different placements of antenna elements (Ant. 1 and Ant. 2, for example), gain values of 4.6/5.15 dB are achieved for the resonators. The efficiencies (radiation and total) of the CPW-fed ring slot resonators are also given in Figure 12. It is evident that high efficiencies with slight variations are achieved within the range of 3.4–4.4 GHz: more than 80% radiation and 70% total efficiencies were observed for the CPW-fed elements of the proposed MIMO design.

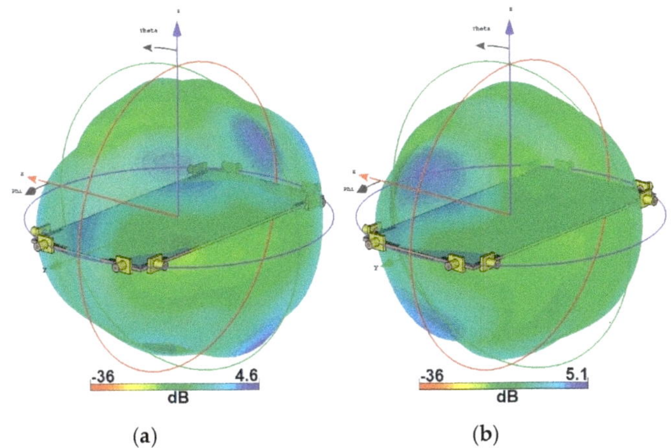

Figure 10. Radiation patterns of the dual-port diversity resonator from (**a**) port 1 and (**b**) port 2.

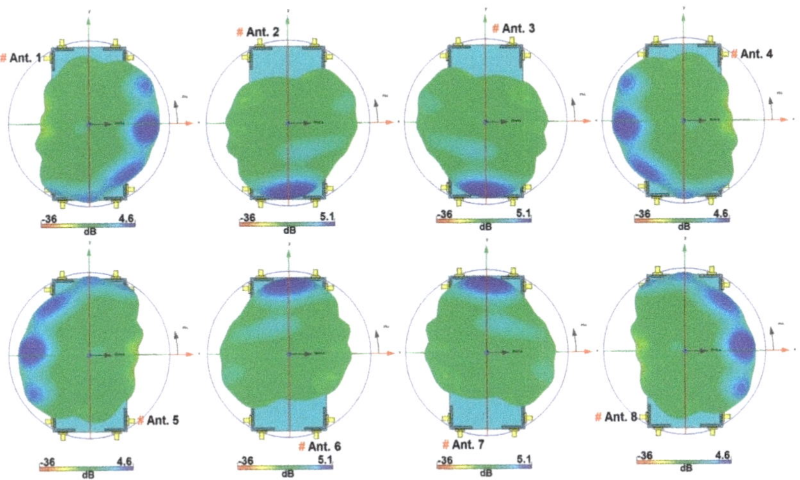

Figure 11. 3D radiation patterns at center frequency (4 GHz).

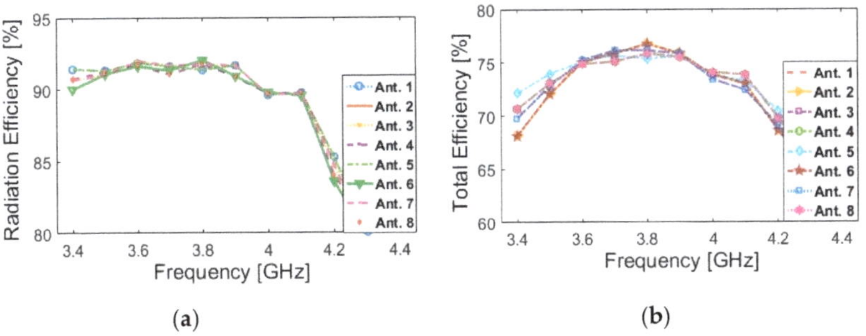

Figure 12. (**a**) Radiation and (**b**) total efficiencies of the antenna elements (Ant. 1–Ant. 8).

A prototype of the 5G smartphone antenna design was fabricated and fed for measurements, as illustrated in Figure 13. In order to verify the simulated results of the smartphone antenna—mentioned above—the S-parameter and radiation patterns were measured. Due, however, to similar placements and also the performances of the CPW-fed antenna pairs, the properties of the smartphone antenna design for port 1 and 2 were measured and compared in the following way. A vector network analyzer and antenna chamber room were used in the measurement process in our research. The feeding mechanism of the design is shown in Figure 14a. During the measurements, in order to avoid unwanted mutual effects, 50-Ω RF loads are installed for the elements not under test. The measured and simulated results of the S-parameters are compared in Figure 14b. As seen, the measured results are in good agreement with the simulated results to cover the required operation band: a quite good impedance bandwidth ($S_{11} < -10$ dB within 3.4–4.4 GHz) and mutual coupling ($S_{21} < -15$ dB) are obtained for the smartphone antenna design.

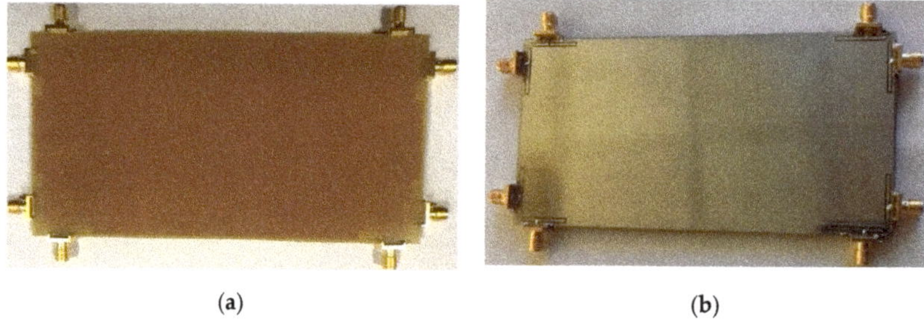

(a)   (b)

Figure 13. (**a**) Front and (**b**) back views of the fabricated sample.

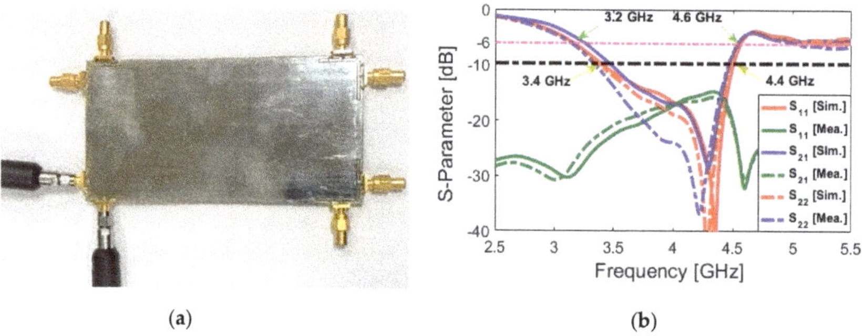

(a)   (b)

Figure 14. (**a**) Feeding mechanism and (**b**) S-parameters of adjacent CPW-fed elements for the 5G smartphone antenna.

Measured and simulated radiation patterns are shown in Figure 15. When measuring radiation patterns, we keep one port excited and another one loaded with a 50-Ω load. In the measurement of radiation patterns, the smartphone MIMO antenna is used as the receiver, and a horn antenna is used as the transmitter. As can be observed from Figure 15 a,b, the sample smartphone antenna prototype offers good quasi-omnidirectional radiation patterns with an acceptable agreement between simulations and measurements [38–40].

Table 2 provides a comparison between the presented smartphone array antenna and another reported smartphone array [11–25]. As can be observed, compared with the recently proposed 5G MIMO smartphone antennas with planar and uniplanar structures, our antenna performs better in terms of impedance match and bandwidth, and its clearance size remains at a satisfactory level, as shown in Table 2. The proposed design achieves not only approximately 1 GHz impedance bandwidth but also sufficient mutual couplings, better than −15 dB. Unlike the reported 5G antenna design, our antenna is implemented in one-side of the smartphone mainboard using CPW-fed technology, which makes it easy to fabricate and integrate with the circuit. It is apparent that all the listed antennas have double-sides or uniplanar configuration. In addition, due to the small clearance of the proposed smartphone antenna, its fundamental radiation properties in data and talk modes are not reduced significantly, as discussed in the following section.

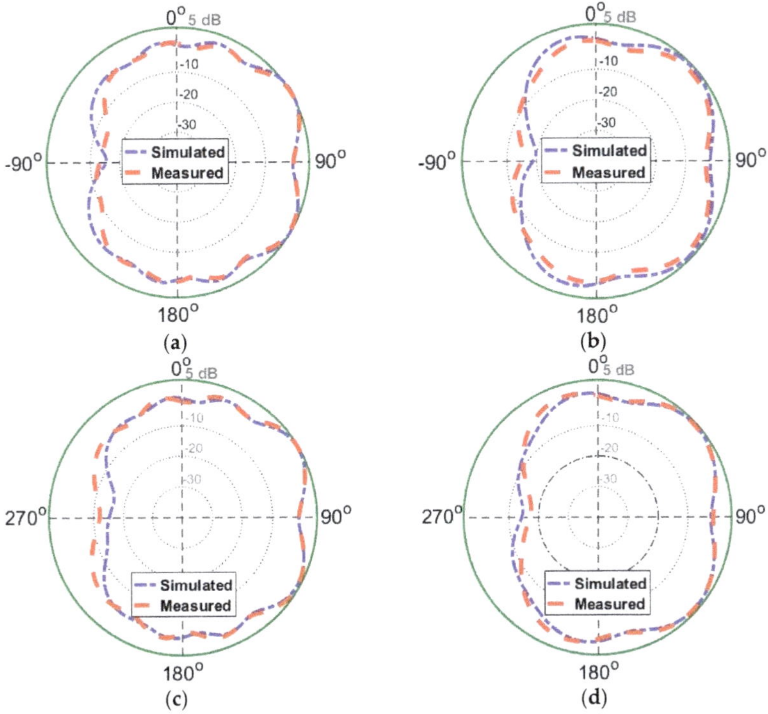

Figure 15. 2D radiation patterns for (**a**) Ant.1 at 3.6 GHz, (**b**) Ant.2 at 3.6 GHz, (**c**) Ant.1 at 4 GHz and (**d**) Ant.2 at 4 GHz.

Table 2. Comparison between our design and the referenced 5G smartphone antennas.

Reference	Design Type	Bandwidth (GHz)	Efficiency (%)	Size (mm²)	Isolation (dB)	ECC
[11]	Gap-Coupled IFA	3.4–3.6	-	150 × 75	15	<0.02
[12]	Inverted-F	3.4–3.6	55–60	100×50	10	-
[13]	Patch-Slot	3.55–3.65	52–76	150 × 75	11	-
[14]	Monopole	3.4–3.6	35–50	150 × 75	11	<0.40
[15]	Spatial-Reuse Antenna	3.4–3.6	40–70	150 × 75	12	<0.2
[16]	Inverted-L Monopole	3.4–3.6	40–60	136 × 68	14	<0.2
[17]	Inverted-F	3.4–3.6	-	120 × 70	20	-
[18]	Ring-Slot	3.4–3.8	60–75	150 × 75	15	<0.01
[19]	Monopole	4.55–4.75	50–70	136 × 68	10	-
[20]	Tightly Arranged Pairs	3.4–3.6	50–70	150× 73	17	<0.07
[21]	Wave-Guide	3.4–3.6	50–80	150 × 75	15	<0.2
[22]	Monopole	3.4–3.6	60–70	150 × 75	18	<0.015
[23]	Diamond-shaped Slot	3.3–3.9	60–80	150 × 75	17	<0.01
[24]	open-end slot	3.4–3.6	50–60	136 × 68	11	0.05
[25]	loop element	3.3–3.6	40	120 × 70	15	0.02
Proposed	CPW-Fed Diversity	3.4–4.4	65–80	150 × 75	16	<0.005

4. User Impacts on the Performance of the CPW-fed Smartphone Antenna Array

For smartphone antennas, it is indispensable to investigate the user effect on the radiation performance of the antenna [41,42]. Different usage postures in data-mode and talk-mode are considered in this section. Figure 16 represents S_{nn} and efficiency results of the 5G antenna in data-mode with different placement modes for right/left hands touching the top/bottom sides of the smartphone. It can be observed that similar characteristics are achieved for different data-mode scenarios. This is mainly due to symmetrical configuration and similar placements of the CPW-fed antenna pairs. In addition, as shown in Figure 16b, the S_{nn} results of all elements are not influenced significantly and still could cover the desired operation band. Some variation is discovered for the elements, which are partially covered with the hand-phantom, due to its absorption. Furthermore, it is evident from Figure 16c that a part of the radiation power of the antenna is absorbed by the medium, which causes some reduction in the efficiencies of the elements. However, the elements still provide around 40% and more total efficiencies within the 3.4–4.4 GHz operation band.

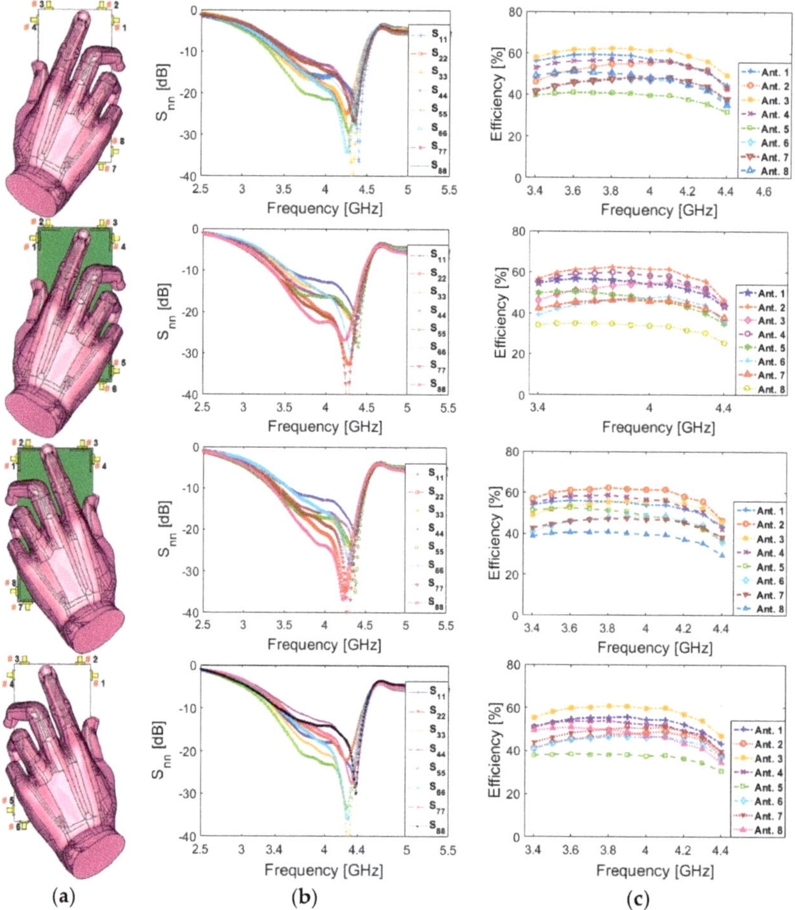

Figure 16. (a) Placement, (b) S_{nn} and (c) total efficiencies for different data-mode scenarios.

Apart from the data-mode, discussed above, the characteristics of the CPW-fed resonators in talk-mode are also investigated and represented in Figures 17–19. It is evident from Figure 17 that the antenna elements work sufficiently and provide good S_{nn} and total efficiency results for different antenna elements. The radiation pattern results for the MIMO smartphone antenna in talk-mode are shown in Figure 18. As can obviously be realized from the simulation, the proposed CPW-fed MIMO antenna offers good radiation patterns in talk-mode. In addition, the gain levels of the CPW-fed antenna resonators vary from 2 to 5 GHz.

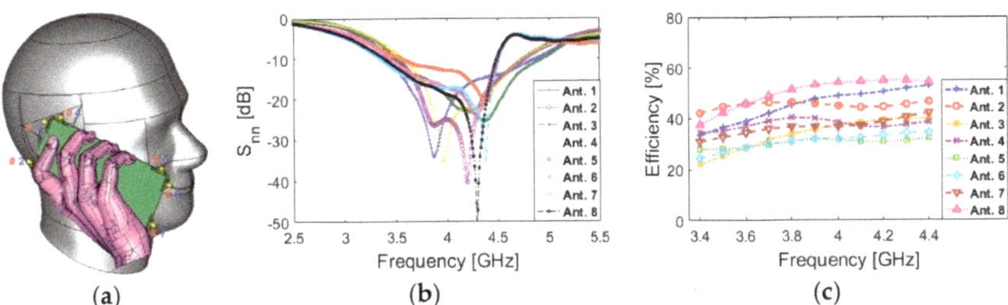

Figure 17. (**a**) Placement, (**b**) S_{nn} and (**c**) total efficiencies for talk-mode.

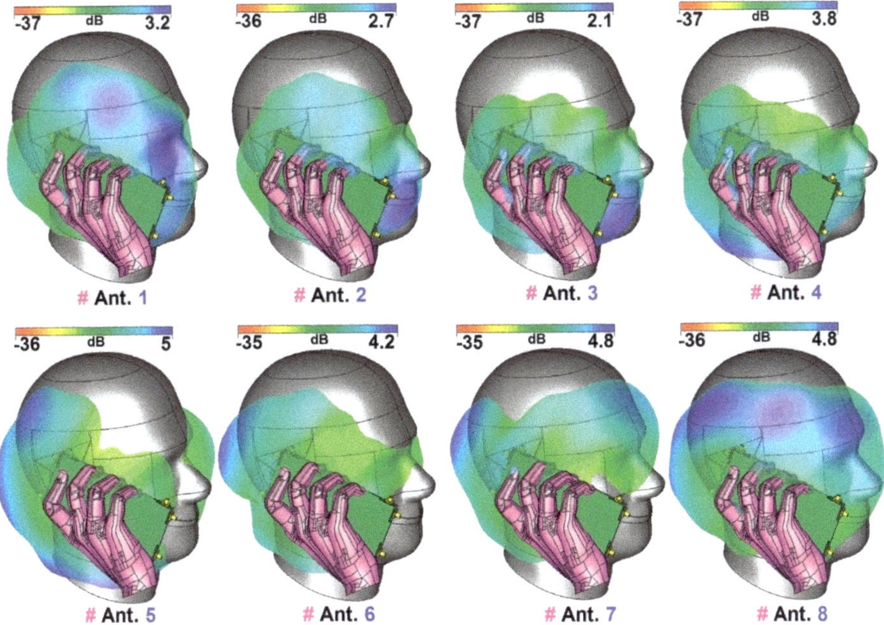

Figure 18. Radiation patterns in Talk-Mode.

Compared with Figure 8, the maximum reduction of antenna gain is observed for the elements closely spaced with the user's head and hand. In general, the closer the distance between the antenna element and the user's hand/head is, the greater the reduction on the gain and the efficiencies [43]. As can be observed, the maximum reductions of the gain levels are discovered for the elements that are located near to the head phantom (Ant. 3). In addition, due to the presence of the head and hand phantoms, the radiation patterns are a bit distorted and become weaker. One can see that antenna elements are touched by different parts of the hand and head phantoms in the presented talk-mode.

The specific absorption rate (SAR) characteristic of the CPW-fed MIMO design is studied and represented in Figure 19. It is shown that Ant. 3 causes the maximum SAR value (2.1), and the minimum SAR value (0.7) is observed from Ant. 7. Therefore, it can be concluded that the closest distance between antenna elements and the user-head leads to a maximum SAR value, and the furthest distance leads to a minimum SAR value [44,45].

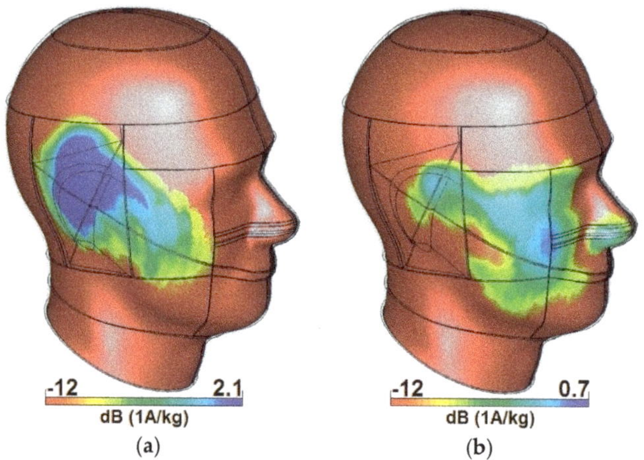

Figure 19. Specific absorption rate (SAR) investigations for (a) Ant. 3 and (b) Ant. 7.

The S_{nn} (S_{11}~S_{88}) and efficiency characteristics of the CPW-fed MIMO smartphone antenna in the presence of battery, speaker, camera, USB connector and LCD screen are investigated and illustrated in Figure 20. It was found that the designed CPW-fed MIMO antenna provides sufficient S_{nn} and efficiencies supporting a 3.4–4.4 GHz band. In addition, as shown in Figure 20c, the array exhibits high efficiencies in the presence of smartphone components.

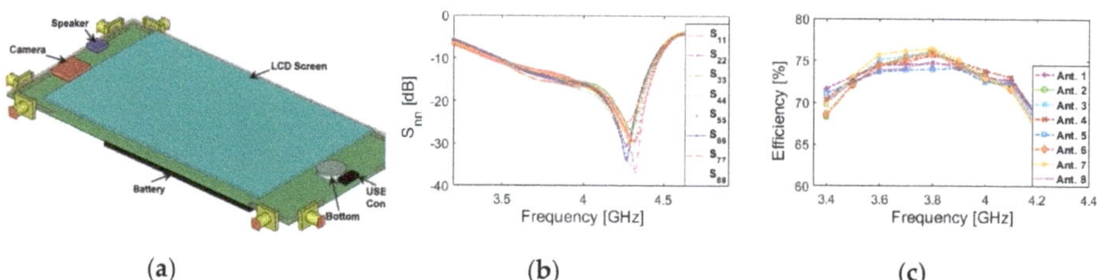

Figure 20. (a) Schematic, (b) S_{nn} and (c) efficiencies of the array in the presence of the smartphone components.

5. Conclusions

A smartphone array antenna design with new double-fed CPW-fed resonators is introduced for sub-6 GHz 5G applications. The structure of the CWP-fed element consists of two closely-spaced modified T-ring radiators operating with a frequency band of 3.3–4.4 GHz. Four pairs of the CPW-fed diversity antennas are placed at four corners of the smartphone board to form an 8 × 8 MIMO antenna. The fundamental characteristics and MIMO performance of the design were studied and sufficient results were achieved. Simulated and experimental results are provided to validate the usefulness of the designed smartphone antenna array for 5G mobile communications.

Author Contributions: Writing—original draft preparation, N.O.P., Y.I.A.A.-Y., A.M.A., M.P. and R.A.A.-A.; writing—review and editing, N.O.P., H.J.B. and R.A.A.-A.; investigation, N.O.P., H.J.B., M.P. and Y.I.A.A.-Y.; resources, N.O.P., M.P. and R.A.A.-A.; for other cases, all authors have participated. All authors have read and agreed to the published version of the manuscript.

Funding: This project has received funding from the European Union's Horizon 2020 research and innovation program under grant agreement H2020-MSCA-ITN-2016 SECRET-722424.

Acknowledgments: The authors wish to express their thanks to the support provided by the innovation program under grant agreement H2020-MSCA-ITN-2016 SECRET-722424.

Conflicts of Interest: The authors declare no conflict of interest.

References

1. Nadeem, Q.U.A.; Kammoun, A.; Debbah, M.; Alouini, S.-M. Design of 5G full dimension massive MIMO systems. *IEEE Trans. Commun.* **2018**, *66*, 726–740. [CrossRef]
2. Ojaroudiparchin, N.; Shen, M.; Pedersen, G.F. Multi-layer 5G mobile phone antenna for multi-user MIMO communications. In Proceedings of the 23rd Telecommunications Forum Telfor (TELFOR), Belgrade, Serbia, 24–26 November 2015; pp. 559–562.
3. Osseiran, A.; Boccardi, F.; Braun, V.; Kusume, K.; Marsch, P.; Maternia, M.; Queseth, O.; Schellmann, M.; Schotten, H.; Taoka, H.; et al. Scenarios for 5G mobile and wireless communications: The vision of the METIS project. *IEEE Commun. Mag.* **2014**, *52*, 26–35. [CrossRef]
4. Yang, H.H.; Quel, Y.Q.S. Massive MIMO Meet Small Cell. *SpringerBriefs Electr. Comput. Eng.* **2017**. [CrossRef]
5. Ojaroudi, N.; Ghadimi, N. Design of CPW-fed slot antenna for MIMO system applications. *Microw. Opt. Technol. Lett.* **2014**, *56*, 1278–1281. [CrossRef]
6. Parchin, N.O.; Basherlou, H.J.; Al-Yasir, Y.I.A.; Abd-Alhameed, R.A.; Abdulkhaleq, A.M.; Noras, J.M. Recent developments of reconfigurable antennas for current and future wireless communication systems. *Electronics* **2019**, *8*, 128. [CrossRef]
7. Hussain, R.; Alreshaid, A.T.; Podilchak, S.K.; Sharawi, M.S. Compact 4G MIMO antenna integrated with a 5G array for current and future mobile handsets. *IET Microw. Antennas Propag.* **2017**, *11*, 271–279. [CrossRef]
8. Ojaroudi, N.; Ojaroudi, H.; Ghadimi, N. Quadband planar inverted-f antenna (PIFA) for wireless communication systems. *Prog. Electromagn. Res. Lett.* **2014**, *45*, 51–56. [CrossRef]
9. Chen, Q.; Lin, H.; Wang, J.; Ge, L.; Li, Y.; Pei, T.; Sim, C.-Y.-D. Single ring slot based antennas for metal-rimmed 4G/5G smartphones. *IEEE Trans. Antennas Propag* **2018**, *67*, 1476–1487. [CrossRef]
10. Ojaroudiparchin, N.; Shen, M.; Pedersen, G.F. Wide-scan phased array antenna fed by coax-to- microstriplines for 5G cell phones. In Proceedings of the 21st International Conference on Microwaves, Radar and Wireless Communications, krakow, Poland, 9–11 May 2016.
11. Liu, Y.; Lu, Y.; Zhang, Y.; Gong, S.-X. MIMO antenna array for 5G smartphone applications. In Proceedings of the 13th European Conference on Antennas and Propagation (EuCAP), Krakow, Poland, 31 March–5 April 2019.
12. Al-Hadi, A.A.; Ilvonen, J.; Valkonen, R.; Viikan, V. Eight-element antenna array for diversity and MIMO mobile terminal in LTE 3500MHz band. *Microw. Opt. Technol. Lett.* **2014**, *56*, 1323–1327. [CrossRef]
13. Parchin, N.O.; Al-Yasir, Y.I.A.; Noras, J.M.; Abd-Alhameed, R.A. Dual-polarized MIMO antenna array design using miniaturized self-complementary structures for 5G smartphone applications. In Proceedings of the 13th European Conference on Antennas and Propagation (EuCAP), Krakow, Poland, 31 March–5 April 2019.
14. Wong, K.-L.; Lu, J.-Y.; Chen, L.-Y.; Li, W.Y.; Ban, Y.L. 8-antenna and 16-antenna arrays using the quad-antenna linear array as a building block for the 3.5-GHz LTE MIMO operation in the smartphone. *Microw. Opt. Technol. Lett.* **2016**, *58*, 174–181. [CrossRef]
15. Chang, L.; Yu, Y.; Wei, K.; Wang, H. Polarization-orthogonal co-frequency dual antenna pair suitable for 5G MIMO smartphone with metallic bezels. *IEEE Trans. Antennas Propag.* **2019**, *67*, 5212–5220. [CrossRef]
16. Abdullah, M.; Ban, Y.-L.; Kang, K.; Li, M.-Y.; Amin, M. Eight-element antenna array at 3.5GHz for MIMO wireless application. *Prog. Electromagn. Res. C* **2017**, *78*, 209–217. [CrossRef]
17. Zhao, X.; Yeo, S.P.; Ong, L.C. Decoupling of inverted-F antennas with high-order modes of ground plane for 5G mobile MIMO platform. *IEEE Trans. Antennas Propag.* **2018**, *66*, 4485–4495. [CrossRef]
18. Parchin, N.O.; Al-Yasir, Y.I.A.; Ali, A.H.; Elfergani, I.; Noras, J.M.; Rodriguez, J.; Abd-Alhameed, R.A.; Al-Yasir, Y.I.A. Eight-element dual-polarized MIMO slot antenna system for 5G smartphone applications. *IEEE Access* **2019**, *9*, 15612–15622. [CrossRef]
19. Xu, S.; Zhang, M.; Wen, H.; Wang, J. Deep-subwavelength decoupling for MIMO antennas in mobile handsets with singular medium. *Sci. Rep.* **2017**, *7*, 12162. [CrossRef]

20. Sun, L.; Feng, H.; Li, Y.; Zhang, Z. Compact 5G MIMO mobile phone antennas with tightly arranged orthogonal-mode pairs. *IEEE Trans. Antennas Propag.* **2018**, *66*, 6364–6369. [CrossRef]

21. Li, M.-Y.; Ban, Y.-L.; Xu, Z.-Q.; Guo, J.; Yu, Z.-F. Tri-polarized 12-antenna MIMO array for future 5G smartphone applications. *IEEE Access* **2018**, *6*, 6160–6170. [CrossRef]

22. Zhao, A.; Zhouyou, R. Size reduction of self-isolated MIMO antenna system for 5G mobile phone applications. *IEEE Antennas Wirel. Propag. Lett.* **2019**, *18*, 152–156. [CrossRef]

23. Parchin, N.O.; Basherlou, H.J.; Alibakhshikenari, M.; Parchin, Y.O.; Al-Yasir, Y.I.A.; Abd-Alhameed, R.A.; Limiti, E. Mobile-phone antenna array with diamond-ring slot elements for 5G massive MIMO systems. *Electronics* **2019**, *8*, 521. [CrossRef]

24. Abdullah, M.; Kiani, S.H.; Abdulrazak, L.F.; Iqbal, A.; Bashir, M.A.; Khan, S.; Kim, S. High-performance multiple-input multiple-output antenna system for 5G mobile terminals. *Electronics* **2019**, *8*, 1090. [CrossRef]

25. Jiang, W.; Liu, B.; Cui, Y.; Hu, W. High-isolation Eight-Element MIMO array for 5G smartphone applications. *IEEE Access* **2019**, *7*, 34104–34112. [CrossRef]

26. Statement: Improving Consumer Access to Mobile Services at 3.6 GHz to 3.8 GHz. Available online: https://www.ofcom.org.uk/consultations-and-statements/category-1/future-use-at-3.6-3.8-ghz (accessed on 21 October 2018).

27. Ojaroudi, N.; Ghadimi, N. Dual-band CPW-fed slot antenna for LTE and WiBro applications. *Microw. Opt. Technol. Lett.* **2014**, *56*, 1013–1015. [CrossRef]

28. Ojaroudi, N. Small microstrip-fed slot antenna with frequency band-stop function. In Proceedings of the 21th Telecommunications Forum. TELFOR 2013, Belgrade, Serbia, 27–28 November 2013.

29. Ojaroudi, N. Design of microstrip antenna for 2.4/5.8 GHz RFID applications. In Proceedings of the German Microwave Conference, GeMic 2014, RWTH Aachen University, Aachen, Germany, 10–12 March 2014.

30. Valizade, A.; Ghobadi, C.; Nourinia, J.; Parchin, N.O.; Ojaroudi, M. Band-notch slot antenna with enhanced bandwidth by using Ω-shaped strips protruded inside rectangular slots for UWB applications. *Appl. Comput. Electromagn. Soc. (ACES) J.* **2012**, *27*, 816–822.

31. *CST Microwave Studio*; ver. 2018; CST: Framingham, MA, USA, 2018.

32. Al-Nuaimi, M.K.T.; Whittow, W.G. Performance investigation of a dual element ifa array at 3 ghz for mimo terminals. In Proceedings of the Antennas and Propagation Conference (LAPC), Loughborough, UK, 14–15 November 2011.

33. Parchin, N.O.; Basherlou, H.J.; Abd-Alhameed, R.A.; Noras, J.M. Dual-band monopole antenna for RFID applications. *Future Internet* **2019**, *11*, 31. [CrossRef]

34. Sharawi, M.S. Printed multi-band MIMO antenna systems and their performance metrics [wireless corner]. *IEEE Antennas Propag. Mag.* **2013**, *55*, 218–232.

35. Ojaroudiparchin, N.; Shen, M.; Pedersen, G.F. Small-size tapered slot antenna (TSA) design for use in 5G phased array applications. *Appl. Comput. Electromagn. Soc. J.* **2018**, *32*, 193–202.

36. Mazloum, J.; Ghorashi, A.; Ojaroudi, M.; Ojaroudi, N. Compact triple-band S-shaped monopole diversity antenna for MIMO applications. *Appl. Comput. Electromagn. Soc. J.* **2015**, *30*, 975–980.

37. Ojaroudiparchin, N.; Shen, M.; Zhang, S.; Pedersen, G.F. A switchable 3-D-coverage-phased array antenna package for 5G mobile terminals. *IEEE Antennas Wireless Propag. Lett.* **2016**, *15*, 1747–1750. [CrossRef]

38. Parchin, N.O.; Alibakhshikenari, M.; Basherlou, H.J.; Abd-Alhameed, R.A.; Rodriguez, J.; Limiti, E. MM-wave phased array quasi-yagi antenna for the upcoming 5G cellular communications. *Appl. Sci.* **2019**, *9*, 978. [CrossRef]

39. Elfergani, I.T.E.; Hussaini, A.S.; Rodriguez, J.; Abd-Alhameed, R. *Antenna Fundamentals for Legacy Mobile Applications and Beyond*; Springer: Switzerland, 2017; pp. 1–659.

40. Ojaroudi, N.; Ojaroudi, Y.; Ojaroudi, S. Compact ultra-wideband monopole antenna with enhanced bandwidth and dual band-stop properties. *Int. J. RF Microw. Comput. Aided Eng.* **2014**, *25*, 346–357. [CrossRef]

41. Ojaroudi, N.; Yazdanifard, S.; Ojaroudi, N.; Naser-Moghaddasi, M. Enhanced bandwidth of small square monopole antenna by using inverted Ushaped slot and conductor-backed plane. *Appl. Comput. Electromagn. Soc. (ACES) J.* **2012**, *27*, 685–690.

42. Syrytsin, I.; Zhang, S.; Pedersen, G.F. Performance investigation of a mobile terminal phased array with user effects at 3.5 GHz for LTE advanced. *IEEE Antennas Wirel. Propag. Lett.* **2017**, *16*, 1847–1850. [CrossRef]

43. Ojaroudiparchin, N.; Shen, M.; Pedersen, G.F. Design of Vivaldi antenna array with end-fire beam steering function for 5G mobile terminals. In Proceedings of the 23rd Telecommunications Forum Telfor (TELFOR), Belgrade, Serbia, 24–26 November 2015; pp. 587–590.
44. Hussain, R.; Sharawi, M.S.; Shamim, A. 4-element concentric pentagonal slot-line-based ultra-wide tuning frequency reconfigurable MIMO antenna system. *IEEE Trans. Antennas Propag.* **2018**, *66*, 4282–4287. [CrossRef]
45. Parchin, N.O.; Ullah, A.; Asharaa, A.S.; Al-Yasir, Y.I.A.; Jahanbakhsh, H.; Nagala, M.; Abd-Alhameed, R.; Noras, J. 8×8 MIMO antenna system with coupled-fed elements for 5G handsets. In Proceedings of the IET Antennas and Propagation Conference, Birmingham, UK, 11–12 November 2019.

MDPI

St. Alban-Anlage 66

4052 Basel

Switzerland

Tel. +41 61 683 77 34

Fax +41 61 302 89 18

www.mdpi.com

Electronics Editorial Office

E-mail: electronics@mdpi.com

www.mdpi.com/journal/electronics